Chemical Processes for a Sustainable Future

Chemical Processes for a Sustainable Future

Contributors

_____

M.A. Martin-Luengo, M. Yates et al.

AURIS
Reference

www.aurisreference.com

Chemical Processes for a Sustainable Future

Contributors: M.A. Martin-Luengo, M. Yates et al.

Published by Auris Reference Limited

www.aurisreference.com

United Kingdom

Chemical Processes for a Sustainable Future

ISBN: 978-1-78154-893-6

British Library Cataloguing in Publication Data
A CIP record for this book is available from the British Library

Printed in the United Kingdom

Exclusively distributed by CBS Publishers & Distributors Pvt. Ltd.

Sales & Distribution Rights only for India, Pakistan, Bangladesh, Sri Lanka, Nepal and Bhutan. This book is not to be sold outside these territories.

Contents

List of Abbreviations

BAF	Biological Aerated Filters
CLE	Cross Linked Enzyme Aggregates
CRL	Candida Rugosa Lipase
CSA	Chemical Safety Assessment
CSR	Chemical Safety Report
DE	Deep Eutectic Solvents
DMC	Dimethyl Carbonate
DO	Dissolved Oxygen
EC	Enzyme Classification
EIA	Environmental Impact Assessment
ESG	Environmental, Social, and Governance
FAME	Fatty Acid Methyl Esters
FBB	Fibrous Bed Bioreactor
GHG	Greenhouse Gas
GHSV	Gas Hourly Space Velocity
GRI	Global Reporting Initiative
HPLC	High Performance Liquid Chromatography
HRT	Hydraulic Retention Time
ILCE	Ionic Liquid Coated Enzyme
MCA	Multicriteria Assessment
MFC	Microbial Fuel Cells
MLSS	Mixed Liquor Suspended Solids
OP	Organophosphates
ORP	Oxidation Reduction Potential
PAH	Polycyclic Aromatic Hydrocarbons
PBR	Packed Bed Reactor
PIL	Protic Ionic Liquids
PPL	Porcine Pancreatic Lipase
QSAR	Quantitative Structure Activity Relationship
RBC	Rotating Biological Contactors
RHA	Rice Husk Ash
SCF	Supercritical Fluids
SDS	Safety Data Sheets
SLM	Supported Liquid Membranes
SVI	Sludge Volume Index
TPP	Three Phase Partitioning
UV	Ultraviolet
VOC	Volatile Organic Compound
WWTP	Wastewater Treatment Plants

List of Contributors

M.A. Martin-Luengo
Department of New Architectures, Institute of Materials Science of Madrid, CSIC, Spain

E. Sáez Rojo
Department of New Architectures, Institute of Materials Science of Madrid, CSIC, Spain
Institute of Catalysis and Petroleochemistry, CSIC, Spain

A.M. Martínez Serrano
Department of New Architectures, Institute of Materials Science of Madrid, CSIC, Spain

M. Diaz
Department of New Architectures, Institute of Materials Science of Madrid, CSIC, Spain
Institute of Catalysis and Petroleochemistry, CSIC, Spain

L Vega Argomaniz
Department of New Architectures, Institute of Materials Science of Madrid, CSIC, Spain
Institute of Catalysis and Petroleochemistry, CSIC, Spain

L. Medina Trujillo
Department of New Architectures, Institute of Materials Science of Madrid, CSIC, Spain

S. Nogales
Department of New Architectures, Institute of Materials Science of Madrid, CSIC, Spain
Institute of Catalysis and Petroleochemistry, CSIC, Spain

M. Yates
Institute of Catalysis and Petroleochemistry, CSIC, Spain

F. Plou
Institute of Catalysis and Petroleochemistry, CSIC, Spain

R. Lozano Pirrongelli
Institute of Catalysis and Petroleochemistry, CSIC, Spain

M. Ramos
Centre for Biomedical Technology, Polytechnical University of Madrid, Spain

J.L. Salgado
AIZCE Technical Committee, Interprofesional Asociation of Juices and Citric Concentrates,

J.L. Lacomb
Institute of Biofunctional Studies, Complutense University of Madrid, Madrid, Spain

A. Civantos
Institute of Biofunctional Studies, Complutense University of Madrid, Madrid, Spain

G. Reilly
Department of Materials Science and Engineering, Kroto Research Institute, Broad Lane, University of Sheffield, Sheffield, United Kingdom

C. Vervaet
Laboratory of Pharmaceutical Technology, Gent, Belgium

Paula Saavalainen
Environmental and Chemical Engineering, Faculty of Technology, University of Oulu, 90014 Oulu, Finland

Satish Kabra
Department of Chemical Engineering, Institute of Chemical Technology (ICT), Matunga, Mumbai 400019, India

Esa Turpeinen
Environmental and Chemical Engineering, Faculty of Technology, University of Oulu, 90014 Oulu, Finland

Kati Oravisjärvi
Environmental and Chemical Engineering, Faculty of Technology, University of Oulu, 90014 Oulu, Finland

Ganapati D. Yadav
Department of Chemical Engineering, Institute of Chemical Technology (ICT), Matunga, Mumbai 400019, India

Riitta L. Keiski
Environmental and Chemical Engineering, Faculty of Technology, University of Oulu, 90014 Oulu, Finland

Eva Pongrácz
Thule Institute, NorTech Oulu, University of Oulu, 90014 Oulu, Finland

Mohamed Samer
Cairo University, Faculty of Agriculture, Department of Agricultural Engineering, Egypt

Mahesh K. Potdar
Centre for Green Chemistry, School of Chemistry, Monash University, Melbourne, Victoria 3800, Australia

Geoffrey F. Kelso
Centre for Green Chemistry, School of Chemistry, Monash University, Melbourne, Victoria 3800, Australia

Lachlan Schwarz
Centre for Green Chemistry, School of Chemistry, Monash University, Melbourne, Victoria 3800, Australia

Chunfang Zhang
Centre for Green Chemistry, School of Chemistry, Monash University, Melbourne, Victoria 3800, Australia

Milton T. W. Hearn
Centre for Green Chemistry, School of Chemistry, Monash University, Melbourne, Victoria 3800, Australia

Rinkoo Devi Gupta
Faculty of Life Sciences and Biotechnology, South Asian University, New Delhi 110021, India

Luiz J Visioli
Department of Chemical Engineering, Federal University of Santa Maria, 97105-900 Santa Maria, Brazil

Heveline Enzweiler
Department of Chemical Engineering, Federal University of Santa Maria, 97105-900 Santa Maria, Brazil

Raquel C Kuhn
Department of Chemical Engineering, Federal University of Santa Maria, 97105-900 Santa Maria, Brazil

Marcio Schwaab
Department of Chemical Engineering, Federal University of Santa Maria, 97105-900 Santa Maria, Brazil

Marcio A Mazutt
Department of Chemical Engineering, Federal University of Santa Maria, 97105-900 Santa Maria, Brazil

Anil C Mali
Department of Process Research and Development, Megafine Pharma (P) Ltd., 201, Lakhmapur, Dindori, Nashik 422 202, Maharashtra, India

Dattatray G Deshmukh
Department of Process Research and Development, Megafine Pharma (P) Ltd., 201, Lakhmapur, Dindori, Nashik 422 202, Maharashtra, India

Divyesh R Joshi
Department of Process Research and Development, Megafine Pharma (P) Ltd., 201, Lakhmapur, Dindori, Nashik 422 202, Maharashtra, India

Hitesh D Lad
Department of Process Research and Development, Megafine Pharma (P) Ltd., 201, Lakhmapur, Dindori, Nashik 422 202, Maharashtra, India

Priyank I Patel
Department of Process Research and Development, Megafine Pharma (P) Ltd., 201, Lakhmapur, Dindori, Nashik 422 202, Maharashtra, India

Vijay J Medhane
Department of Chemistry, Organic Chemistry Research Center, K. T. H. M College, Nashik 422 002, Maharashtra, India

Vijayavitthal T Mathad
Department of Process Research and Development, Megafine Pharma (P) Ltd., 201, Lakhmapur, Dindori, Nashik 422 202, Maharashtra, India

Mukesh Kumar
Department of Biophysics, All India Institute of Medical Sciences, New Delhi, India

Joyeeta Mukherjee2
Chemistry Department, Indian Institute of Technology Delhi, New Delhi, India

Mau Sinha
Department of Biophysics, All India Institute of Medical Sciences, New Delhi, India

Punit Kaur
Department of Biophysics, All India Institute of Medical Sciences, New Delhi, India

Sujata Sharma
Department of Biophysics, All India Institute of Medical Sciences, New Delhi, India

Munishwar Nath Gupta
Department of Biochemical Engineering and Biotechnology, Indian Institute of Technology Delhi, New Delhi, India.

Tej Pal Singh
Department of Biophysics, All India Institute of Medical Sciences, New Delhi, India

Joshua E. S. J. Reid
York Structural Biology Laboratory, Department of Chemistry, University of York, Heslington, York YO10 5DD, UK
TWI Ltd., Granta Park, Great Abington, Cambridge CB21 6AL, UK

Neil Sullivan
The Durham Genome Centre, Park House, Station Road, Lanchester DH7 0EX, UK

Lorna Swift
St. Saviour's and St. Olave's School, London SE1 4AN, UK

Guy A. Hembury
University of Hull, Kingston-upon-Hull HU6 7RX, UK.

Seishi Shimizu
York Structural Biology Laboratory, Department of Chemistry, University of York, Heslington, York YO10 5DD, UK

Adam J. Walker
TWI Ltd., Granta Park, Great Abington, Cambridge CB21 6AL, UK

Preface

In a scientific sense, a chemical process is a method or means of somehow changing one or more chemicals or chemical compounds. Such a chemical process can occur by itself or be caused by an outside force, and involves a chemical reaction of some sort. In an "engineering" sense, a chemical process is a method intended to be used in manufacturing or on an industrial scale to change the composition of chemical(s) or material(s), usually using technology similar or related to that used in chemical plants or the chemical industry. The text *Chemical Processes for a Sustainable Future* approaches sustainability from two directions, the reduction of pollution and the maintaining of existing resources, both of which are addressed in a thorough examination of the main chemical processes and their impact. First chapter focuses on sustainable materials and biorefinery chemicals from agriwastes. Second chapter suggests multicriteria based evaluation tool to assess the sustainability of three different reaction routes to dimethyl carbonate: direct synthesis from carbon dioxide and methanol, transesterification of methanol and propylene carbonate, and oxidative carbonylation of methanol. Third chapter elucidates the technologies of biological and chemical wastewater treatment processes. Fourth chapter provides an overview of recent developments in this field with special emphasis on the application of more sustainable enzyme-catalyzed reactions and separation processes employing ionic liquids, driven by advances in fundamental knowledge, process optimization and industrial deployment. The aim of fifth chapter is to provide recent developments on the understanding of the mechanism of catalytic promiscuity, evolvability of promiscuous functions and the applications of enzyme promiscuity in the designing of enhanced or new functional biocatalysts. Sixth chapter reviews the technical and economic feasibility of the main technologies available to produce biobutanol. An efficient and high yielding process for the production of impurity free rivaroxaban, an anti-coagulant agent using alternate synthon is reported in seventh chapter. Enhancement of stability of a lipase by subjecting to three phase partitioning is presented in eighth chapter. Last chapter evaluate the mutagenicity of protic ionic liquids using the mini ames test.

Chapter 1

SUSTAINABLE MATERIALS AND BIOREFINERY CHEMICALS FROM AGRIWASTES

M.A. Martin-Luengo[1], E. Sáez Rojo[1,2], A.M. Martínez Serrano[1,3], M. Diaz[1,2], L Vega Argomaniz[1,2], L. Medina Trujillo[1], S. Nogales[1,2], M. Yates[2], F. Plou[2], R. Lozano Pirrongelli[2], M. Ramos[3], J.L. Salgado[4], J.L. Lacomb[5], A. Civantos[5], G. Reilly[6] and C. Vervaet[7]

[1] Department of New Architectures, Institute of Materials Science of Madrid, CSIC, Spain

[2] Institute of Catalysis and Petroleochemistry, CSIC, Spain

[3] Centre for Biomedical Technology, Polytechnical University of Madrid, Spain

[4] AIZCE Technical Committee, Interprofesional Asociation of Juices and Citric Concentrates,

[5] Institute of Biofunctional Studies, Complutense University of Madrid, Madrid, Spain

[6] Department of Materials Science and Engineering, Kroto Research Institute, Broad Lane, University of Sheffield, Sheffield, United Kingdom

[7] Laboratory of Pharmaceutical Technology, Gent, Belgium

INTRODUCTION

Countries with economies based on agriculture generate vast amounts of low or null value wastes which may even represent an environmental hazard. In our group, agricultural industrial wastes have been converted into value added liquid substances and materials with several aims: decreasing pollution, giving added value to wastes and working in a sustainable manner in which the wastes of an industry can be used as the raw materials of the same or others, as the "cradle to cradle" philosophy states [1].

Sub-products from the agricultural food industry are being employed as renewable low cost raw materials in the preparation of Ecomaterials, designed for use in a number of industrial processes of great interest. Given their origin, these materials may compete with conventional ones since with this process a sustainable cycle is closed, in which the residues of one industry are used as raw materials in the same or other industries [2].

With regards to the composition of the residues produced from agriculture, the pH of soil is of great importance, since plants can only absorb the minerals that are dissolved in water and pH is mandatory for the physical, chemical and biological properties of soil and the main cause of many agronomic questions related to nutrient assimilation [3-5]. Variations of pH modify the solubility of most elements necessary for the development of crops and also influence the microbian activity of soil, which will affect the transformation of elements that are liberated to the soil and can be assimilated to form crops or not [3]. For example at pH lower than 6 or higher than 8 bacterian activities are lowered, the oxidation of nitrogen to nitrate is reduced and the amount of nitrogen available for plant food is decreased. However Al, Fe and manganese are more soluble at low pHs, reaching even toxic concentrations. Potassium and sulphur are easily adsorbed at pH higher than 6, calcium and magnesium between 7 and 8.5 and iron at pH lower than 6. For alkaline pH in soil, the availability of $H_2PO_4^-$ can be reduced through precipitation of phosphorous containing salts with cations such as calcium Ca^{2+} or magnesium Mg^{2+}. However when soils have acid pH other compounds with HPO_4^{2-} and iron (Fe^{2+}), aluminium (Al^{3+}) and manganese (Mn^{2+}) can form, with increased solubility.

The main factors that influence soil pH are the mineral composition and how it meteorizes, the decomposition of organic matter, how nutrients are partitioned among the solution and aggregates and of course the pluviometry of the zone and atmospheric contamination. Lower pHs are found in places with high pluviometry, with high organic matter decomposition, young soils developed on acid substrates, and places with high atmospheric contamination (acid rain).

Depending on the species, crops can benefit from calcareous soils with high calcium carbonate content such as alfalfa, but other plants prefer soils with acid pH such as potatoes, coffee or tobacco. It is clear that different seasons will produce plants with a varying composition depending on the atmospheric conditions and therefore the materials derived from them need to be characterized and analyzed to determine their possible uses.

Variation in content of components in fruits and vegetables depends upon both genetics and environment, including growing conditions, harvest and storage, processing and preparation. For example in broccoli, low soil water content during plant growth and post harvest cold storage were conditions that, combined, gave higher amounts of l-ascorbic acid [6,7]. Higher polyphenol contents are found in organic tomato juices compared to non-organic ones due to a higher phosphorus uptake and limited nitrogen availability in the first case [8]. Therefore, thorough characterization of the residue composition is a key step before determination of the possible uses of a given residue.

Among the applications being developed in our group, Bio-refinery processes (preparation of sustainable *p*-cymene and hydrogen, avoiding the use of petroleum derivatives and synthesis of pharmaceutical and fine chemical intermediates), design of structured materials capable of effluent decontamination and preparation of biomaterials to act as scaffolds for cell growth towards development of prostheses and implants, will be considered here.

Given its multidisciplinary approach, this work is being carried out through the collaboration among national (Institute of Materials Science of Madrid (ICMM, CSIC), Institute of Catalysis (ICP, CSIC), Centre of Molecular Biology Severo Ochoa (UAM-CSIC), Polytechnic University of Madrid (UPM), University at distance (UNED), University Complutense of Madrid (UPM) and international (University of Sheffield and University of Ghent) research groups, in addition to various industries interested in the transformation of their residues and or sub-products into "value added materials", with whom various research projects have been and are being sponsored by the MICINN and CDTI.

PRODUCTION OF SUSTAINABLE *P*-CYMENE AND HYDROGEN

Environmental problems pose a great challenge, particularly in countries such as Spain, where water use, residues desertic and contaminated soils have become a matter of the utmost concern. Efforts from academia, industry and government are mainly based on technological changes that improve chemical processes to avoid negative environmental consequences. New renewable and sustainable chemicals are now being obtained from agricultural wastes, applying "biorefinery technologies", which reduce the need for non-renewable fossil fuel resources to help solve many environmental problems. Spain is the third producer of citrus fruits in the world [9]. Limonene is a six membered ring terpene, present in agricultural wastes derived from citrus peels, with a purity of more than 95% in orange peel oil. Sustainable *p*-cymene and hydrogen were prepared form limonene, comparing commercial and agricultural waste derived catalysts and conventional and non-conventional activation paths. These studies show interesting results through the development of solids that activate certain kinds of reactions shown in Figure 1.

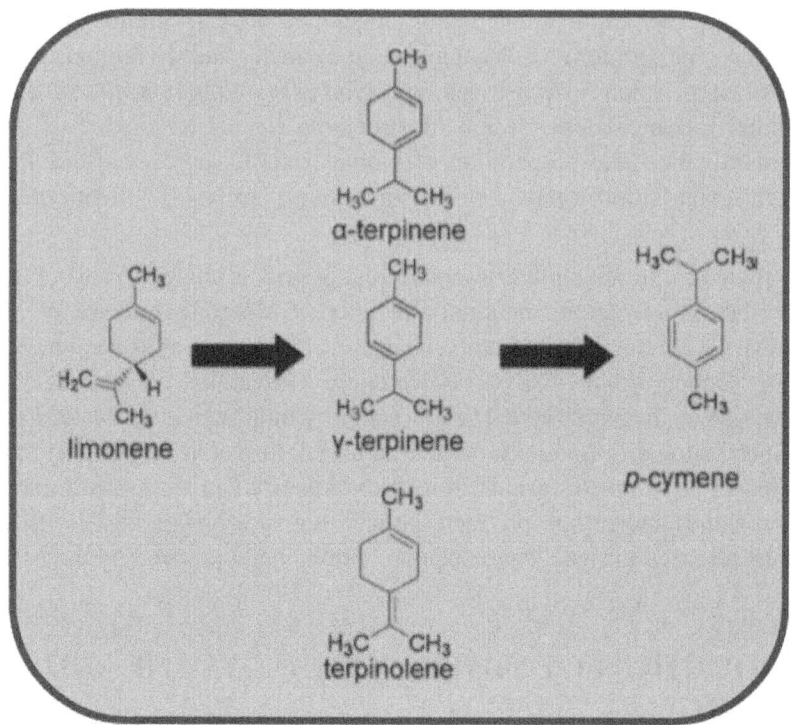

Figure 1. Conversion of limonene to *p*-cymene and reaction intermediates.

This work is based on designing a clean process to transform limonene, into *p*-cymene, in a rapid, simple and economic way. Limonene, a cyclic terpene with empiric formula $C_{10}H_{16}$, is extracted with a purity of *ca.* 95% from citrus fruit processing. Although initially used in pharmaceutical industries, food is the main means for human exposure, although due to its low toxicity [10] it does not constitute a risk to human health at the actual exposure levels. This compound has various uses, as a biodegradable solvent for resins (replacing organic solvents such as mineral oils, methyl-ethyl ketone, toluene, glycolic ethers, CFCs, or as an additive in pigments, inks and adhesives. However, new uses are being sought to give added value to this subproduct, thus increasing the income of these industries, but also with an obvious benefit to society. From the chemical structure of limonene, shown in Figure 1, it may be appreciated that having a six membered ring this substance has the potential to obtain compounds that are usually produced employing petroleum derivatives, with the added bonus of it being a less toxic renewable intermediate. In this study the production of *p*-cymene, an important intermediate in industrial fine chemicals syntheses, for fragrances, flavourings, herbicides, pharaceuticals, *p*-cresol production, syntheses of not nitrated musk's (i.e. tonalide), etc. [11] will be described.

The preparation of *p*-cymene is usually carried out by Friedel-Crafts alkylation of toxic benzene or toluene (from petroleum), with $AlCl_3$ as catalyst and the respective halides or propanol [12]. In these processes mixtures of *ortho* and *para* isomers, are found and therefore further separation processes are required. Furthermore, the use of benzene or toluene and $AlCl_3$ are restricted by environmental legislation in industrialised countries. In this study a series of commercial silica–alumina mixed oxides, supplied by Sasol, with silica contents ranging from 1 to 40 wt.% designated as SIRAL 1 to SIRAL 40, accordingly, were employed. The textural and acid characteristics of these materials were determined and related with their catalytic activities in the transformation of limonene to *p*-cymene using microwave irradiation as a rapid and efficient energy source, alternative to conventional heating, where apart from the advantage of substantially reduced reaction times the product selectivity was also enhanced [10].

The texture of the mixed oxides: surface areas and pore volumes were analysed by nitrogen adsorption/desorption isotherms at $-196°C$, in a Tristar apparatus from Micromeritics, on samples previously outgassed at $300°C$ to a vacuum of less than 10^{-4}torr to ensure that they were clean, dry and free from any loosely adsorbed species. The BET method was used to determine the specific surface areas (S_{BET}) from the adsorption data in the relative pressure range of $0.05–0.30$ $p/p°$ and the mesopore size distributions were calculated from the desorption branch of the nitrogen isotherm using the Kelvin equation and the BJH method with the parameters for the thickness of the adsorbed layers from the Harkins–Jura equation, chosen since this employed a metal oxide as the non-porous standard [13].

The acidities of the solids were analyzed from their ammonia adsorption capacities, determined on a Micromeritics ASAP 2010 device, after being outgassed overnight at $300°C$ to a vacuum of $>10^{-4}$torr, then the ammonia adsorption isotherms at $30°C$ were determined up to a pressure of about 350 torr, obtaining the total chemisorption plus physisorption capacities. Subsequently the sample was outgassing at $30°C$ and a second adsorption isotherm at $30°C$ determined to measure the physisorption capacity of the sample. The chemisorption capacity was calculated from the differences in the ammonia uptakes between the first and second isotherms. The total acidities were subsequently calculated assuming that each molecule of ammonia reacted with one acid site [14].

Pyridine adsorption coupled with infrared analysis was used to qualitatively determine the Lewis (Al^{3+}) and Brønsted (Al-OH) acid sites of the materials, these being capable of catalyzing the conversion of limonene to *p*-cymene.

The catalysts were pressed as self-supporting discs, outgassed under high vacuum, contacted with pyridine vapour and outgassed to obtain the infrared spectra corresponding to Lewis and Brønsted acid sites with adsorption peaks at 1455 and 1545 cm^{-1}, respectively [15].The microwave reactions were carried out with a programmable focalised monomodal microwave apparatus, Synthewave 402 from Prolabo, under both dry media and reflux conditions. Dry media microwave induced reactions can be considered a clean technology, since no solvents are used. In those experiments 50 µl of limonene were physically mixed with 200 mg of solid, placed in a glass reactor and irradiated at 300 W for 5, 10 or 20 min. In reflux conditions 500 mg of solid was mixed with 5 ml of limonene and heated with a reflux column to 165°C for 10 or 20 min, avoiding overheating of the reaction mixture. The final temperature was chosen to be slightly lower than the boiling points of the reactant and products (limonene 175°C, p-cymene 177°C) in order to control the reaction. At the end of the experiments the reaction mixtures were cooled, extracted and dissolved in ethanol and analysed by gas chromatography coupled to a mass spectrometer (GC–MS). Following the extraction of the reaction products with ethanol, the catalytic activities of the samples were redetermined to ascertain if the materials had suffered any change in their activities or selectivities.

It has been previously proposed that the available surface area coupled with the accessibility to the active acid sites play important roles in controlling the catalytic process in this reaction [16]. The specific surface areas, pore volumes and average mesopore diameters of the solids are summarised inTable 1. The nitrogen adsorption/desorption isotherm for SIRAL 1 was of type IV with a well-defined plateau at high relative pressures with a type H1 hysteresis loop, characteristic of a mesoporous solid with a narrow well defined mesopore size distribution [13]. On increasing the silica content the specific surface area, mesopore volume and average mesopore diameters increased progressively as evidenced by a widening of the hysteresis loop between the adsorption and desorption branches and with the highest silica contents the plateau at high relative pressure became less well defined as the pore sizes shifted into the narrow macropore range. From the isotherms obtained for SIRAL 1 and SIRAL 40 presented in Figure 2 the change in the hysteresis loop due to the widening of the pores may be appreciated while the upward displacement of the curve with the increased silica content was indicative of the greater specific surface area. The other samples with intermediate silica contents, not shown here for clarity, gave rise to adsorption/desorption curves that lay between these two extremes.

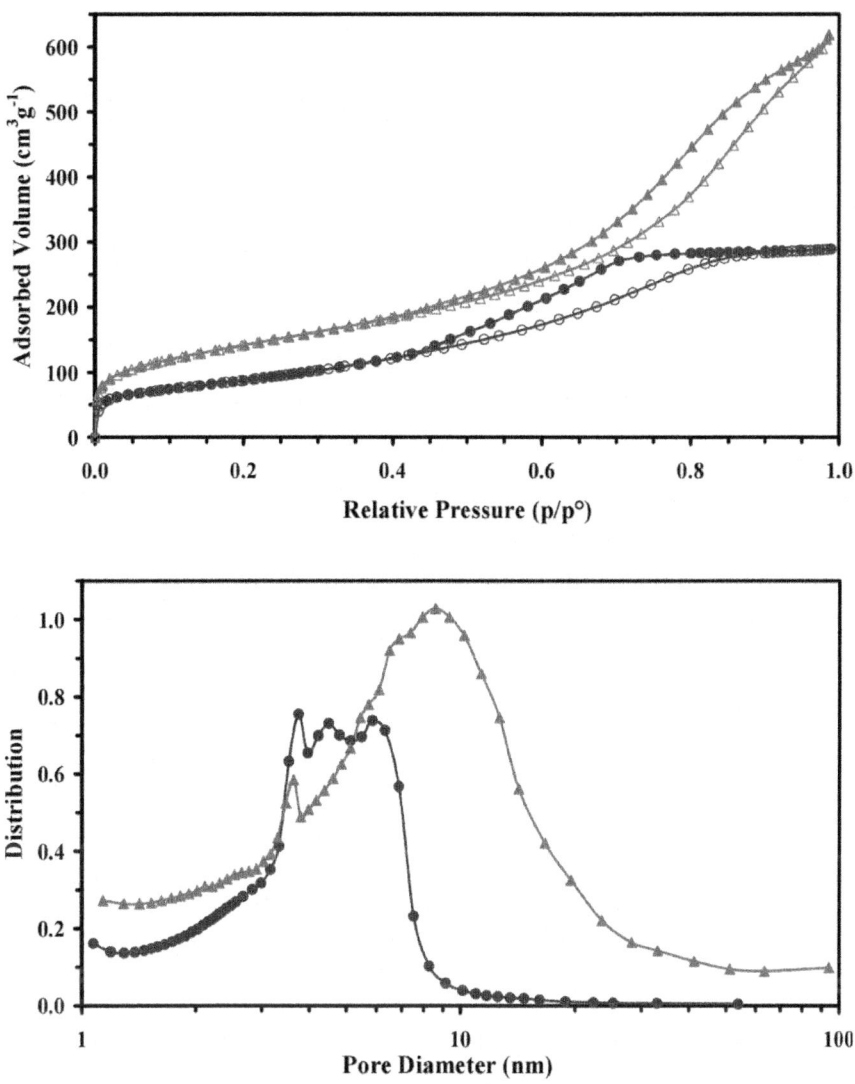

Figure 2. Textural characteristics of mixed oxides SIRAL 1○● and SIRAL 40Δ▲.

Since the reaction is catalyzed by the acid sites, the higher activity as the number and accessibility of the surface acid sites rose with increased silica content was to be expected. Greater activities were found as the total acidity was increased (measured by ammonia adsorption, shown in Figure 3).

Table 1. Textural properties of the mixed oxides

Sample	Surface area (m^2g^{-1})	Pore volume (cm^3g^{-1})	Average mesopore diameters (nm)	
SIRAL 1	321	0.45	4.5	5.8
SIRAL 5	364	0.57	4.8	5.9
SIRAL 10	422	0.59	4.8	6.8
SIRAL 20	432	0.68	5.4	9.4
SIRAL 40	506	0.95	5.8	12.7

The catalytic activity results obtained with microwave irradiation of limonene are presented in Table 2.

Table 2. Limonene conversions and selectivities under dry media conditions

			Selectivity			
Sample	Time (min)	Conversion (%)	α-terpinene (%)	Γ-terpinene (%)	α –terpinolene (%)	p-cymene (%)
	5	15	85	0	15	0
SIRAL 1	10	56	52	14	0	34
	20	100	0	0	0	100
	5	42	60	12	21	7
SIRAL 5	10	63	27	25	0	48
	20	100	0	0	0	100
	5	65	48	11	31	10
SIRAL 10	10	69	15	0	14	71
	20	100	0	0	0	100
	5	88	31	6	15	48
SIRAL 20	10	100	0	0	0	100
	20	100	0	0	0	100
	5	93	0	0	3	87
SIRAL 40	10	100	0	0	0	100
	20	100	0	0	0	100

During the reactions (SIRAL 1 shown as example in Figure 4), only α-terpinene, γ-terpinene, γ-terpinolene (from isomerisation) and p-cymene (from dehydrogenation) were found as products, with no other compounds such as menthanes or menthenes, produced by disproportionation or polymerisation, detected. In Table 2 it should be noted that the selectivity to p-cymene increased with both longer irradiation times and increased silica contents. For the two samples with the highest silica contents reaction times of just 10 min gave selectivities to p-cymene of greater than 90%. Since these samples had the most acid centres, calculated on both per gram or per m^2g^{-1} basis, it would appear that the selectivity to *p*-cymene was due to the rapid aromatisation of the intermediates produced by isomerisation over the acid centres. Although conversion of limonene to *p*-cymene over Lewis acid sites with conventional heating is known, the much longer reaction times necessary to increase the overall conversion (3 h) leads to reduced selectivities due to the formation of undesirable mentanes, etc. [17].

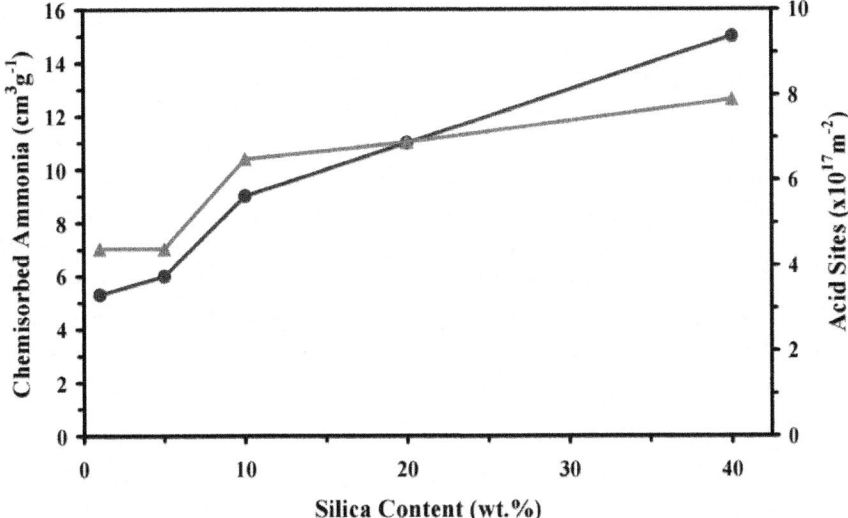

Figure 3. Ammonia chemisorption cm^3g^{-1} ● and number of acid sites m^{-2} ▲ vs. silica content of the mixed oxides.

In order to carry out the reaction in conditions similar to those used industrially the two samples with the highest activities in dry media were chosen for further study in reflux conditions, with the ratio between the reactants and the solid: 0.00025 and 0.01 cm^3g^{-1}, respectively. From Table 3 it may be observed that the intermediate isomerisation product α-terpinene was only found for SIRAL 20 and at a very low level (4%) and 100% conversion of limonene to *p*-cymene was achieved with Siral 40 after 10 min. The speed of the reaction when heated by microwave irradiation is possibly the reason for the high selectivities, since the short reaction time necessary to attain these high conversions and selectivity to *p*-cymene avoid the formation of undesirable by-products such as mentanes (products of disproportionation) or polymers, that are found with the longer reaction times employed with conventional heating [18]. Thus, microwave irradiation favoured the production of *p*-cymene from limonene, avoided the use of highly toxic benzene, toluene and aluminium trichloride and allowed high conversions and excellent selectivities towards the desired product due to the accelerated reaction rates. As the reaction is governed by the number and accessibility of the acid sites greater activities were achieved with increased silica contents, which led to higher specific surface areas, pore volumes and average pore sizes in addition to an increased number of acid sites.

Modified Clays

Sepiolite modified with sodium hydroxide and impregnated with either iron or manganese salts were also used as catalytic supports for conversion of limonene to *p*-cymene. The use of an inexpensive natural clay for the catalyst preparations reduces costs and the need for commercial synthesised solids is avoided. The sepiolite used was from Tolsa SA (hydrated magnesium silicate of the philosilicate type 2:1 with a layer of magnesium between two layers of silica tetrahedra. The octahedral sheet is composed mainly of Mg^{2+}, mainly composed of SiO_2, 62%, MgO 25%, Al_2O_3 1.2%, Fe_2O_3 0.5%). This natural clay of high abundance in Madrid (Spain), has a specific surface area (S_{BET}) close to 300 m^2g^{-1}, of which 150 m^2g^{-1} is external (pores with diameters >2 nm ø) and the rest is due to the micropores of the material (< 2nm). It has a high density of –SiOH groups originated at the edges due to breakage of Si-O-Si bonds at *ca.* 0.5 nm intervals, having a density of *ca.* 2.2 groups/10 nm^2[19].

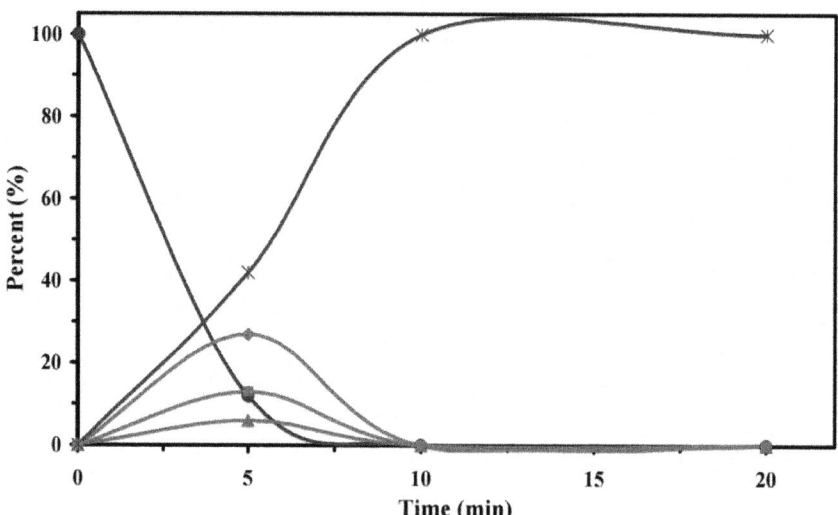

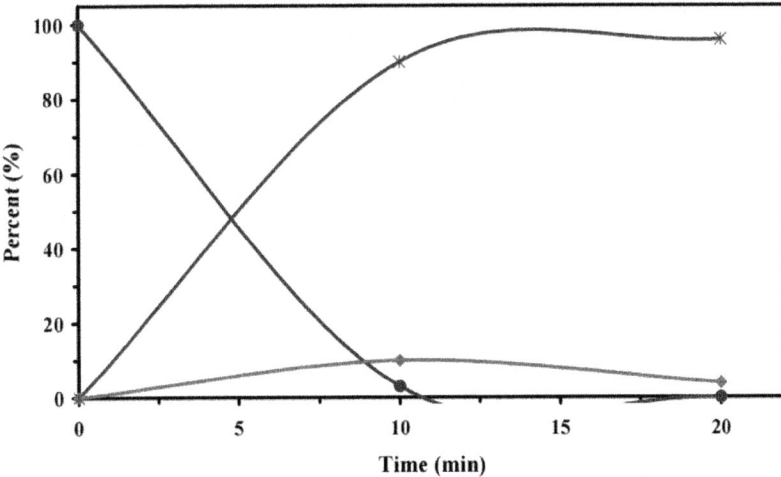

Figure 4. Reactant and product distributions vs. time for (a) Siral 1 in dry media and (b) Siral 20 under reflux: limonene (●), α-terpinene (♦), γ-terpine (▲), α-terpinolene (■) and *p*-cymene (Ú).

The activity of the parent sepiolite was modified by adding 5.75 wt.% of iron or manganese, in the form of their nitrate or acetate, respectively, since their oxides efficiently absorb microwave radiation and have low toxicity compared to other metals commonly used as catalysts (*i.e.*Pd, Cd, Cr etc.), or price (Au or Ag). A further sample was pretreated with sodium hydroxide to introduce a similar amount of sodium.

The best procedure to decompose the impregnated compounds to their corresponding supported oxides was determined with TG-DTA analyses in air flow of the precursors in a Stanton model STA 781 thermogravimetric analyser up to 1000°C. The thermograms obtained show 8% loss of adsorbed water in the interval 20-120°C, 5% at 120-250°C loss of sepiolite (zeolitic) water, and 8% loss for decomposition of the anions at *ca.* 400°C for iron nitrate and at ca. 300°C for the manganese acetate. In accordance with these findings all the precursor solids were calcined at 400°C for 4 h in a 50 cm^{3}min^{-1}air flow [20].

The nitrogen adsorption isotherms for the parent sepiolite heat treated at 400°C gave a mixed type I/II form, due to the presence of both micropores (0-2 nm) and mesopores (2-50 nm) that extend into the macropore region (> 50 nm). For the SepFe (Figure 5), SepMn and SepNa there was a loss in the specific surface area due to the collapse of the microporous structure of sepiolite by folding due to the thermal treatment. For all samples the hysteresis loops were of type H3, typical for solids with slit-shaped pores, commonly found with clay materials.

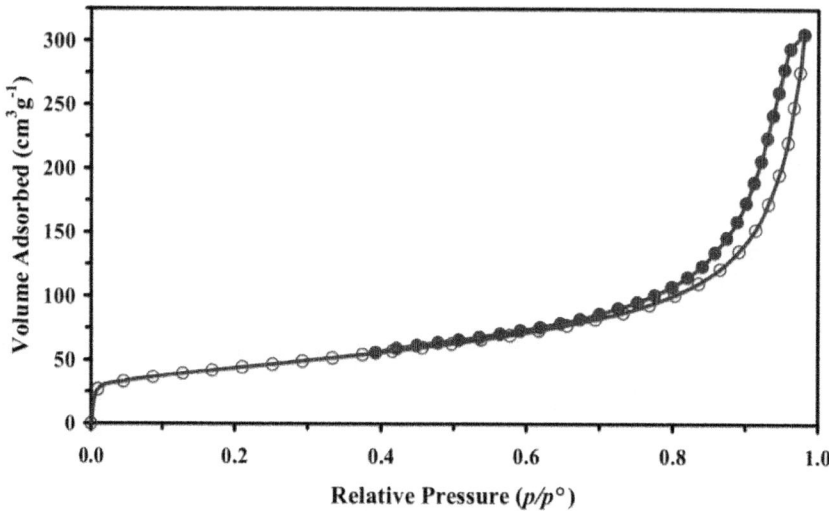

Figure 5. Nitrogen isotherm for SepFe.

Sample SepNa had the lowest pore volume and surface area, probably due to blocking of pores with sodium species. The textural data from the corresponding isotherms are presented in Table 3.

Table 3. Textural characteristics of the solids

Solid	S_{BET}	S_{EXT}	V_{mic}	V_{mes}	V_t
Sepiolite	298	149	0.038	0.417	0.455
SepNa	100	93	0.002	0.362	0.364
SepFe	155	149	0.002	0.416	0.418
SepMn	153	138	0.008	0.431	0.439

A JEOL model FXII electron microscope operating at 200kV was used to study the structure of the materials. The basic sites of the solids were quantified by the adsorption-desorption of carbon dioxide in a Coulter Omnisorp 100 apparatus on solids previously outgassed to clean the surface at different temperatures to quantify the amount and strength of basic sites. The results obtained by this procedure are included in Figure 5 [21].Transmission electron microscopy showed the oxide particles present in the structures of SepFe and SepMn with sizes of 4-5 nm but with no other observable alteration to the fibrilarsepiolite structure and a more homogeneous distribution of the oxide particles for SepFe compared to SepMn, which reduce the exposed oxide area during reaction compared to SepFe. Homogeneous particle sizes have been claimed to be a positive effect for the reactivity in these kinds of reactions.

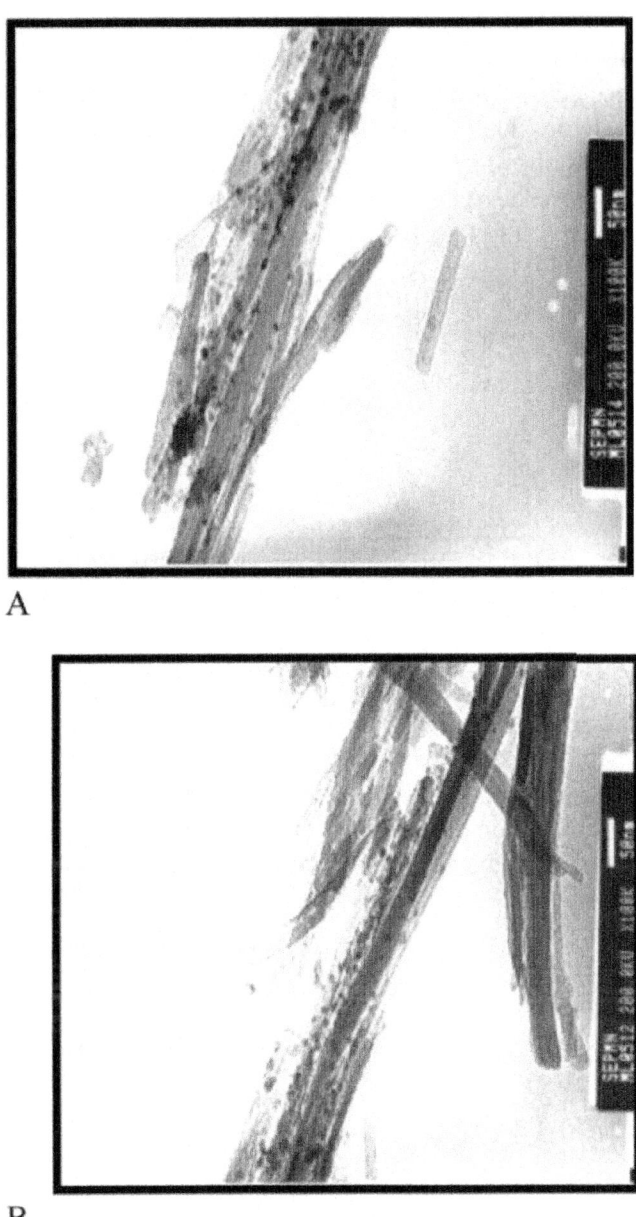

Figure 6. Transmission electron micrographs of (a) SepMn and (b)SepFe.

The amount and strength of acid sites determined by TPD of pyridine (Figure 7) showed the measured acidities followed the order: SepFe>SepMn> Sep. As expected no pyridine adsorption was found for the sample pretreated with sodium hydroxide (SepNa). The increase in the amount and strength of acid sites by addition of iron or manganese oxides to these supports was mainly related to the presence of Fe^{x+} and Mn^{x+} ions(Lewis acidity) and to surface OH groups(Brönsted acidity) [22].

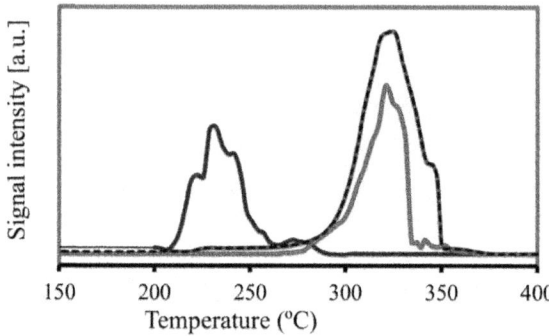

Figure 7. Pyridine determination of acid sites on Sepiolite, SepMn and SepFe.

From the adsorption-desorption of carbon dioxide shown in Figure 8 the basicity of Sepiolite (due to the existence of basic sites mainly Mg^{2+} ions and oxygen on the surface) is low, while temperatures in excess of 400°C were needed to desorb carbon dioxide from the other three materials. The solid with greater amount of stronger basic sites was SepNa. Samples SepFe and SepMn were more basic than Sepiolite and had greater total amounts of basic sites of lower strength.

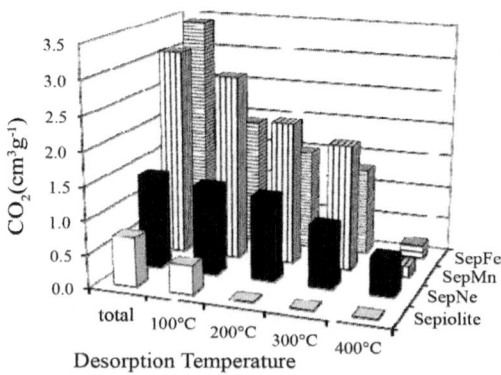

Figure 8. Basicity of the catalysts by CO_2 adsorption-desorption.

For the reactions under dry media (200 mg of catalyst mixed with 0.057ml of limonene) and solvent free conditions, after irradiation the system was cooled and eluted with 2 ml ethanol and filtered to extract the reactants and products and subsequently analysed as described in section 1.1.In the solvent free reactions, a temperature of 165°C, was chosen to transform limonene, based on the results obtained under dry media conditions. In this case 5 ml of limonene and 500 mg of catalyst were used and a reflux attachment included in the microwave oven. At the end of the experiment 0.057 ml of liquid were dissolved in 2 ml of ethanol to follow the analysis procedure as described above. The results for the dry media reactions are shown in Table 4.

Table 4. Conversions and selectivities under dry media conditions

Sample	Reaction Time (Min)	Conversion (%)	Product Distribution (%)			
			α-terpinene	γ-terpinene	α-terpinolene	p-cymene
	5	35	25	8	31	36
Sepiolite	10	73	26	0	0	74
	20	100	0	0	0	100
	5	75	0	0	0	100
SepFe	10	100	0	0	0	100
	20	100	0	0	0	100
	5	89	0	0	0	100
SepMn	10	100	0	0	0	100
	20	100	0	0	0	100

SepNa did not show activity, with high amounts of basic sites and practically non-existent acid sites, even though it reached the highest temperature during microwave heating. With sepiolite, selectivity to p-cymene rose, reaching 100% after 20 min and for SepFe and SepMn after only 10 min due to iron or manganese that are microwave adsorbing centres compared to the parent sepiolite, but not to the final temperature attained (Figure 9).

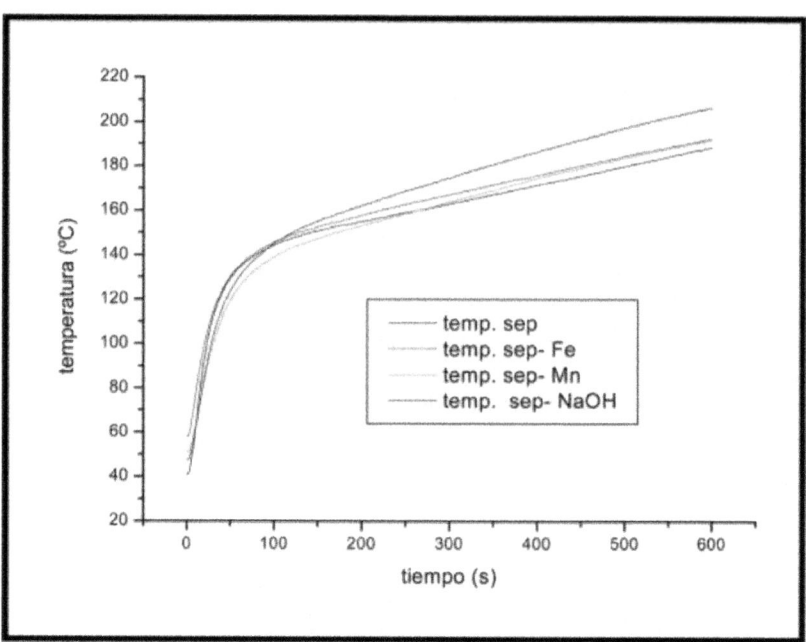

Figure 9. Temperatures reached under dry media conditions.

The solids that presented activity under dry media conditions (Sepiolite, SepFe and SepMn) were tested in the liquid phase reactions that allow higher liquid to solid ratios (Table 5) and the temperatures reached are included in Figure 10.

Table 5. Conversions and selectivities in liquid phase

Sample	Reaction Time (Min)	Conversion (%)	Product Distribution (%)			
			α-terpinene	γ-terpinene	α-terpinolene	p-cymene
Sepiolite	10	100	0	0	0	100
	20	100	0	0	0	100
SepFe	10	85	32	19	35	14
	20	100	9	3	0	88
SepMn	10	10	34	8	58	0
	20	75	33	17	33	17

As in dry media, the selectivity to p-cymene increased with longer reaction times and higher temperatures, in agreement with the fact that the formation of this product is an endothermic reaction. Under these reaction conditions Sepiolite gave the best results for both activity and selectivity, reaching

100% conversion and selectivity after only 10 minutes. The differences in the temperature profiles for these three materials during the reaction, were not so great as to explain such differences in their catalytic activities and selectivities.

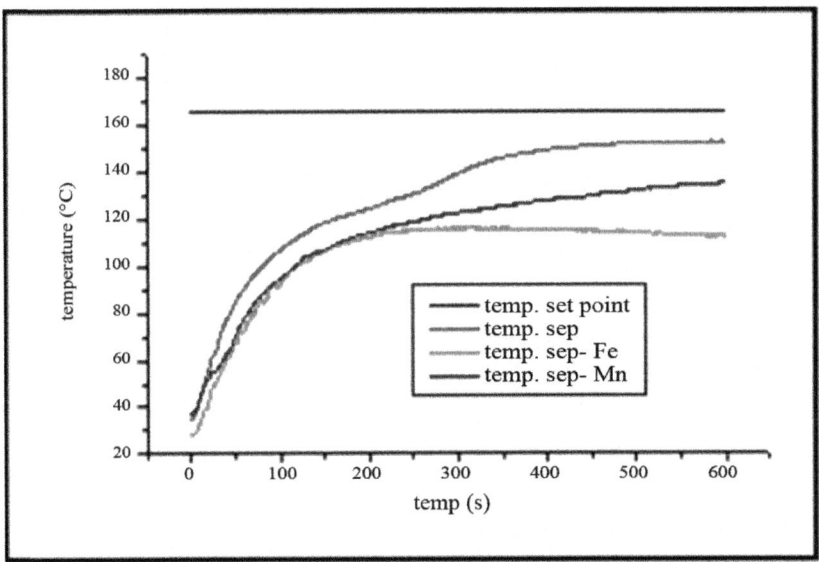

Figure 10. Temperatures reached under solvent free conditions.

The strong basic sites present in SepNa, SepMn and SepFe may be capable of adsorbing limonene to such an extent that it was not readily available for reaction under the reaction conditions, this being the reason why Sepiolite is the most active solid for the dehydrogenation of limonene to *p*-cymene. Previous work by the authors showed that the acid sites of silica–alumina mixed oxides could dehydrogenate limonene in conditions similar to those used here [10]. Thus, the acid sites present on the Sepiolite surface, although weaker than those found with iron or manganese containing solids, were of the right strength to convert limonene to *p*-cymene. Further experiments showed that 300 mg of Sepiolite were enough to convert 5 ml of limonene to *p*-cymene in 10 min with a 100% conversion and selectivity. Similar conversions and selectivities to those found in this study have recently been reported for the transformation of limonene, although dangerously high hydrogen pressures, greater than atmospheric and the use of toxic and expensive palladium catalysts were required [23].

Conventional heating of limonene using acid solids as catalysts in liquid-solid conditions leads to lower activities and selectivities than those found in this work. We believe the short reaction times required using dielectric heating were responsible for the higher conversions and selectivities, due to the increased activity of these paramagnetic absorbing centres, thus avoiding other products such as dimers and polymers, that are formed during the longer reaction times necessary with conventional heating [17].

SUSTAINABLE FINE CHEMICALS FROM AGRI WASTES

A vast amount of work has been carried out regarding the preparation of fine chemicals, with oxidation being one of the main paths followed. Renewable value added chemicals were prepared in this work using solid and liquid agricultural industrial wastes from rice and citrus production, as renewable raw materials, avoiding the use of substances toxic to the environment and achieving a reduction in energy expenditure. The whole process is consistent with a sustainable development. The present investigation has demonstrated how the transformation of a low value subproduct (limonene) into high value materials (carvona, carveol and limonene epoxides) can be achieved with similar conversions and selectivities to those found in the literature for catalysts that have higher toxicity and are less environmentally friendly.

After a thorough review of literature data, careful design of oxidation conditions of limonene allows the production of carvone and carveol through allylic oxidation and limonene oxides through double bond epoxidation (Figure 11). These products have interesting pharmaceutical properties and also are chemical intermediates with prices from 5 to 10 times higher than limonene [9].

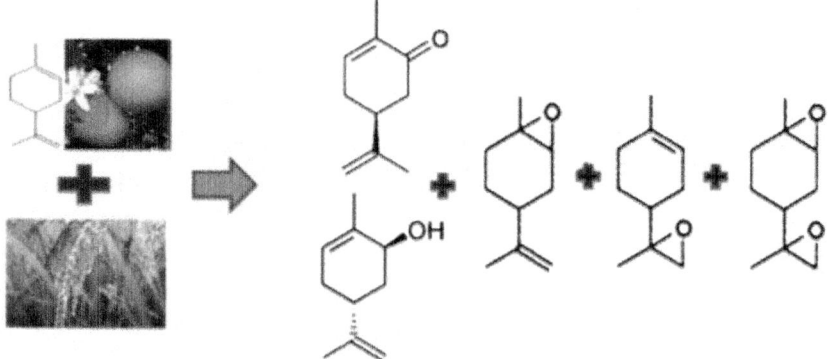

Figure 11. Renewable carvone, carveol and epoxides prepared from limonene. Source of photographs: http://www.finecooking.com/assets/uploads/posts/5552/ ING-oranges_sql.jpg   http://1.bp.blogspot.com/_siEMqPzFQAY/TJO2Pt4tX3I/ AAAAAAAABW0/gWJ_-7IP9co/s1600/arroz9.jpg

A clean process for this transformation was developed here by using iron oxide supported on silica (RHS) from rice husk (RH) to catalyse the reaction. RH is produced in huge amounts annually and its ash contains *ca.* 94% silica, with the added bonus of zero CO_2 energy being produced during the calcination of RH to produce RHS.

The organic reactions carried out in this work were activated with dielectric heating that allows higher yields under mild reaction conditions, avoiding thermal decomposition of sensitive products or reagents, and therefore affecting selectivity [10]. Tert-butyl hydroperoxide (tBHP) was used as oxidant due to its easy handling characteristics and high thermal stability. The tert-butanol and its by-products on oxidation, can be distilled and recycled, a particularly important step when industrial amounts are required. Silica from rice husk (RHS) was used as support and catalyst. It was prepared by calcining rice husk in air at 500°C for 4 h. For comparison reasons, silica AerosilEvonik was used since it had similar textural properties to RHS (surface area of 91 m^2g^{-1}). The supported catalysts were prepared by dry impregnation of RHS with iron nitrate ($Fe(NO_3)_3 \cdot 9H_2O$ (> 99% purity, Sigma-Aldrich) solutions, the impregnated precursors were air dried overnight at 100°C, prior to TG-DTA and FTIR studies of the calcination procedures necessary to produce FeOx/RHS catalysts (400 or 600°C, 4 h in air). Since treating silica with low pH iron nitrate solutions can produce structural changes, fresh RHS was treated with nitric acid solutions of the impregnating pH (1.4) and then calcined at 400 or 600°C, for comparison purposes.

The basicity of the solids was measured by decomposition of acetic acid, in a gas analysis system with quadrupole mass spectrometry, model M3 QMS200 Thermostar coupled to a thermogravimetric/differential thermal analysis equipment, Stanton STA model 781. Increasing the temperature of the spiked solids under nitrogen flow at a heating rate of 5°Cmin^{-1} up to 700°C, recording the evolution of mass 44, ascribed to CO_2. The amount and temperature of evolution of the CO_2 signal gave an indication of the strength and amount of basic sites. The CO_2 signal was calibrated from the decomposition of a known amount of calcium oxalate.

The catalytic oxidations of limonene were studied in a programmable focussed microwave oven Syn402 from Prolabo, described above in section 1.1. Preliminary experiments were carried out in dry media conditions on the Fe-containing catalysts, to determine the best conditions to use in further liquid/solid experiments, choosing the conversion of limonene and selectivity to carvone as parameters for comparison of the effects caused by the various conditions studied. Studied parameters were the limonene: tBHP volume ratio, temperature and reaction time. Based on these preliminary experiments, the conditions chosen for the liquid phase work were 300W microwave power, with a reflux attachment, reaction temperature (150°C) and reaction time up to 120 min, using 4 ml limonene, 14 ml of TBHP solution and 150 mg of catalyst. Conventional heating results were carried out for comparison with similar amounts of reactants and catalyst, starting the reaction time from the moment the reaction temperature was reached and the catalyst added. The effects of reaction temperature, amount of catalyst and oxidant on both the reactivity and selectivity were studied using the most promising catalyst. The reactants and products were analysed in a GC-MS system, as described above.

The chemical composition of the RHS used in this work was 6% S, 2% K, 1% Cl, P and Ca, 0.1% Mn and Fe and 0.04% Zn. The iron containing precursors were calcined at 400 or 600°C for 4 h, according to TG-DTA and FTIR data. The iron containing catalysts (FeO$_x$/RHS) had 4.8% or 8.9% Fe. XRD patterns indicated the amorphous nature of RHS and the FeOx/RHS show crystalline iron oxide particles, more common on increasing the amount of iron and calcination temperature, due to sintering, corresponding to maghemite (γ-Fe$_2$O$_3$: 36°, 44°, 54°, 58° and 63°) and hematite (α-Fe$_2$O$_3$: 24°, 33°, 36°, 50° and 62.5°).

Textural analyses gave rise to the results included in Table 6, showing nitrogen isotherms of type IIb with hysteresis loops type H3, typical of samples with slit-shaped mesopores that extended into the macropore region, caused by the spaces between the plates of material and with no microporosity according to the t-plot analyses [13]. It may be observed that higher calcination temperatures caused a fall in the surface area and a shift in the hysteresis loop to higher relative pressure/wider pores, due to sintering of the samples, in agreement with XRD results.

Table 6. Textural analyses of catalysts

Solid calc. T ⁰C	S$_{BET}$/m^{2}g^{-1}	V$_{mes}$/mLg^{-1}
RHS	85	0.172
RHS HNO₃ 400	93	0.203
RHS HNO₃ 600	93	0.210
4.8%Fe/RHS 400	118	0.200
4.8%Fe/RHS 600	92	0.185
8.9%Fe/RHS 400	122	0.181
8.9%Fe/RHS 600	87	0.163

Most silicas are almost transparent under the electron beam of the transmission microscope and therefore easy to distinguish from oxide particles deposited on their surfaces [24]. However, the RHS particles prepared in this work showed an unexpected dense structure under study by TEM and SEM (Figure 12), formed by lamellar entities, in agreement with the shapes of the nitrogen isotherms, explained above, that made it difficult to distinguish the iron oxide particles.

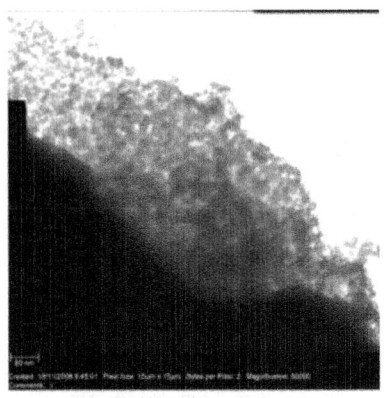

(a) Scale bar: 80 nm

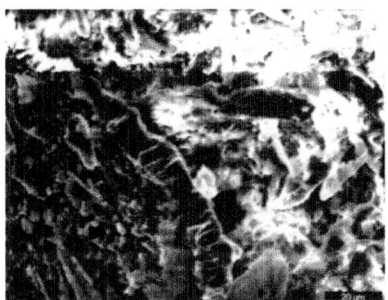

(b) Scale bar: 20 μm

Figure 12. Electron micrographs for 8.9% Fe/RHS 400a. TEM, b. SEM.

The adsorption of acetic acid used to characterize the catalysts basic sites produces mainly carbonate, bidentate or bridged acetate species, that decompose to CO_2, giving rise to bands due to the interaction with basic centres of higher basicity with increasing temperature, i.e. T< 150°C for the desorption of physically adsorbed CO_2, those at T = 150-250°C corresponding to molecules of CO_2 evolved from basic sites of low to medium strength, whilst those at T = 300-400°C correspond to the interaction with basic sites of medium to high strength (Figure 13)[25].

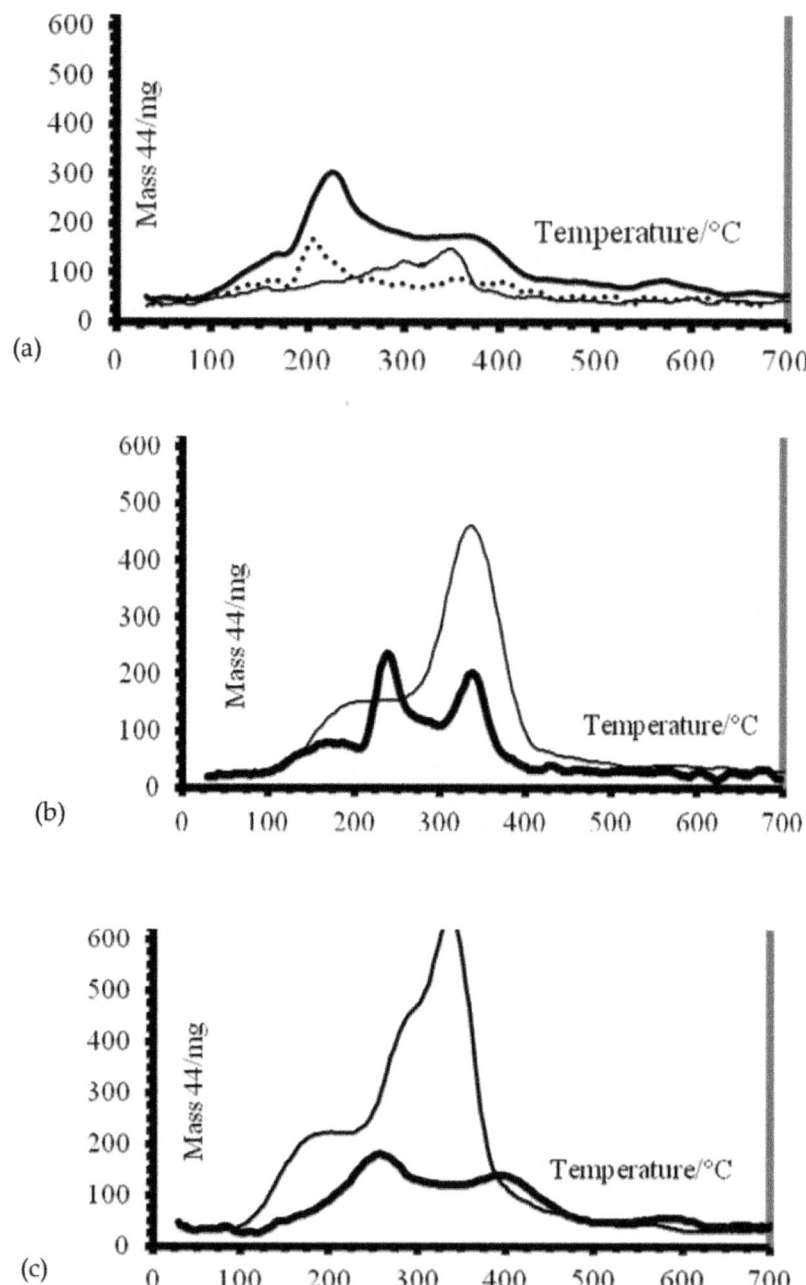

Figure 13. Mass 44/mg evolved from acetic acid decomposition on:4a. Original and modified RHS (dotted line: RHS, thick line RHS HNO$_3$ 400, thin line RHS

HNO_3 600)4b. 8.9%Fe/RHS (thick line calcined at 600°C, thin line calcined at 400 °C)4c. 4.8%Fe/RHS (thick line calcined at 600°C thin line calcined at 400 °C).

From these data it was concluded that the materials employed had the following order of basicity:

Silica Evonik<<< RHS HNO_3 600 < RHS < RHS HNO_3 400

The traces obtained with silica Evonik were within the noise level of the technique, indicating that the acetic acid was only physisorbed, due to its low basicity. These results can be explained taking into account that the basicity, related to both the strength and number of basic sites of RHS was mainly due to the presence of basic oxides. On treating RHS with nitric acid, particles are disaggregated leading to an increase of basic sites available on the surface, but at higher calcination temperatures sintering of the particles occurs decreasing the amount of surface basic centres. The Fe-containing catalysts present higher basicities than the RHS due to the presence of iron oxide, but those calcined at 600°C have lower basicity than their analogues calcined at 400°C, in agreement with the sintering occurring in the particles of iron oxide, XRD, textural and published data [26].

Epoxidation and allylic oxidation reaction pathways of limonene are competitive, with valuable epoxides, carvone and carveol being produced. The selective oxidation of alkenes in the presence of peroxides and basic site containing materials is advantageous, since epoxides undergo breakage on acid sites, which decreases the selectivity of the processes. A thorough analysis of literature on limonene oxidative transformations data provide an insight into the conditions needed to convert limonene into value added and renewable limonene oxides (mono or diepoxides) from epoxidation through electrophilic attack at the double bonds, and allylic oxidation when hydrogen abstraction is the dominant reaction towards carvone and carveol, both interesting molecules, since they retain the olefinic functionality, which allows further useful transformations [27, 28].

The catalytic activity was first measured under dry media conditions, reacting 150 mg of catalysts and 48 μL of limonene:tBHP solution, with different limonene:tBHP volume ratios and temperatures, in order to determine optimal conditions for further liquid phase reactions, that allow treating higher amounts of limonene. Experiments with no control of solid temperature and times over 20 min produced evaporation of reactants, giving rise to irreproducible results, which were avoided by maintaining the temperature of the solid constant throughout the reaction, leaving the microwave power to vary accordingly and limiting the reaction times to 20 min. Under these conditions, the experiments were carried out in duplicate and the results obtained with the Fe-containing catalysts under dry media conditions at 120°C and 20 min irradiation time, with limonene:tBHP solution= 1:1 (volume ratios) are included in Table 7, showing that under equal conditions, the best catalyst for conversion and selectivity to carvone was 8.9%Fe/RHS 400. The catalysts with higher iron content and lower calcination temperatures present higher activities although the selectivities were similar for catalysts with similar iron contents.

With other reaction variables maintained constant, Table 8 shows that by increasing the amount of oxidant the conversion of limonene increases, but that at high amounts the selectivity to carvone starts to decrease and the more oxidised product carvacrol was found. From these results an optimum limonene:tBHP solution=1:3.5 (volume ratio) was chosen.

Table 7. Conversions and selectivities of Fe-containing catalysts under dry media conditions (limonene:tBHP ratio=1:1, 120°C, 20 min, free microwave power, 0.15 mg catalyst)

Catalyst	Conversion (%)	$S_{carvone}$ (%)
No catalyst	0	
4.8%Fe/RHS 400	35	35
4.8%Fe/RHS 600	25	36
8.9%Fe /RHS 400	48	45
8.9%Fe/RHS 600	39	48

S = selectivity

Table 8. Conversions and selectivities of 8.9%Fe/RHS 400 catalyst at 120°C under dry media conditions with different Limonene:tBHP ratios (20 min reaction, free microwave power, 0.15mgs catalyst)

Limonene:tBHP solution	Conversion (%)	$S_{carvone}$ (%)
1:1	48	45
1:2	55	37
1:4	63	23
1:3.5	60	36

With regards to the reaction temperature, from Table 9, a value of 150°C was favoured, since higher temperatures produced over oxidation of carvone, decreasing the selectivity to this compound and at lower temperatures lower conversions were attained.

Table 9. Conversions and selectivities of 8.9%Fe 400 catalyst at different temperatures under dry media conditions (limonene:tBHP solution ratio=1:3.5, 20 min reaction time, free microwave power, 0.15mg catalyst)

Temperature (°C)	Conversion (%)	$S_{carvone}$ (%)
140	45	38
150	60	36
160	70	21

Based on these results, the parameters chosen for the experiments carried out under liquid conditions were a reaction temperature of 150°C, reaction times up to 2 h (with reflux attachment), limonene:tBHP solution=1:3.5 (volume ratio), 18 ml total volume, 0.15 g of catalyst. The experimental results obtained under these conditions are presented in Table 10. The main reaction products were the epoxides, carvone and carveol in different amounts. The epoxide found was mainly the endo, with ratios endo/exo+diepoxide close to 2/1+1.

Repeat experiments with fresh catalysts showed a 1-2% difference in conversions and 2-3% in selectivities in both the dry media and liquid phase conditions. It can be seen that in the absence of catalyst, a conversion of ca. 5% after 30 min of reaction was reached and maintained throughout the experiment, with a selectivity of 91% to limonene oxides and 9% carvone. RHS gave a 12% conversion after 30 min, reaching 22% at 90 min with no further increase by the final reaction time of 2 h. The selectivity to epoxides, after 30 min was higher than 60%, decreasing to *ca* 45% by the end of the reaction, with increasing amounts of carvone and carveol. This reactivity was as expected, due to the intrinsic basicity of the oxides contained in this silica (see Figure 14).

Table 10. Conversions and selectivities under liquid conditions. (Limonene:tBHP ratio=1:3.5, 0.15g catalyst, 150°C, reflux

Catalyst/T_{calc} (°C)	Reaction time min	Conversion %	$S_{epoxides}$ %	$S_{carvone}$ %	$S_{carveol}$ %
Blank 14mL decane	30	5	91	9	0
	60	5	86	14	0
	90	6	85	15	0
	120	6	84	16	0
Silica Evonik	120	6	81	19	0
	30	12	62	24	14
RHS	60	16	50	27	23
	90	22	47	29	24
	120	22	45	31	24
	30	16	74	15	11
RHS HNO₃ 400	60	23	59	19	22
	90	33	55	24	21
	120	35	49	30	21
	30	16	69	31	0
RHS HNO₃ 600	60	18	55	31	14
	90	24	50	34	16
	120	25	43	35	22
	30	17	76	14	10
4.8%Fe/RHS 400	60	23	72	16	12
	90	35	67	18	15
	120	37	55	26	19
	30	16	77	13	10
4.8%Fe/RHS 600	60	20	69	18	13
	90	29	61	22	17
	120	30	54	27	19
	30	16	88	12	0
8.9%Fe/RHS 400	60	25	68	19	13
	90	41	64	21	15
	120	43	58	23	19
	30	19	73	16	11
8.9%Fe/RHS 600	60	21	68	18	14
	90	31	63	20	17
	120	33	54	24	22
8.9%Fe/RHS 400ᶜ	30	5	75	20	5
	60	9	70	21	9
	90	20	65	22	10
	480*	22	55	20	10

ᶜ Conventional heating (* Other products found, mainly carvacrol)

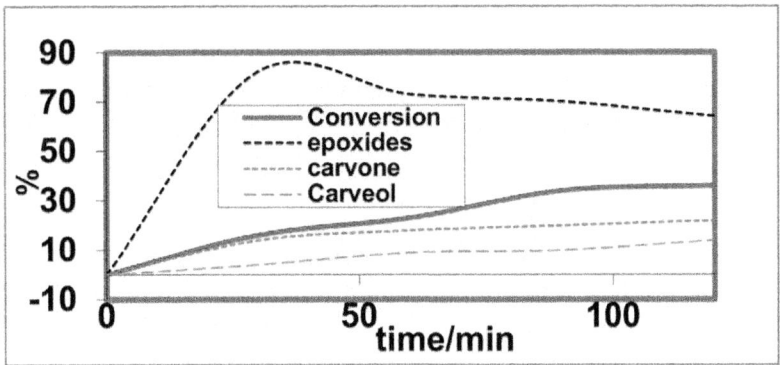

Figure 14. Conversions and selectivities of 8.9%Fe/RHS with orange peel oil, under the conditions shown in Table 10.

The reactivity of RHS HNO$_3$ 400 was higher than that of RHS, reaching *ca* 34% after 90 min, in agreement with the higher basicity, surface area and pore volume of the former. The reactivity of RHS HNO$_3$ 400 was higher than that of RHS HNO$_3$ 600, for the same reasons. The activity of commercial silica AerosilEvonik, chosen for its similar textural properties (91 m^{2}g^{-1}), to RHS but with no basic oxides content was practically that observed with no catalyst, showing the importance of basic sites on the reaction. With regards to the catalytic activity of the Fe-containing catalysts, higher conversions were reached with those catalysts prepared with greater amounts of iron and lower calcination temperatures, as seen in the dry media results, in agreement with their higher amounts of basic sites (Figure 13) and better accessibility, as shown in the study of their textural properties (Table 7).

The low reaction times needed in this work due to optimization of the catalysts and dielectric activation avoids the formation of undesired products, such as polymers sometimes found in the oxidation of limonene at longer reaction times [24]. These workers find one of the technological advantages to be that, since the final mixtures contain high concentrations of valuable oxygenated products of interesting organoleptic properties, they can be used directly, for example in fragrance compositions where separation is unnecessary.

As in Menini´s work [29] at ca. 40% conversion, the reactions become stagnant. Addition of fresh catalyst to the mixture of the substrate and the products does not promote further conversion of the substrate, which seems to indicate that the products accumulated are probably acting as radical scavengers. No iron leaching occurred under the reaction conditions and the catalysts could be recovered by thorough washing with decane and reused at least three times with no appreciable loss of activity. Under the conditions used in Table 10, the behavior of catalyst 8.9%Fe 400 was determined:

- adding a small amount (70 µg) of radical scavenger hydroquinone
- with conventional heating
- with orange oil, instead of limonene

Addition of hydroquinone greatly decreased the activity, down to that found with no catalyst, indicating the radical nature of the reaction. Under conventional heating after 2 h refluxing, only 22% conversion was reached with 55% selectivity to the epoxides, 20% to carvone, 10% to carveol and 15% to carvacrol, the latter compound due to the extended oxidation under the conditions used. When orange oil was used (Figure 14), the compounds present other than limonene (myrcene, alpha pinene, beta pinene, linalool and decanal) were recovered unreacted and thus the results were promising, especially bearing in mind its lower price, compared to commercial limonene.

Comparing the results obtained here with some from the literature, it can be seen that optimised mixed oxides of iron, cobalt and manganese manage to convert 0.5 ml of limonene after 7 h with an oxygen atmosphere, whereas in this work 1.6 ml limonene were converted with tBHP in 1.5 h. As in our work, these authors obtained as products limonene epoxides, carvone and carveol, with selectivity towards limonene oxide, although the presence of cobalt in the composition of those catalysts had a negative influence on their environmental impact [30,31]. Similar conversions were found with V_2O_5/TiO_2 and tBHP. Again, vanadimpentoxide is more toxic than the iron oxide used in our work. Furthermore, limonene glycol and polymers decrease the selectivities to the desired compounds in that work [32]. Laborious and time consuming preparation and optimisation of synthetic hidrotalcites with immobilised palladium and copper in their compositions achieve similar conversions and selectivities to those found here, however reaction times up to 6 h were required in that case and palladium and copper, although quite reactive, are not very environmentally friendly [33].

The main advantage of our process is that the renewable raw materials used for the preparation of the catalysts, with adequate design, can give rise to catalysts that can compete with synthetic ones, prepared usually in more expensive and time consuming ways and therefore less environmentally friendly [34].

THE USE OF AGRICULTURAL RESIDUES TO IMPROVE THE TEXTURAL CHARACTERISTICS OF STRUCTURED SOLIDS TO DECONTAMINATE EFFLUENTS

Atmospheric contamination from industrial processes that contain VOCs which lead to the formation of photochemical smog is of great important with respect to air quality. The most usual remedy for this type of pollution are the so called "end of pipe" methods in which catalysts, adsorption beds or a combination of the two are employed. When large volumes of gas are to be treated, the conformation of the catalysts into open channel monoliths or extrudates is a necessity to avoid or reduce the pressure drop across the catalytic bed. However, if incorporated catalysts are used due to their robust nature and abrasion resistance a reduced activity compared to powder catalysts of similar composition is generally encountered due to diffusion limitations of the gas to be treated into the conformed catalyst. This limitation can be reduced by controlling the texture, surface area and porosity, of the incorporated catalyst.

Toluene was chosen as a typical aromatic VOC to be eliminated by its total oxidation from a spiked air stream. The subproducts from the rice industry chosen to modify the physical characteristics of the conformed catalysts were rice husk ash (RHA), from burning the husks to produce electricity, with the added advantage of drastically reducing the volume of this residue, and rice bran (Bran) that is separated from the rice grain during the whitening process. The Bran is a Pore Generating Agent (PGA) that during the firing of the extruded green body at 500°C to form the conformed ceramic is burnt out with the subsequent increase in the overall porosity of the structure. Whilst the RHA was used to both modify the porosity of the structure and hopefully increase the mechanical strength of the ceramic due to its high silica content [35].

Different compositions and textural properties can be achieved by careful design of the structured solids prepared using rice production wastes(rice husk ash and rice bran). Optimization of these processes lowers the temperature and therefore the economic expenditure for decontamination of effluents spiked with toluene, chosen as standard for comparison purposes and for being a classical example of a VOC. The textural developement of the solids is a key point to change their behaviour, where the use of rice wastes are a convenient, cheap and ecologic way to improve the activities of the final materials.

The clay used as the agglomerating agent and as the final support for the catalyst was α-sepiolite (Sep), due to its exceptional rheological properties that allow extrusion of the paste produced by mixing with water [36] and the formation of a stable ceramic body with an acceptable strength when heat treated at temperatures above 330°C. Iron nitrate was used as the precursor to the catalyst due to its low cost, toxicity and environmental impact [37]. Four incorporated catalysts [38] were prepared in which the iron to sepiolite ratio was maintained constant but different amounts of Bran and RHA were also incorporated into the original paste before mixing, extrusion and firing at 500°C. Heat treatment at 500°C was chosen since at this temperature the sepiolite forms a ceramic structure that binds all of the other components in the extruded paste. Also this temperature is sufficient for the decomposition of the iron salt to iron oxide which is complete by 200°C, and the thermal decomposition of the Bran. The catalytic activity studies were carried out by passing an air flow spiked so as to contain 100 ppm of toluene through the catalyst bed with a Gas Hourly Space Velocity (GHSV) similar to that expected for the decontamination of an industrial effluent stream. The temperature of the catalyst bed was raised from ambient to 500°C in a number of steps, maintaining each for a period of 20 min to ensure that the result was in equilibrium and monitoring the destruction of the organic by the use of a flame ionisation detector. The gas balance was confirmed by the amount of CO_2 produced, monitored by infra red spectroscopy.

The original compositions of the green bodies and the textural, mechanical and catalytic activities of these materials after heat treatment at 500°C to produce the conformed ceramic catalyst as a solid extrudate with a diameter of 3 mm are collated in Table11.

Table 11. Compositional and Textural Characteristics of the catalysts heat treated at 500°C

Sample	Composition Sep/Fe/Bran/RHA Parts by Weight	Crushing Strength MPa	Surface Area m^2g^{-1}	Pore Volume cm^3g^{-1}	Catalytic Activity T_{50} °C
1	17/1/0/0	18.3	143	0.576	382
2	17/1/2/0	13.8	140	0.734	350
3	17/1/0/2	19.0	125	0.656	377
4	17/1/2/2	15.8	124	0.797	343

From the results presented in Table 11 it may be observed that the mechanical strength of the materials was reduced by the incorporation of Bran into the original paste before extrusion and its subsequent removal by heat treatment at 500°C comparing the result obtained with samples 1 and 2 and that of samples 3 and 4. This was to be expected since with the removal of the PGA on firing at 500°C there is a significant increase in the total pore volume. The mechanical strength of brittle ceramics is related to both the total pore volume and the pore size distribution[39] where an increase in the pore volume or the size of the pores is detrimental to the strength development. From the porosimetry curves presented in Figure 15 it may be seen how the thermal decomposition of the Bran led to both an increase in the total pore volume and a shift to wider pores. Thus, sample 1 had a monomodal pore size distribution curve with practically the whole pore volume located in pores of less than 200 nm. When Bran was incorporated and subsequently burnt out of the conformed material there was a 27% increase in the total pore volume located in the pores between 100 nm and 1 μm.

With the incorporation of RHA the most obvious change in the porosity of sample 3 compared to that of sample 1 was the development of a bimodal curve with the pores close to 900 nm being due to the particle size of the RHA. The slight reduction in the porosity in pores up to about 100 nm was due to the reduction in the amount of Sep per gramme of material. The effects of the incorporation of both RHA and Bran may be seen by comparing the curves obtained with sample 3 with that of sample 4. The curves were similar up to about 100 nm then their was a gradual increase in the pore volume that reached about 21% and a shift in the diameter of the wider pores to approximately 1.5 μm.

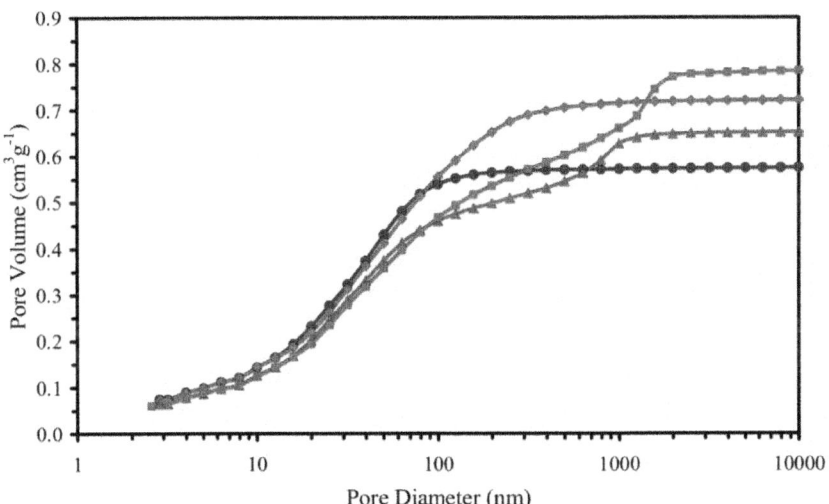

Figure 15. Cumulative pore volume curves for samples heat treated at 500°C: 1 ●, 2 ♦, 3 ▲ and 4 ■.

The catalytic oxidation activity of the four samples is presented in Figure 16 where it may be seen that the least active was that based on sample 1 which had the lowest total pore volume and narrowest pores. With the introduction of Bran into the original composition and its subsequent removal by heat treatment at 500°C which led to a significant increase in both the pore volume and the connectivity the temperature at which 50% of the toluene was decomposed (T_{50}) was reduced by 32°C. For sample 3, although the pore volume and width of the largest pores was increased, which should have lead to less diffusion limitation, the T_{50} was only reduced by 5°C. The highest catalytic activity was found with sample 4 which had the advantage of the highest pore volume and the widest pore diameters due to the presence of both Bran and RHA in the original composition. The T_{50} was reduced by 39°C and the material also had a respectable mechanical strength due to the incorporation of the RHA.

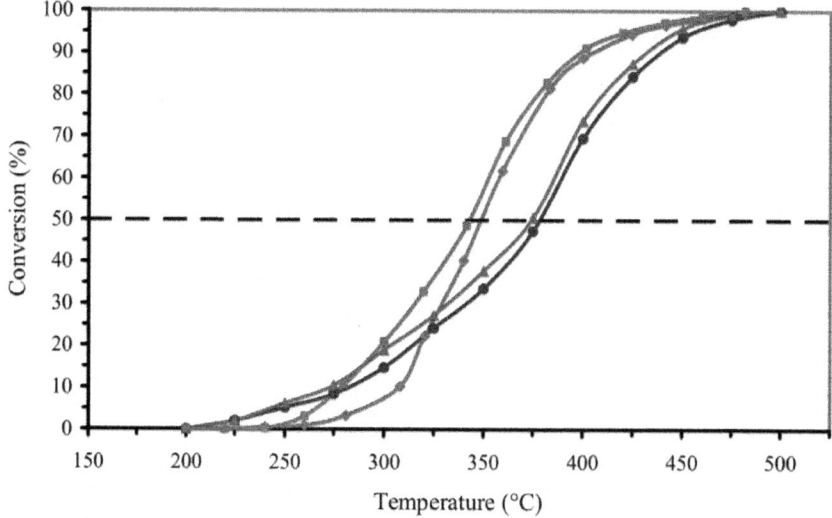

Figure 16. Catalytic activity curves for samples heat treated at 500°C: 1 ●, 2 ♦, 3 ▲ and 4 ■.

PREPARATION OF RENEWABLE BIOCOMPATIBLE SCAFFOLDS FOR BONE REPLACEMENT THERAPY

Under this heading we present how agricultural wastes have been analyzed, modified and changed in order to have properties similar to bone growth scaffolds. SEM micrographs (Figure 17) allow the observation of the porous nature of the inorganic skeleton that remains after the treatment of the waste, and its similarity to bone.

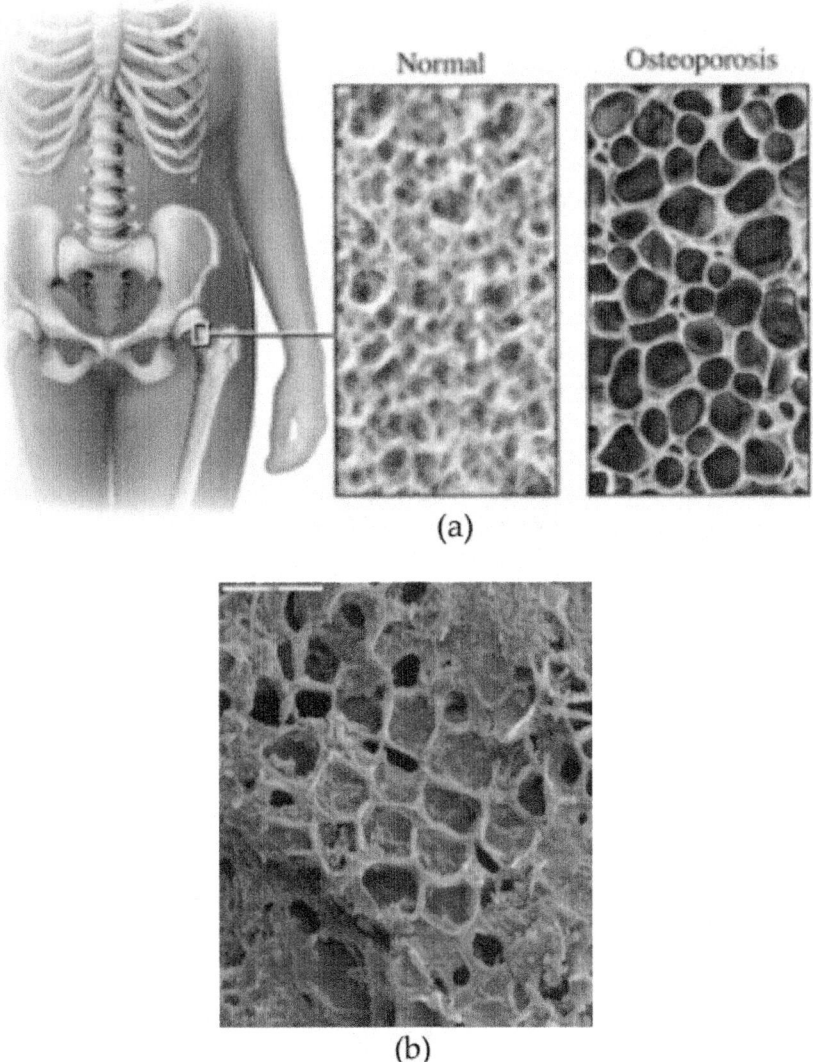

(a)

(b)

Figure 17. Scanning electron micrograph of treated beer bagasse (scale bar 20 μm) and similarity with bone structure.

XRD analysis of these materials show tricalcium phosphate and calcium silicate of structure and composition similar to the synthetic materials presently used in bone and tooth replacement (Figure 18). These materials have proven to be biocompatible and capable of promoting bone growth as confirmed by similar growth rates in vitro of osteoblasts to those of hydroxyapatite used as standard.

It is estimated that approximately a million bone grafts are performed each year to treat bone defects resulting from trauma and diseases in the United States. Various strategies have been used to solve this problem. Autografts are used to treat these defects, but available bone can be inadequate and difficult to shape and obtain. Allografts and xenografts must be processed to eliminate the risk of transmission of live viruses [40]. These difficulties have been the impetus for research into a variety of bone grafting materials.

Bone tissue engineering techniques have become an expanding research area in regenerative medicine. Bone and tooth replacement require materials that act as scaffolds or artificial extra-cellular matrix, directing tissue formation and allowing the transport of biological nutrients to restore the structure and function of damaged tissues. These materials, which require tailored porosity, surface chemistry, and mechanical strength, are typically produced from animal bone, organic oil-derived polymers, inorganic materials, or complex mixtures of all the above [41]. Figure 19 includes the mercury intrusion porosimetry results for a beer bagasse derived bioecomaterial.

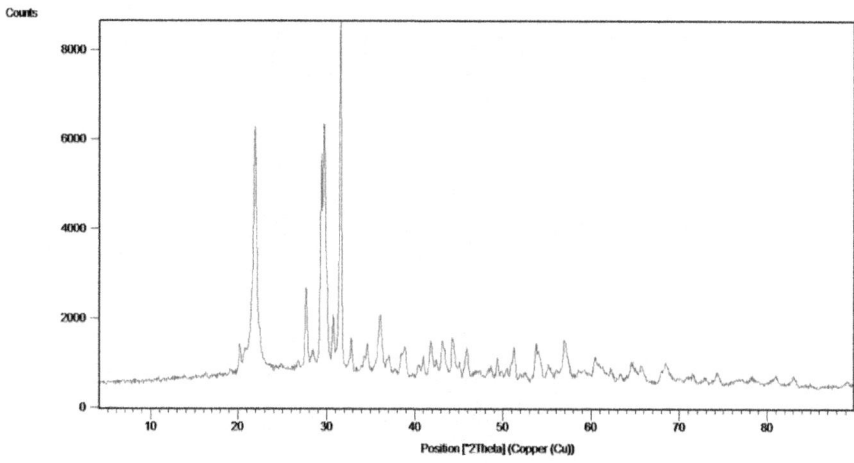

Figure 18. X-ray diffraction of beer bagasse based Bioecomaterial.

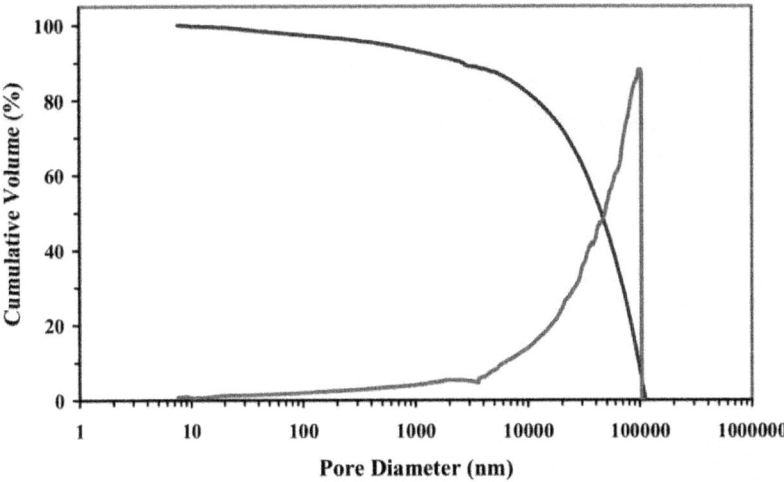

Figure 19. Cumulative Pore Volume (blue line) and pore size distribution(red line) of beer bagasse based bioecomaterial.

The SEM image of the treated beer bagasse (Figure 17b) displayed pores of about 10μm. However, from the mercury intrusion porosimetry curve presented in Figure 19 it may be appreciated that a continuously increasing curve was observed from the largest pores down to those at the physical limit of this technique (4 nm). Thus, although the cumulative intruded volume in the largest pores was due to filling of the intraparticulate space, while at diameters below 10 μm the porosity must be due to the interparticulate porosity. However, what is most important from this figure is that the porosity between the particles and within them formes a continuous network that favours cell growth when used as a scaffold for bone regeneration.

Materials used as bone substitutes need to embrace several important requirements:

i.   biocompatibility;

ii.   osteoinduction and osteopromotion,

iii.   porosity,

iv.   stability under stress,

v.   resorbability and degradability,

vi.   plasticity,

vii.   sterility and

viii.   stable and long-term integration of implants [42].

All these requirements serve as a basis for effective long-term tolerance in bone replacement therapies [43].

The first approach in the selection of a suitable scaffold for bone regeneration is the evaluation of its in vitro cytotoxicity using appropriate cell lines in culture, such as osteoblasts. Thereby, we can determine the biocompatibility of each scaffold under study. A good biocompatibility could be expected from materials whose physicochemical properties promote cell adhesion and differentiation into mature osteoblasts.

In the present study we aimed to develop scaffolds for bone regeneration obtained from agricultural wastes from several crops. These materials have proved to be good replacement candidates for use as biomaterials for the growth of osteoblasts and could be used in bone replacement therapies.

In this study, beer bagasse, a low-cost waste material from beer production, was employed as a renewable raw material that was tailored for use as biocompatible scaffolds for osteoblast growth, given its structure and composition [44-46]. To the best of our knowledge this is the first time that agricultural waste has been modified to produce solids that can act as scaffolds for tissue regeneration.

To study cell adhesion to the materials, their biocompatibility and efficiency for bone growth, MC3T3-E1, an osteoblast-like cell line, was chosen because they are well characterised for modelling endogenous osteoblasts [47]. A commercial synthetic scaffold, hydroxyapatite (HA) nanopowder (B200 nm) from Sigma Aldrich, was used for comparison purposes.

The osteoblastic cells were seeded onto material-coated plates in DMEM-10%FBS and allowed to adhere for 2 h at 37°C in a 5% CO_2 atmosphere. Plates were then washed with phosphate-buffered based saline solution (PBS) to remove non-adherent cells. Adherent live cells were quantified using the live/dead viability assay kit (Molecular Probes [40]). The number of live adhered cells was evaluated 2h after seeding by fluorescence microscopy, counting at least 10 representative fields per well. The cell adhesion process was studied after 2h of incubation at 37°C. The adhesion of MC3T3E1 cells to the various materials evaluated is shown in Figure 20. As can be seen, the number of osteoblast cells which adhered to the materials derived from beer bagasse was comparable to that of the commercial material used as control (HA). The cell adhesion rate was even higher in the case of sample BB47. However, the adhesion of MC3T3E1 cells to the materials was slightly lower than that on plastic, especially for cells seeded on HA and BB48. Assuming 100% as the rate of cell adhesion for the commercial material (HA), those for BB47, BB48 and BB410 were found to be 141.5%+/- 6.1, 92.5%+/-7.4 and 118.1%+/-5.8, respectively. Thus, confirming similar adhesion rates for the studied materials, demonstrating that they possess an appropriate porosity to allow cell adhesion.

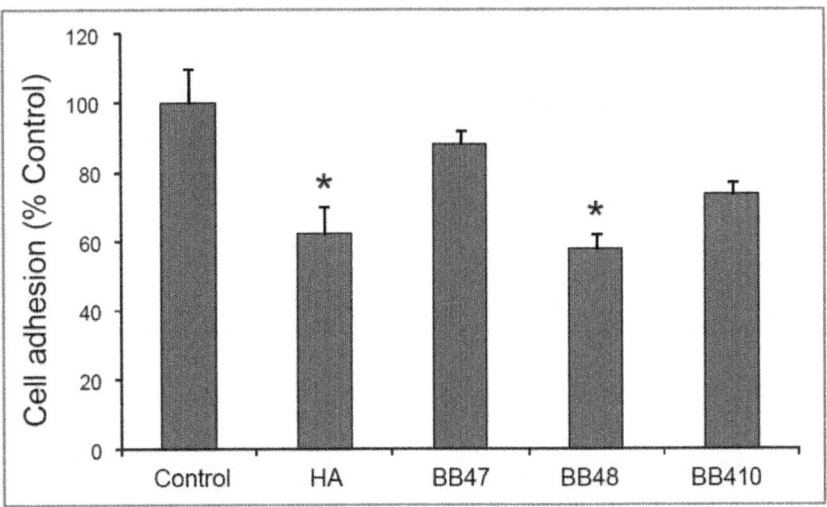

Figure 20. Adhesion of MC3T3E1 cells. Cells were plated on a plastic control dish (Control) and on plates coated with the control material (HA) and with materials derived from beer bagasse (BB47, BB48, BB410). Cells were incubated for 2h at 37°C. Then, adherent live cells were quantified as described in the text. Results are expressed as a percentage of the control (cells seeded on plastic plates). Data represent mean+/-S.E.M. of three independent experiments. ($p < 0.05$, ANOVA, post-hoc Tukey HSD test, * vs. Control).

The biocompatibility of the materials derived from beer bagasse, was assessed with the same MC3T3-E1 cells after different periods of exposition, comparing the results to those obtained with the commercial material HA and to regular tissue culture polystyrene plates, used as controls. The viability of MC3T3-E1 cells growing on plastic plates, BB47, BB48, BB410 and the commercial material HA was determined at one, three and seven days after seeding by the live/dead viability assay [48], to distinguish dead cells (red) from live ones (green), as observed by inverted fluorescence microscopy (Figures 21 and 22).

Calculating the percentage of live cells over total cells (live cells plus dead cells), a high cell viability rate was observed for all scaffold materials produced from beer bagasse, similar to that of the commercial product HA and the control cell culture plastic plates (Figure 21).

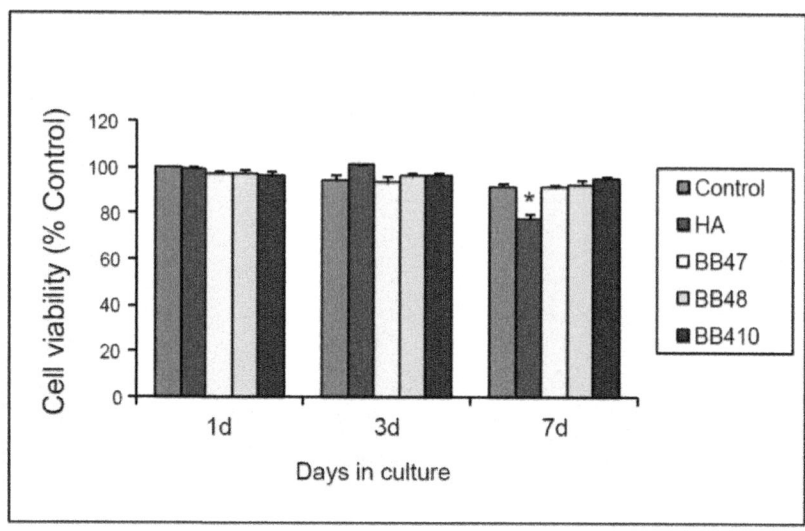

Figure 21. MC3T3-E1 cell viability after 1, 3 and 7 days growing on plastic plates or on plates coated with HA, BB47, BB48 and BB410. Cell viability was evaluated by live/dead viability assay kit. Data represent mean+/-S.E.M. of three independent experiments. ($p<0.05$, ANOVA, post-hoc TukeyHSD test, * vs. Control).

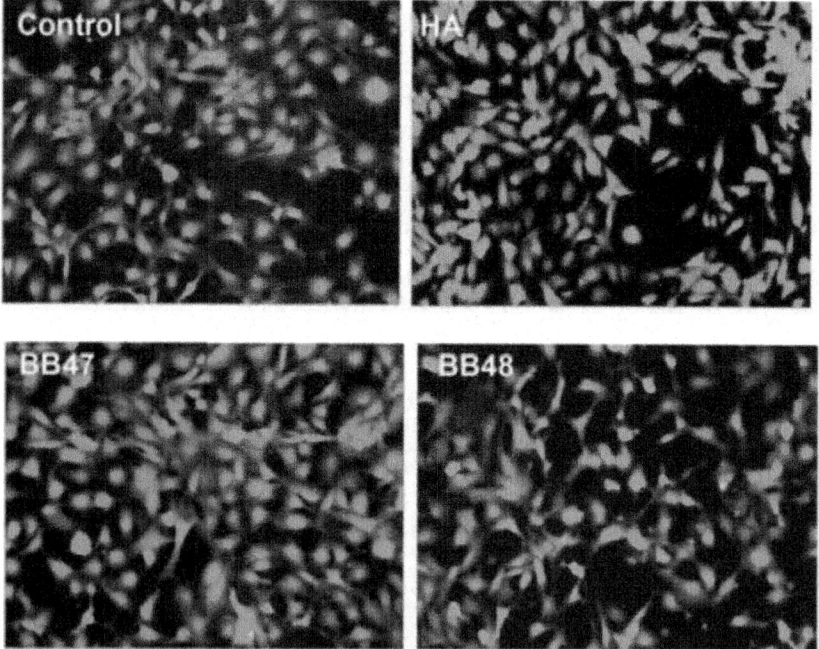

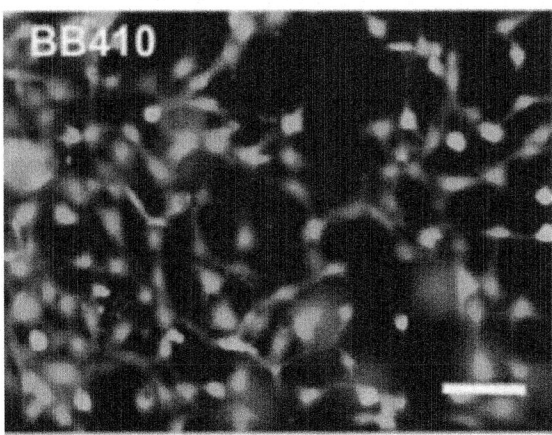

Figure 22. Fluorescence microscopy images showing MC3T3-E1 cells growing on different materials 3 days after seeding and stained with Live (green)-Dead (red) viability assay kit. Scale bar: 100µm.

None of the materials analysed resulted in a reduction of cell viability at any time points, more than 90% of the cells growing on these materials remained viable over the whole evaluation period. Only cells growing on HA for 7 days showed a significant reduction in the rate of cell viability.

Since bone formation is a lengthy process, the most relevant aspect is that even after seven days of culture, all materials displayed similar biocompatibilities rates to those observed on the commercial sample and the plastic plates.

In summary, murine-derived osteoblastic cells (MC3T3-E1) actively adhereto the beer bagasse-derived materials, with no significant changes in cell viability of cells growing on these materials compared to those observed with either the commercial material or the cells seeded directly onto plastic plates. Such characteristics are desired in dental and orthopaedic prostheses [49].

We conclude that our renewable raw material (RRM) are scaffolds that have the right characteristics to support adhesion and survival of the osteoblast-derived cell line MC3T3E1, strongly suggesting that they could be employed in oral and/or bone tissue regeneration.

CONCLUSIONS

1. Agro-industrial wastes are a renewable source of many solid and liquid substances and furthermore, their use leads to a reduction in environmental hazards. These wastes are an inexhaustible and sustainable source of materials and substances in a biorefinery based economy that

is constantly gaining interest in industrial sectors, especially because the actual petroleum based will come to an end.

2. Some agriculturalwastes, can be used as catalysts for production of fine chemicals. For example, silica prepared from rice husk, with a lamellar structure, basicity due to oxides from its natural origin and unusual high density, can act both as support and catalyst for the oxidation of limonene. Also, sustainable chemicals can be produced from agricultural wastes.

3. The use of agricultural wastes allows the production of extrudates that can be employed to optimise cleaning of contaminated effluents. The catalytic activity of conformed catalysts may be significantly enhanced by the use of low cost sub-products and residues from the rice production industry. Rice bran is a useful PGA that improves the connectivity of the porous matrix in these incorporated catalysts while the RHA both improves the strength development of the ceramic body while also improving the texture and thus reducing the diffusion limitations of the gas to be treated into the porous structure of the catalyst.

4. Some waste derived solids, after the appropriate pretreatment, can be used for bone growth. The biocompatibility of some modified agro-industrial wastes has shown to be interesting in their use in fields like release of medicines, supports for enzymes and food supplements. Recent studies "in vivo" are showing promising results for these materials as economical sustainable scaffolds to be used in tissue engineering.

REFERENCES

1. K. Kem-Laurin, 2012Approaches to A Sustainable User ExperienceUser Experience in the Age of Sustainability, 3168

2. K. Mahesh, Ravindra. M. Utkarsh, M. Adinpunya, 2012Rapid separation of carotenes and evaluation of their in vitro antioxidant properties from ripened fruit waste of Areca catechu- A plantation crop of agro-industrial importance. Ind. Crops and Products, 40204209

3. Brady NC, Weil, RR1999The Nature and Properties of Soils (12th ed), Prentice Hall, Upper Saddle River, New Jersey.

4. B. Notarnicola, H. Kiyotada, Huisingh. D. Curran, 2012Progress in working towards a more sustainable agri-food industry.J. of Cleaner Production, 2818

5. Karlen DL, Ditzler CA, Andrews SS2003Soil quality: why and how?Geoderma, 114145156

6. S. L. P. Cogo, F. C. Chaves, Zambiazi. R. C. Schirmer, L. Nora, J. A. Silva, C. V. Rombaldi, 2011Low soil water content during growth contributes to preservation of green colour and bioactive compounds of cold-stored broccoli (Brassica oleraceae L.) florets.Postharvest Biology and Technology, 60158163

7. S. Pérez-Balibrea, D. A. Moreno, C. García-Viguera, 2011Improving the phytochemical composition of broccoli sprouts by elicitation.Food Chemistry, 1293544

8. A. Vallverdú-Queralt, A. Medina-Remón, I. Casals-Ribes, R. M. Lamuela-Raventos, 2012Is there any difference between the phenolic content of organic and conventional tomato juices? Food Chemistry, 130222227

9. Salgado JL, AIZCE Technical Committee (Interprofessional association of Citric Juices and concentrates of Spain).

10. Yates. M. Martin-Luengo, Domingo. Martínez, B. Casal, M. Iglesias, M. Esteban, E. Ruiz-Hitzky, (200, 2008Synthesis of p-cymene from limonene, a renewable feedstock.Appl. Catal. B: Env. 81218224

11. S. M. Aschmann, J. Arey, R. Atkinson, 2011Formation of p-cymeme from OH + y-terpinene: H-atom abstraction from the cyclohexadiene ring structure.Atmospheric Env. 4544084411

12. Yadav GD, Purandare SA2007Vapor phase alkylation of toluene with 2-propanol to cymenes with a novel mesoporous solid acid UDCaT-4Microp. and Mesop. Materials, 103363372

13. F. Rouquerol, J. Rouquerol, K. S. W. Sing, 1999Adsorption by Powders and Porous Solids: Principles, Methodology and Applications.

14. Lavalley JC1996Infrared spectrometric studies of the surface basicity of metal oxides and zeolites using adsorbed probe moleculesCatal. Today 27377401

15. C. Leyva, J. Ancheyta, A. Travert, F. Maugé, L. Mariey, J. Ramírez, (201. Rana, 2012Activity and surface properties of NiMo/SiO2Al2O3 catalysts for hydroprocessing of heavy oils.Appl. Catal. A: Gen. 425-426: 1-12.

16. C. Fernandes, C. Catrinescu, P. Castilho, P. A. Russo, M. R. Carrott, C. Breen, 2007Catalytic conversion of limonene over acid activated Serra de Dentro (SD) bentoniteAppl. Catal. A: Gen. 318108120

17. C. Catrinescu, C. Fernandes, P. Castilho, C. Breen, 2006Influence of exchange cations on the catalytic conversion of limonene over Serra de Dentro (SD) and SAz-1 clays: Correlations between acidity and catalytic activity/selectivity. Appl. Catal. A: Gen 311172184

18. R. C. S. Schneider, V. Z. Baldissarelli, M. Martinelli, Holleben. von, E. B. Carama, E. B. , 2003Determination of the disproportionation products of limonene used for the catalytic hydrogenation of castor oil.J. Chromatogr. A 985313319

19. Martin-Luengo MA, Pajares JA, González TejucaL1985Particle size determination of palladium supported on sepiolite and aluminum phosphate.J. of Colloid and Interface Sci. 107540546

20. Oh S-T, Kim M-S, Choa Y-H, Kim KH, Lee S-K2012Preparation of Fe-50 wt.% Co nanopowders by calcination and hydrogen reduction of nitrate powders. Microelectronic Eng.899799

21. W. Xu, X. Liu, J. Ren, H. Liu, Y. Wang, Y. Lu, G. , 2011Synthesis of nanosizedmesoporous Co-Al spinel and its application as solid base catalyst. Microp. andMesop. Materials 142251257

22. C. Gervasini, P. Messi, A. Carniti, N. Ponti, F. Ravasio, F. Zaccheria, (200, 2009Insight into the properties of Fe oxide present in high concentrations on mesoporous silicaJ. of Catal. 262224234

23. C. Zhao, W. Gan, X. Fan, Z. Cai, P. J. Dyson, Y. Kou, 2008Aqueous-phase biphasic dehydroaromatization of bio-derived limonene into p-cymene by soluble Pdnanocluster catalysts.J.Catal. 254244250

24. Lee OC, Oh Y-G. and S-G2009Synthesis and characterization of hollow silica microspheres functionalized with magnetic particles using W/O emulsion methodColl. and Surf. A: Physicochem. and Eng. Aspects 337208212

25. M. X. Baraton, X. Chen, K. E. Gonsalves, 1997Ftir study of a nanostructured aluminum nitride powder surface: Determination of the acidic/basic sites by CO, CO2 and acetic acid adsorptions.Nanostr. Mater. 8435445

26. P. F. Ross, G. Busca, V. Lorenzelli, M. Lion, C. Lavalley, 1988Characterization of the surface basicity of oxides by means of microcalorimetry and fourier transform infrared spectroscopy of adsorbed hexafluoroisopropanol., J. Catal. 109378386

27. P. Oliveira, A. Machado, A. M. Ramos, I. Fonseca, F. M. B. Fernandes, Rego. A. M. Botelho, J. Vital, 2009MCM-41 anchored manganese salen complexes as catalysts for limonene oxidation Microp. andMesop. Materials 120432440

28. Yates. M. Martin-Luengo, M. Diaz, Rojo. E. Saez, Gil. L. Gonzalez, 2011Renewable fine chemicals from rice and citric subproducts: EcomaterialsAppl. Catal. B: Env. 106488493

29. L. Menini, M. C. Pereira, L. A. Parreira, Gusevskaya. E. V. Fabris, 2008Cobalt- and manganese-substituted ferrites as efficient single-site heterogeneous catalysts for aerobic oxidation of monoterpenic alkenes under solvent-free.J. Catal. 254: 355364

30. Menini L,Pereira MC, Parreira LA, Fabris JD,Gusevskaya EV2008Cobalt- and manganese-substituted ferrites as efficient single-site heterogeneous catalysts for aerobic oxidation of monoterpenic alkenes under solvent-free conditionsJ. Catal. 254355364

31. Robles-Dutenhefner PA, da Silva MJ, Sales LS, Sousa EMB, Gusevskaya EV2004Solvent-free liquid-phase autoxidation of monoterpenes catalyzed by sol-gel Co/SiO2J. Mol. Catal. A: Chem. 217139144

32. P. Oliveira, M. L. Rojas-Cervantes, A. M. Ramos, I. M. Fonseca, Rego. A. M. Botelho, J. Vital, 2006Limonene oxidation over 2O5TiO2 catalysts. Catal. Today 118: 307-314.

33. J. Bussi, A. López, F. Peña, P. Timbal, D. Paz, D. Lorenzo, E. Dellacasa, (200, 2003Liquid phase oxidation of limonene catalyzed by palladium supported on hydrotalcitesAppl. Catal. A: Gen. 253177189

34. M. Guidotti, N. Ravasio, R. Psaro, G. Ferraris, G. Moretti, 2003Epoxidation on titanium-containing silicates: do structural features really affect the catalytic performance?J. Catal. 214242250

35. Jimmy Nelson Appaturi, Farook Adam, ZakiaKhanam2012A comparative study of the regioselective ring opening of styrene oxide with aniline over several types of mesoporous silica materialsMicroporous and Mesoporous Materials1561621

36. Murray HH1991Overview- clay mineral applications. Applied Clay Science 5379395

37. Kim SC, Shim WG2008Influence of physicochemical treatments on iron-based spent catalyst for catalytic oxidation of tolueneJournal of Hazardous Materials154310316

38. Cybulsky Aand Moulin JA1998Structurated Catalysts and Reactors. Ed. Marcell and Dekker, Inc New York.

39. M. Yates, 2006Application of mercury porosimetry to predict the porosity and strength of ceramic catalyst supportsParticle & Particle Systems Characterization2394100

40. C. G. Simon, W. F. Guthrie, F. W. Wang, 2004Cell Seeding into Calcium Phosphate Cement.J. Biomed. Mater. Res. Part A 68A: 628639

41. H. Cao, N. A. Kuboyama, 2010Biodegradable Porous Composite Scaffold of PGA/b-TCP for Bone Tissue Engineering. Bone386395

42. A. Kolk, J. Handschel, W. Drescher, D. Rothamel, F. Kloss, M. Blessmann, M. Heiland, K. D. Wolff, R. Smeets, 2012Currenttrends and futureperspectives of bone substitutematerials- From space holders to innovative biomaterials. J Craniomaxillofac Surg. (in press, corrective proof, available online 30th Jan)

43. H. H. Horch, R. Sader, C. Pautke, A. Neff, H. Deppe, A. Kolk, 2006Synthetic, pure-phase beta-tricalcium phosphate ceramic granules (Cerasorb) for bone regeneration in the reconstructive surgery of the jaws. Int. J. Oral MaxillofacSurg, 35708713

44. Prashanth. K. V. Harish, R. N. Tharanathan, 2007Chitin/Chitosan: Modifications and Their Unlimited Application Potential-an Overview. Trends Food Sci. Technol. 18117126

45. Seadi. T. Al, B. Holm-Nielsen, 2004III. 2 Agricultural Wastes. Waste Management Series 4207215

46. Luengo. M. A. Martin, M. Yates, B. Casal, 2010Preparation of Biocompatible Materials from Beer Production and Their Uses. Spanish patent PCT/ES2009/070475.

47. F. Chen, H. Ouyang, X. Feng, Z. Gao, Y. Yang, X. Zou, T. Liu, G. Zhao, T. Mao, Anchoring Dental Implant in Tissue-Engineered Bone Using Composite Scaffold: A PreliminaryStudy in Nude Mouse Model. J. Oral Maxillofac. Surg. 2005200563586595

48. Y. F. Chou, W. Huang, J. C. Dunn, T. A. Miller, Wu, 2005The Effect of Biomimetic Apatite Structure on Osteoblast Viability, Proliferation, and Gene Expression.Biomaterials285 EOF95 EOF

49. S. Bertazzo, W. F. Zambuzzi, Silva. H. A. da, C. V. Ferreira, C. A. Bertran, Bioactivation of Alumina by Surface Modification: A Possibility for Improving the Applicability of Alumina in Bone and Oral RepairClin. Oral Implants Res. 2009288 EOF293 EOF

Chapter 2

SUSTAINABILITY ASSESSMENT OF CHEMICAL PROCESSES: EVALUATION OF THREE SYNTHESIS ROUTES OF DMC

Paula Saavalainen[1], Satish Kabra[2], Esa Turpeinen[1], Kati Oravisjärvi[1], Ganapati D. Yadav[2], Riitta L. Keiski[1] and Eva Pongrácz[3]

[1]Environmental and Chemical Engineering, Faculty of Technology, University of Oulu, 90014 Oulu, Finland

[2]Department of Chemical Engineering, Institute of Chemical Technology (ICT), Matunga, Mumbai 400019, India

[3]Thule Institute, NorTech Oulu, University of Oulu, 90014 Oulu, Finland

ABSTRACT

This paper suggested multicriteria based evaluation tool to assess the sustainability of three different reaction routes to dimethyl carbonate: direct synthesis from carbon dioxide and methanol, transesterification of methanol and propylene carbonate, and oxidative carbonylation of methanol. The first two routes are CO_2-based and in a research and development phase, whereas the last one is a commercial process. The set of environmental, social, and economic indicators selected were renewability of feedstock, energy intensity, waste generation, CO_2 balance, yield, feedstock price, process costs, health and safety issues of feedstock, process conditions, and innovation potential. The performance in these indicators was evaluated with the normalized scores from 0 to +1; 0 for detrimental and 1 for favorable impacts. The assessment showed that the transesterification route had the best potential toward sustainability, although there is still much development needed to improve yield. Further, the assessment gave clear understanding of the main benefits of each reaction route, as well as the major challenges to sustainability, which can further aid in orienting development efforts to key issues that need improvement. Finally, it was concluded that a multicriteria analysis such as the one presented in this paper was a viable method to be used in the process design stage.

INTRODUCTION

In the last decades, sustainable development has become the cornerstone of environmental policy and a leading principle for resource management. The widely used definition of sustainable development is that of the United Nations' Brundtland Commission [1]: "Development that meets the needs of the present without compromising the ability of future generations to meet their own needs." In corporate terms, sustainability can be summarized as the "triple bottom line" (TBL) success [2], which implies that firms have to maintain and grow their economic, social, and environmental capital base, while actively contributing to sustainability in the political domain [3, 4].

One of the key challenges of sustainable development is that it demands new and innovative choices and ways of thinking. Innovations in technology are challenging organizations to make new choices in their operations, products, and activities that impact the earth and people as well as economics [5]. There is, however, no standard method for measuring the triple bottom line success of technological innovations at the design phase and the principles to achieve sustainability by themselves are insufficient to create the right framework for design towards sustainability [6]. It would be useful to have a screening tool to assess how a new product or process under development would perform in terms of sustainability, or compared with a commercial process. Although there are various international efforts to measure sustainability, only a few of them have an integral approach taking into account environmental, economic, and social aspects. In most cases, the focus is on one of the three aspects [7]. For example, Life Cycle Assessment (LCA) is used to evaluate the environmental performance of products, but it concentrates on environmental impacts only [8]. As well, environmental impact assessment (EIA), a procedural tool for the design phase, only evaluates the environmental implications of decisions [9].

In order to fully evaluate the sustainability of new process routes, there is a need for a comprehensive evaluation of the environmental, economic, and social impacts of these new routes at an early process design stage. The paper suggests using multicriteria assessment for sustainability assessment and demonstrates its use in assessing a novel carbon dioxide-based reaction route to dimethyl carbonate (DMC).

SUSTAINABILITY ASSESSMENT METHODOLOGIES

There are a number of sustainability assessment methodologies evaluating the performance of industrial facilities. The World Business Council for Sustainable Development [10], the Global Reporting Initiative [5], and development of standards [11] are key drivers for adopting sustainability management in industries.

The most extensive work in terms of sustainability assessment has been done by the Global Reporting Initiative (GRI). GRI is a nongovernmental organization that aims at driving sustainability and has developed an environmental, social, and governance (ESG) reporting framework to be used worldwide. GRI version 4 on Sustainability Reporting Guidelines defines the principles and indicators that organizations can use to measure and report their economic, environmental, and social performance. Many companies use these indicators while publishing their annual or environmental reports. GRI is committed to continuously improve and increase the use of the guidelines which are available to the public [5].

The American Institute of Chemical Engineers (AIChE) has defined the AIChE Sustainability Index (SI) to measure the sustainability performance of representative companies in chemical industry [12]. The AIChE SI uses publicly available data on the companies' strategic commitment, sustainability innovation, environmental performance, safety performance, product stewardship, social responsibility, and value chain management to measure their sustainability performance. Metrics to measure the "greenness" of the companies' chemistry have been developed by the American Institute of Chemical Engineers' Center for Waste Reduction Technologies (AIChE/CWRT) assessing material intensity, energy intensity, water consumption, toxic release, and pollutant effects. The metrics developed are simple, understandable, easy to reproduce, and comparable [13]. They take into notice also the social aspects of sustainability by considering the health effects the chemicals used/produced have. However, they are developed for companies and are adjusted for existing process improvements rather than for a new process design.

Similarly, the Institution of Chemical Engineers (IChemE) has developed a set of metrics to enable process industry companies to measure and report progress along the path of sustainable development [14]. The Sustainable Development Progress Metrics are intended to help companies to set targets and develop internal standards and to monitor their progress in time [15]. The IChemE metrics are divided into environmental, economic, and social indicators. The environmental indicators are concentrating on resource use by considering how much energy, material, and water are consumed and land is used. Also atmospheric, aquatic impacts, and impacts on land caused by emissions, effluents, and waste are taken into notice. The economic indicators are concentrating on the profit gained, value added and taxes paid, and investments made by the company. The social indicators are considering the employment situation, health and safety at work, and also impacts to society. Not all the metrics proposed are valid in every case and it is up to the companies to decide which of the metrics are relevant for them. Key indicators have to be chosen from each of the aspects of sustainability to give a balanced view of the sustainability performance [15]. Whilst the IChemE metrics account for all three aspects of sustainability, they are meant as a sustainability management tool for companies, aiming at enhancing their sustainability performance, and are not suitable for assessing processes under development.

In terms of sustainability guidance for chemicals and chemical process design, Green Chemistry was developed to reduce or eliminate negative environmental impacts [16, 17]. The 12 Principles of Green Chemistry have been a cornerstone of environmentally conscious chemical process design since the late 1990s. Green Chemistry had been suggested to be used as a pollution prevention tool as it applies innovative scientific solutions to real-world environmental situations [18]. However, the assessment range of Green Chemistry does not cover the full depth of sustainability. As it was its original purpose, its emphasis is on reducing the toxicity of chemical products and driving inherently safer chemistry.

Protection of human health and the environment from chemicals and associated risks is also the goal of the European REACH (The Registration, Evaluation, Authorization and Restriction of Chemicals) regulation, which came into force in 2007. It renewed and upgraded the previous chemicals regulatory framework of the European Union (EU) [19], in order to ensure that there is free circulation of substances on the internal market and to enhance competitiveness and innovation. REACH confirms that industries are responsible for both assessing and managing the risks associated with chemicals, giving suitable safety information of chemicals to users, and promoting alternative testing methods [20]. About 143 000 chemicals marketed in the EU were preregistered by the December 1, 2008, deadline in REACH. The registration document of chemicals under REACH includes general information, safety data sheets (SDS), chemical safety report (CSR), and chemical safety assessment (CSA). Testing for health hazards under REACH includes acute toxicity, skin corrosion and irritation, serious eye damage and irritation, skin or respiratory sensitizer effect, mutagenic or carcinogenic impacts, toxicity for reproduction, specific target organ toxin in single exposure, specific target organ toxin in repeated exposure, and aspiration hazard [21, 22].

Table 1 summarizes some of the main evaluation guidelines or indicators used in the mentioned assessment processes. All methods outlined in Table 1 take into notice some key aspects of sustainability and clearly intend to evaluate triple bottom line success. Indicators that are possible to assess in the design phase and would give a good signal of sustainability performance are bolded. In terms of sustainability assessment of chemical processes in the design phase, Green Chemistry is the most thorough; however, should it be used to assess sustainability, it is recommended to extend it with social and economic indicators of GRI, AIChE, and IChemE to give a comprehensive measure of sustainability.

Table 1: Main principles/evaluation guidelines of reviewed assessment methods [5,12–16]

	GRI	AIChE	IChemE	Green Chemistry
Environmental performance	Materials Energy Water Biodiversity Emissions Effluents and waste Products and services Compliance Transport Suppliers	(i) Resource use (a) Energy (b) Materials (c) Renewables (d) Water GHG emissions Waste, wastewater Compliance management Value chain management	Resource usage (i) Energy (ii) Material (iii) Water (iv) Land Emissions, effluents, and waste	Prevent waste Use renewable feedstock Avoid chemical derivatives Catalysts Product degradability
Economic performance	Economic performance Market presence Procurement practices	(i) Sustainability innovation (ii) Strategic commitment to sustainability	Profit, value, tax (i) Investments	Maximise atom economy Increase energy efficiency
Social performance	Labor practices (i) Employment (ii) Health and safety (iii) Innovation and knowledge potential (iv) Diversity and equality society (i) Acceptability and social dialogue Human rights	Social responsibility (i) Stakeholder partnership (ii) Social investment (iii) Image in the community Product stewardship (i) assurance system (ii) risk communication (iii) legal proceedings	Workplace (i) Employment (ii) Health and safety society	Less hazardous chemical syntheses Safer chemicals, products, solvents, and reactions Accident prevention and real time analysis

Following the recommendation of IChemE, we selected key indicators from each of the aspects of sustainability to give a balanced view of the sustainability performance. The suggested indicators are as follows:

(i) environmental indicators:
   a) feedstock renewability,

   b) energy intensity,

   c) waste generation,

   d) CO_2 balance,

(ii) economic indicators:
   a) yield,

   b) feedstock price,

   c) process costs,

(iii) social indicators:
   a) process conditions,

   b) chemicals safety,

   c) innovation potential.

These indicators were selected as they can be assessed based on reaction routes as well as laboratory scale experiments and thermodynamic simulations. We propose that these 10 indicators are a necessary and sufficient set of meters for screening purposes at the design phase and give a balanced view of chemical process sustainability.

ASSESSMENT OF DMC PRODUCTION ROUTES

Dimethyl carbonate (DMC, $(CH_3)_2CO$) is an important chemical intermediate that can be used as a fuel additive and a polar solvent in the chemical industry. The production of DMC has received increasing attention over the least years [23–28]. There are several methods for the synthesis of DMC, such as phosgenation of methanol, oxidative carbonylation of methanol, transesterification method, and esterification of carbon dioxide with methanol [29–31]. In this paper, three reaction routes for DMC synthesis are evaluated. The reaction routes are outlined in Table 2.

Table 2: The assessed reaction routes for DMC production

Route A: direct synthesis from carbon dioxide and methanol
$CO_2 + 2CH_3OH \ (CH_3O)_2CO + H_2O$

Route B: transesterification of methanol and propylene carbonate using ionic liquid (IL) as a catalyst
$C_3H_6O + CO_2$ $C_4H_6O_3$
$C_4H_6O_3 + 2CH_3OH$ $(CH_3O)_2CO + C_3H_8O_2$
Route C: oxidative carbonylation of methanol (ENiChem)
$2CH_3OH + 1/2O_2 + CO$ $(CH_3O)_2CO + H_2O$

All three routes provide a safer alternative for the primary synthesis pathway, the "phosgene route" $COCl_2 + 2CH_3OH$ ™ $(CH_3O)_2CO + 2HCl$. The use of phosgene route is phased out from the commercial processes, as phosgene is one of the most acutely toxic substances used in industrial scale. As this route presents inherent hazards and potential environmental problems in handling and waste disposal [23], it is crucial that it is replaced by a more sustainable method.

Route A is currently in academic research phase. This route is particularly attractive for being CO_2-based. Generally speaking, carbon dioxide (CO_2) can be considered as an environmentally friendly and widely available feedstock, available as a waste emission of industrial processes. Chemical utilization of CO_2 for DMC manufacture would be a means to turn this waste into a nonwaste, allowing us to view CO_2 as a useful resource. It has been reported earlier that CO_2-based synthesis processes are meeting many of the provisions for environmental, economic, and social sustainability [32]. Therefore, much academic research has concentrated on the search for benign by design synthesis involving CO_2 as a raw material [33]. The synthesis of carbonic esters is one example [34]. The expectation is that the CO_2-based DMC production routes have significant potentials toward sustainable production. However, there are also numerous challenges of CO_2 utilization [32]; therefore, long-term research efforts for acquiring the necessary knowledge in its chemical reactivity are needed.

Route B is also an attractive "carbon-friendly" route, due to using CO_2 as a reaction feedstock. However, the complexity of a two-step process, the use of toxic propylene oxide, and the coproduction of propylene glycol make this process demanding. The challenge in both Routes A and B is that scale-up of the production would not be economically feasible at the moment.

The commercial route (ENiChem), Route C, is based on the catalytic oxidative carbonylation of methanol. It offers operational and environmental advantages, for example, fewer side products, better atom economy, and safer production comparing to the phosgene route [23, 33] but it is not responding to the current demand of DMC.

In terms of "measuring" innovation potential, we performed a literature review using a simple keyword search in Science Direct to evaluate the volume of publications and calculated the percentage of recent publications (2012 or later) of the 50 most relevant publications. Our reasoning is that the volume of publications is indicative of the level of knowledge potential, and the high percentage of recent publications indicates intensified academic interest, which will contribute to the renewal of science and is more likely to drive innovation.

PROCESS SIMULATIONS

Mass and energy balances of process routes were calculated by Aspen Plus simulations. Process flow sheets are presented in Figures 1 and 2. In order to make comparison of processes simple and appropriate the flow sheets were designed as similar as possible. Processes A and C consist of a reactor, gas separation unit, flash separator, and two columns. Process B is composed of two reactors and three columns. The reactors used were modeled as stoichiometric reactors based on known fractional conversion of a certain component. Radfrac model was used in separation units. The routes were assumed to be ideal (no mass and heat losses and no pressure drops and ideal component properties). The foundation for the calculation was stoichiometric, based on reaction equation, 1 kmol of each. The process conditions for the inlet stream were as follows: temperature 20°C and pressure 1 bar. The outlet stream temperature was set at 20°C and pressure at 1 bar. Concentration of DMC after purification was adjusted to 85 vol-%.

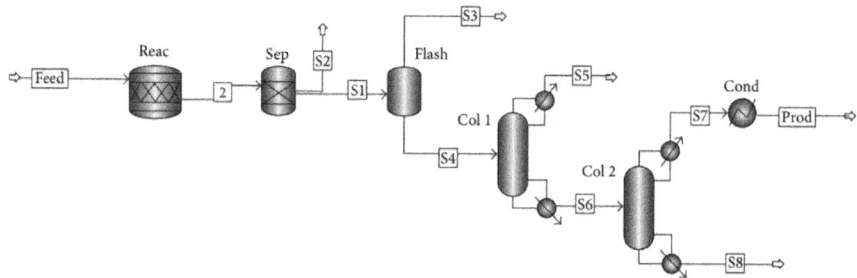

Figure 1: Process flow sheet for Routes A and C.

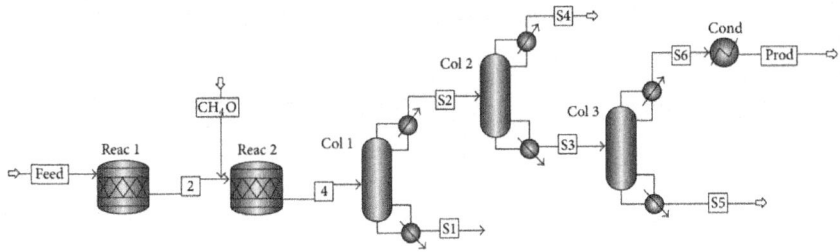

Figure 2: Process flow sheet for Route B.

Detailed descriptions of the process units and conditions are presented in Tables 3–5.

Table 3: Process description for Route A

Process unit	Type	Conditions	Notes
Reac	Stoichiometric reactor	T = 50°C, P = 150 bar	Conversion of CH_3OH = 8.3%
Sep	Component separator	Split fraction of CO_2 = 100%	Separation of unreacted CO_2
Flash	Flash separator	T = 97°C, P = 3 bar	
Col 1	RadFrac column	15 stages, distillate rate = 1.421, reflux ratio = 5	Separation of methanol
Col 2	RadFrac column	15 stages, distillate rate = 0.07, reflux ratio = 9	Concentration of DMC after distillation = 85.5%
Cond	Cooler	T = 20°C, P = 1 bar	Cooling of DMC

Table 4: Process description for Route B

Process unit	Type	Conditions	Notes
Reac 1	Stoichiometric reactor	T = 100°C, P = 140 bar	Conversion of propylene oxide = 100%
Reac 2	Stoichiometric reactor	T = 150°C, P = 1 bar	Conversion of methanol = 5.25%
Col 1	RadFrac column	15 stages, distillate rate = 2, reflux ratio = 5	Separation of propylene carbonate
Col 2	RadFrac column	15 stages, distillate rate = 1.895, reflux ratio = 5	Separation of methanol

| Col 3 | RadFrac column | 15 stages, distillate rate = 0.06, reflux ratio = 5 | Concentration of DMC after distillation = 85.9% |
| Cond | Cooler | T = 20°C, P = 1 bar | Cooling of DMC |

Table 5: Process description for Route C

Process unit	Type	Conditions	Notes
Reac	Stoichio-metric reactor	T = 120°C, P = 27 bar	Conversion of CH_3OH = 16.49%
Sep	Com-ponent separator	Split fraction of CO and O_2 = 100%	Separation of unreacted CO and O_2
Flash	Flash separator	T = 99.1°C, P = 3 bar	
Col 1	RadFrac column	15 stages, distillate rate = 1.307, reflux ratio = 5	Separation of methanol
Col 2	RadFrac column	15 stages, distillate rate = 0.148, reflux ratio = 10	Concentration of DMC after distillation = 85.4%
Cond	Cooler	T = 20°C, P = 1 bar	Cooling of DMC

ASSESSMENT PROCESS FOR REACTION ROUTES

The assumptions for all 3 reaction routes used in the assessment are summarized in Table 6 and the simulation results are gathered in Table 7. Process details (reactants, products, solvents, wastes, catalyst, temperature, pressure, conversion, and selectivity) were taken from the articles or/and academic theses [25, 35–39]. In addition, in Table 6 also the data for the literature review was included.

Table 6: Facts and assumptions regarding the three reaction routes

	Route A: direct synthesis from carbon dioxide and methanol	Route B: transesterification of methanol and propylene carbonate	Route C: oxidative carbon-ylation of metha-nol (ENiChem)
Reaction route (stoichio-metric feed [kmol])	CO_2 + $2CH_3OH$ ™$(CH_3O)_2CO$ + H_2O	$C_3H_6O + CO_2$ ™$C_4H_6O_3$ $C_4H_6O_3$ + $2CH_3OH$ ™ $(CH_3O)_2CO$ + $C_3H_8O_2$	$2CH_3OH + 1/2O_2$ + CO ™$(CH_3O)_2CO$ + H_2O

Atom economy [%] (theoretical)	83.3	60.8	83.3
Raw materials	CO_2 and CH_3OH	CH_3OH, CO_2, $C_3H_6O_3$ Intermediate: $C_4H_6O_3$	CH_3OH, O_2 and CO
Supply chain	CH_3OH from natural gas CO_2 separated from flue gas by absorption (MEA)	CH_3OH from syngas CO_2 separated from flue gas by absorption (MEA) Propylene oxide from H_2O_2 and propene	CH_3OH from natural gas O_2 from air (distillation) CO from natural gas
Solvents and auxiliary chemicals	IL101 as a promoter	No solvents or auxiliary chemicals	No solvents or auxiliary chemicals
Catalyst	CHT-HMS	First step: ion exchange resin D201 Second step: IL 103	$CuCl_2$
By-products and coproducts	H_2O unreacted CH_3OH	Propylene glycol, unreacted CH_3OH, and propylene carbonate	H_2O, unreacted CH_3OH, O_2, CO, and H_2O
Waste and emissions	Methylformate, unreacted CO_2	No wastes	No wastes
Process conditions	Pressure 150 bar Temp. 50°C Supercritical CO_2	First step: Pressure 140 bar Temp. 100°C Supercritical CO_2 Second step: Pressure 1.01325 bar Temp. 150°C	Pressure 27 bar Temp. 120°C
Health and safety issues	Methyl formate (i) is extremely flammable (ii) is harmful if swallowed or inhaled (iii) causes serious eye irritation (iv) may cause respiratory irritation	Propylene oxide (i) is extremely flammable (ii) is harmful if swallowed, inhaled, or came in contact with skin (iii) may cause respiratory irritation (iv) may cause genetic defects and cancer	CO is (i) flammable (ii) toxic for human CO and O_2 must be fed at a carefully controlled rate to avoid the risk of explosion
Volume of articles	542	129	76
Percentage of recent publications	25%	50%	8%

Table 7: Simulation results and cost calculations of reaction routes

	Route A	Route B	Route C
Conversion of MeOH [%]	9.16	(1) Step: 100 (2) Step: 10.5	17
Selectivity to DMC [%]	90.56	(1) Step 100 (2) Step 50	97
Yield [%]	5.99	5.15	12.64
Atom economy [%] (Real)	7.19	8.47	15.17
Amount of DMC (kmol)	0.06	0.051	0.126
Concentration of DMC (vol-%)	85.6	85.9	85.4
CO_2 emissions [kmol]	0.92	0	—
CO_2 consumption [kmol]	0.08	1.00	—
CO_2 balance	0.84	1	0
Energy consumption (specific) [MJ/DMC produced]	1152.8	−746.3	131.5
Energy consumption (Aspen) [MJ]	69.17	−38.06	16.57
Costs of feedstock (€/ kmol)	CO_2: 24.8	CH_3OH: 28.9	CH_3OH: 28.9
	CH_3OH: 28.9	CO_2: 24.8	O_2: 13.4
		C_3H_6O: 72.6	CO: 199
	tot. 53.7	tot. 126.3	tot. 241.3
Operational costs (feed-stock + process) [€]	53.7 + 18.7 = 72.4	126.3 − 10.3 = 116	241.3 + 4.5 = 245.8
Operational costs (feed-stock + process)/DMC produced €/kmol	1206.7	2274.5	1950.8
Treatment cost/waste disposal cost	High disposal cost of methylformate Water can be dis-charged to drain	By-product can be sold	Water can be dis-charged to drain

In reaction Route A, fossil fuel based raw materials are used, where it is assumed that methanol is produced with carbon monoxide (CO), CO_2, and hydrogen (H_2). Reaction Route B uses oil refinery products and CO_2 as raw material. In reaction Routes A and B, commercial catalytic materials are under development. The academic research toward reaction Route A uses calcined hydrotalcite on hexagonal mesoporous silica (CHT-HMS) as a catalyst with an IL promoter. In reaction Route B, ion exchange resin and ionic liquid (IL) are used as catalyst material. Finally, in Route C, commercial copper chloride catalysts are used. Catalytic materials for routes under development should be chosen for the assessment in order to minimize the environmental impact of catalyst materials, that is, enhancement of reaction activity and selectivity and stability of the catalyst, as well as environmentally benign catalytic materials.

In reaction Routes A and B, optimal reaction conditions such as temperature and pressure are not yet resolved as these routes are still under development; however, they are expected to be rather high and supercritical CO_2 is used. Reaction conditions should further be developed so that temperature and pressure are optimized at a lower level to minimize risks and environmental impacts. Reactions in Route C are using lower pressure, but higher temperature. The environmental benefit of this is to be highlighted, when compared with routes under research.

In reaction Routes A and C, only water is produced as a by-product. In Route A also small amounts of methylformate are produced. In Route B, toxic propylene glycol is produced. Propylene glycol is valuable from a commercial point of view; however its possible utilization needs to be considered at the design phase.

Atom economy is the best for the methanol-based reaction Routes A and C. However, it needs to be assessed if the atom economy benefit overweighs other impacts of the reactions. Reaction route C uses CO as a raw material, the production of which is rather energy demanding. In reaction Routes A and B, the yield is very low because of low conversion of methanol. This highlights the need for research for more efficient catalyst materials.

Prices of feedstock were acquired from chemical suppliers. Total operating costs were calculated by summing prices of feedstock and energy consumption of the process. Energy consumption of the process (MJ) was converted to euros by rate of 0.27 e/MJ (Eurostat). Capital costs were left out of considerations because all the cases are quite similar, and thus they were assumed to be equal in capital costs.

SUSTAINABILITY ASSESSMENT OF DMC ROUTES

Sustainability assessments are multicriteria based evaluations, which necessitate the inclusion of a wide variety of data typology with various certainty degrees. In this paper, we use multicriteria assessment (MCA) to perform the evaluation of the three DMC routes. Various multicriteria decision analysis methods have been put forward as an excellent candidate to perform sustainability assessment recently, and a variety of applications have emerged [40]. MCA is formal approach that takes into account multiple criteria in order to help making decisions that matter [41]. MCA stands in contrast to single goal optimization and approaches which, when using "unifying units," may offset poor performance of one criterion by good performances of another criterion, therefore allowing for substitution and compensability between criteria [42].

MCA methods require data to be normalized in order to obtain comparable scales. A common method is the ratio normalization that attributes value 1 to the best performance on a criterion and a proportional value to the other performances [43]. The objective of this method is to provide an easy to use screening tool for assessment and comparison in the design phase, in order to point out key aspects that need to be improved on or further explored. In some cases, we have amended this method in a way that the most preferred performance was valued 1 while detrimental performance was valued 0 and, if applicable, the third value normalized in between. In some cases we were reduced to qualitative evaluation, assigning 1 for best, 0 for worst, and 0.5 for medium values.

Table 8 lists the normalized values of selected indicators. Routes B and C use one-third of raw materials from oil refinery products, and 50% of raw materials in Route A are renewable. Values are normalized accordingly. In terms of energy demand, Route B releases 746.3 MJ energy, while reaction Routes A and C consume energy. We assigned Route A (1152.2 MJ/DMC production) a 0 value, to route B the value 1, and normalized the consumption of Route C (131.52 MJ/DMC) to the value 0.54. Only Routes A and B are CO_2-based. Route C has therefore no direct CO_2 implication. Route B consumes CO_2, while Route A generates it. We assigned Route B the value of 1 and Route C value 0 and normalized Route A in between. In case of wastes, Routes B and C produce no wastes, while Route A produces low amounts of methyl formate. Therefore, A is valued 0 while B and C are valued 1.

Table 8: Sustainability indicator values

	Route A	Route B	Route C
Environmental indicators			
Feedstock renewability	1	0.67	0.67
Energy intensity	0	1	0.54
CO_2 balance	0.84	1	0
Wastes	0	1	1
Economic indicators			
Yield	0.48	0.41	1
Feedstock price	1	0.67	0
Process costs	1	0.3	0
Social indicators			

Process conditions	0	0	0.06
Chemicals safety	1	0.5	0
Innovation potential	0.8	1	0.1
	6.12	6.55	3.37

The yield in reaction Route C is the highest (1), as expected from a commercial process. The yields of Routes A and B are moderate; normalized values are 0.48 and 0.41. Both have the potential to enhance the selectivity and yield as well. The yield of DMC in the process Route A can be improved by circumventing the thermodynamic limitations. The water generated in the process can be chemically trapped as discussed by Eta et al. [44] and thus the equilibrium can be shifted in the forward direction for a higher yield of DMC. The feedstock costs of raw materials for Route C are highest, and therefore it is valued 0; for Route A the costs are the lowest, thus valued 1, and route B has a normalized value of 0.67. The real processing costs are difficult to assess for Routes A and B, which are in the design phase; therefore, theoretical figures of operational costs were used. The assessment was based on the composite costs of feedstock and energy, divided by the amount of produced DMC. Based on this, Route C is the most expensive (0), Route A is the cheapest (1), and Route B is moderate and has normalized value 0.3.

For process conditions, the process temperature and pressure were evaluated. Room temperature (21°C) and atmospheric pressure (1 bar) were considered the safest, which would be valued 1. We assigned 0 for highest temperature 150°C (Route B) and the highest pressure 150 bar (Route A) normalized the other values, 50°C in Route A 0.61 and 120°C in Route C 0.07 and 140 bar in Route B 0,07 and 27 bar in Route C 0,92. These values were multiplied for a composite value. Health and safety issues are most severe for reaction Route C (0) due to the use of CO, less severe for Route B 0.5 that is using organic solvents, and benign in the case of Route A (1). Innovation potential was valuated based on the volume of articles published on these production methods and the percentage of recent papers. Most articles were written on subject related to Route A but only 20% of the relevant were recent, indicating a receding interest. In case of Route B, the volume of publications is moderate, but 50% of the most relevant are recent, which indicates this is of rising scientific relevance. The innovation potential of Route B was evaluated highest (value 1) and Route A was normalized to 0.8, while Route C with the fewest volume and least recent publications was valued 0.1.

The results of this comparative assessment are presented in Figure 3. Route B (red line) seems to be the most positive from environmental and social points of view; the only negative issue is the relatively highest safety risk in terms of process conditions, but it performs best in terms of low energy consumption and CO_2 balance as it consumes CO_2. Route A (blue line) seems to have some potential toward economic and social sustainability; however, in terms of environmental sustainability, it has some shortcomings, such as high energy consumption and waste generation. The commercial process (green line) performs best in terms of yield, which is expected from a mature process; however, it has the worst social sustainability performance and it is also based on nonrenewable feedstock. Table 9 summarizes the benefits and challenges of the three routes.

Table 9: Summary of sustainability assessment, benefits, and challenges to sustainability.

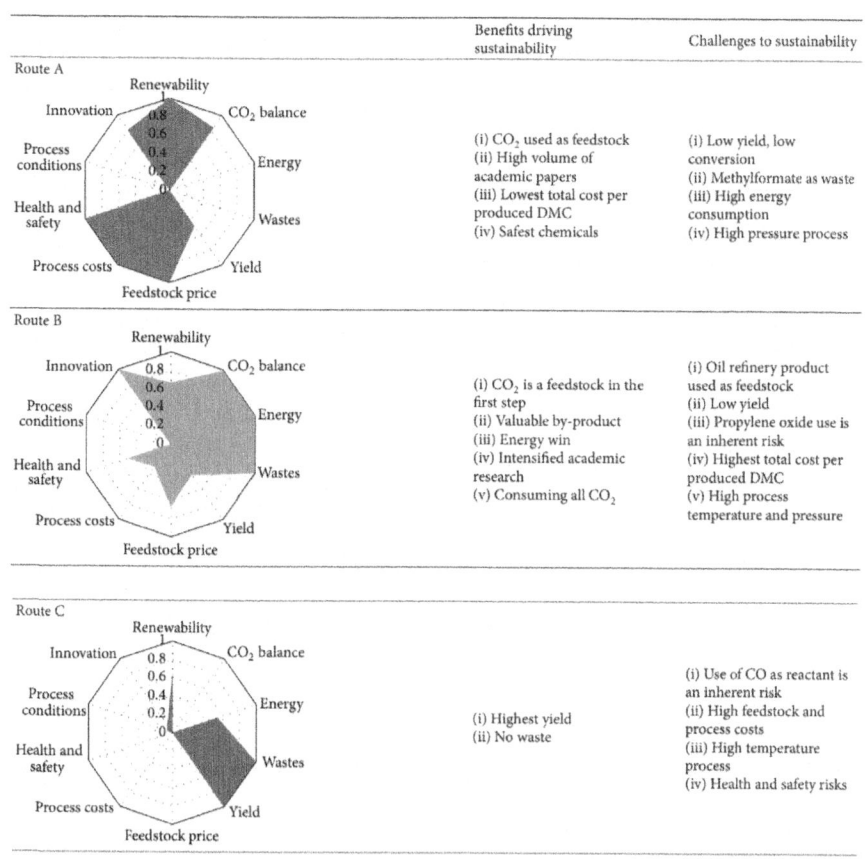

	Benefits driving sustainability	Challenges to sustainability
Route A	(i) CO_2 used as feedstock (ii) High volume of academic papers (iii) Lowest total cost per produced DMC (iv) Safest chemicals	(i) Low yield, low conversion (ii) Methylformate as waste (iii) High energy consumption (iv) High pressure process
Route B	(i) CO_2 is a feedstock in the first step (ii) Valuable by-product (iii) Energy win (iv) Intensified academic research (v) Consuming all CO_2	(i) Oil refinery product used as feedstock (ii) Low yield (iii) Propylene oxide use is an inherent risk (iv) Highest total cost per produced DMC (v) High process temperature and pressure
Route C	(i) Highest yield (ii) No waste	(i) Use of CO as reactant is an inherent risk (ii) High feedstock and process costs (iii) High temperature process (iv) Health and safety risks

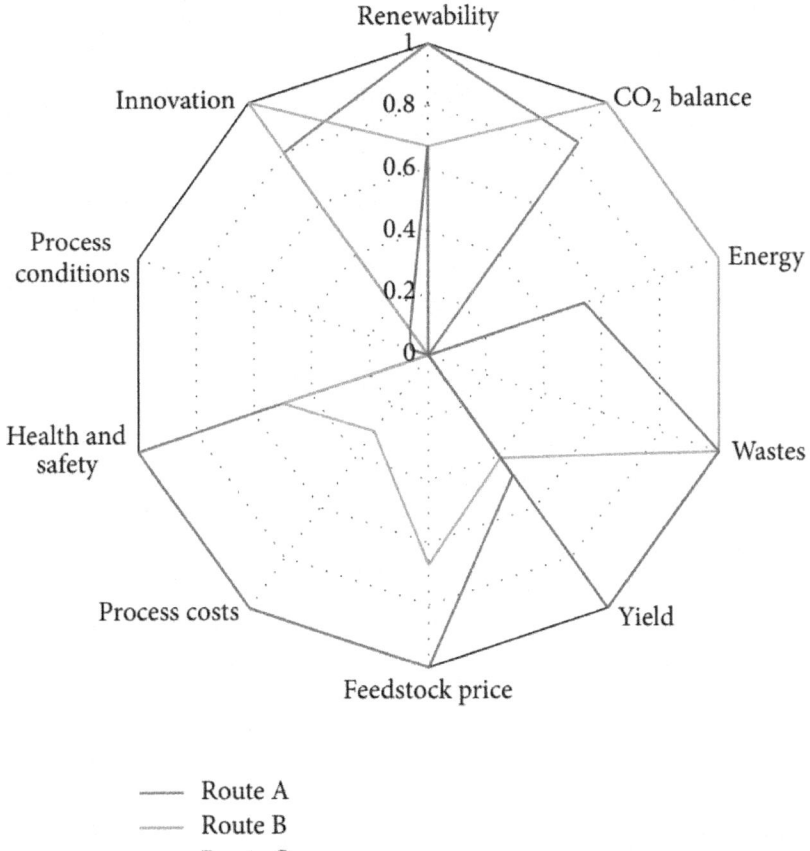

Figure 3: Comparison of the three DMC reaction routes.

In summation, it can be asserted that Route B has the best potential toward sustainability, although there is still much research needed to improve yield and conversion and thus reduce the amounts of wastes. In this case, use of a better catalyst would be further useful and add to sustainability positively. As its shape also indicates, Route A is very conflicting, as it has almost equal amounts of positive and negative factors. Many of the challenges are, however, difficult to overcome, such as the use of nonrenewable feedstock and yield stemming from low theoretical atom economy. In the commercial process (Route C), the toxicity of the reactant and the high feedstock price and production cost are failings that may not be further improved.

CONCLUSIONS

In order to drive sustainability in the chemical industry, there is a need for a methodology capable of assessing the impact of new choices in products, processes, and operations at the design phase. Most sustainability assessment methods are meant to be tools of sustainability management on the corporate level. There are tools available to assess the environmental performance of products, such as Life Cycle Assessment; however they do not take into account economic and social implications. For the assessment tool presented in this paper, the principles of Green Chemistry were used as the basis of evaluation. The objective of Green Chemistry to promote safer chemistry is its strength in terms of driving sustainability, but it also has some limitations. It was meant to provide guidelines for design rather than being an assessment or a screening tool. Sustainability assessments are multicriteria based evaluations; therefore, we used multicriteria assessment (MCA) to perform the evaluation of the three DMC routes. Cross-referencing the Green Chemistry principles with established sustainability assessment and reporting methods (Global Reporting Initiative, AIChE, IChemE, and REACH), this paper suggested a manageable list of factors considered necessary and sufficient to gain an overview of impacts toward sustainability.

The renewable nature of feedstock, energy intensity, CO_2 balance, and waste generation were evaluated as environmental indicators. To assess economic performance, yield, price of feedstock, and process and production costs were selected. In terms of social sustainability, process conditions and chemicals safety were assessed, the latter using the guidelines of REACH. In addition, innovation and knowledge potential was assessed based on the volume and novelty of scientific publications recently published. It was argued that these factors could be assessed based on reaction routes, laboratory scale experiments and results, as well as thermodynamic simulations. As MCA methods require data normalization, we used the common method of 0-1 attribute values, 1 being the best and 0 the worst.

Of the three reaction routes to DMC, two are CO_2-based still in a research phase. The assessment indicated that transesterification has the best potential toward sustainability, although there is still much research needed to improve yield and selectivity. Direct synthesis from CO_2 and methanol has many positive attributes, but an almost equal amount of negative factors. The commercial process, oxidative carbonylation, has performed worst in terms of sustainability, the toxicity of the reactant, and the high feedstock cost providing the major limitations to further improvement.

It can be concluded that the assessment allowed pointing out the main benefits of each reaction route, as well the major challenges to sustainability.

This can further aid in orienting development efforts to key issues that need to be improved. Further, it can be asserted that, of the established sustainability tools our method builds on, Green Chemistry holds the most potential for chemical industry research and development. Green Chemistry is well known and trusted amongst chemical engineers and has practical tools and guidelines developed for process designers. Finally, it is suggested that multicriteria assessment can be used as a sustainability assessment method in the process design stage.

ACKNOWLEDGMENTS

This work was performed within the collaborative project "Sustainable Catalytic Syntheses of Chemicals using Carbon Dioxide as Feedstock (GreenCatCO2)" supported by Department of Science and Technology, Government of India (DST-GOI), and The Academy of Finland. The authors would like to thank the Academy of Finland (Project nos. 129173 (SUSE) and 140122 (GreenCatCO2)) and the Finnish Funding Agency for Technology and Innovation, Tekes (Project no. 40313/09 (Fermet)) for financial support. Ganapati D. Yadav also thanks DST for J.C. Bose National Fellowship and received support from R.T. Mody Distinguished Professor Endowment.

REFERENCES

1.   WCED (World Commission on Environment and Development), "United Nations General Assembly document A/42/427," in Our Common Future, Oxford University Press, Oxford, UK, 1987.

2.   J. Elkington, Cannibals with Forks: The Triple Bottom Line of 21st Century Business, New Society Publishers, Stoney Creek, Conn, USA, 1998.

3.   T. Dyllick and K. Hockerts, "Beyond the business case for corporate sustainability," Business Strategy and the Environment, vol. 11, no. 2, pp. 130–141, 2002.

4.   J. B. Bowell, "Sustainability metrics, indicators, and indices for the process industries," in Sustainable Development in the Process Industries: Cases and Impact, J. Harmsen, Ed., pp. 5–23, John Wiley & Sons, 2010.

5.   The GRI Sustainability Reporting Guidelines, 2013,https://www.globalreporting.org/resourcelibrary/GRIG4-Part1-Reporting-Principles-and-Standard-Disclosures.pdf.

6.   J. García-Serna, L. Pérez-Barrigón, and M. J. Cocero, "New trends for design towards sustainability in chemical engineering: green

engineering," Chemical Engineering Journal, vol. 133, no. 1–3, pp. 7–30, 2007.

7.   R. K. Singh, H. R. Murty, S. K. Gupta, and A. K. Dikshit, "An overview of sustainability assessment methodologies," Ecological Indicators, vol. 15, no. 1, pp. 281–299, 2012.

8.   M. Aresta and M. Galatola, "Life cycle analysis applied to the assessment of the environmental impact of alternative synthetic processes. The dimethylcarbonate case: part 1," Journal of Cleaner Production, vol. 7, no. 3, pp. 181–193, 1999.

9.   Directive 2011/92/EU of the European parliament and of the council of 13 December 2011 on the assessment of the effects of certain public and private projects on the environment, http://eur-lex.europa.eu/LexUriServ/LexUriServ.do?uri=OJ:L:2012:026:0001:01:EN:HTML.

10.   World Business Council for Sustainable Development (WBCSD), Signals of Change: Business Progress Toward Sustainable Development, World Business Council for Sustainable Development (WBCSD), Geneva, Switzerland, 1997.

11.   OECD—Organisation for Economic Co-operation and Development, "An Update of the OECD Composite Leading Indicators Short-Term Economic Statistics Division, Statistics Directorate/OECD," 2002, http://www.oecd.org.

12.   The (AIChE) Sustainability Index: The Factors in Detail, 2009,http://www.aiche.org/resources/publications/cep/2009/january/aiche-sustainability-index-factors-detail.

13.   D. Tanzil, G. Ma, and B. R. Beloff, "Automating the sustainability metrics approach," in Proceedings of the AIChE Spring Meeting, New Orleans, La, USA, April 2004.

14.   M. Wilkinson, "Sustainable development and IChemE," Process Safety and Environmental Protection, vol. 78, no. 4, p. 236, 2000.

15.   IChemE sustainability metrics, sustainable development progress metrics recommended for use in the process industries, 2001,http://nbis.org/nbisresources/metrics/triple_bottom_line_indicators_process_industries.pdf.

16.   P. T. Anastas and J. C. Warner, Green Chemistry: Theory and Practice, Oxford University Press, New York, NY, USA, 1998.

17.   P. T. Anastas and M. M. Kirchhoff, "Origins, current status, and future challenges of green chemistry,"Accounts of Chemical Research, vol. 35, no. 9, pp. 686–694, 2002.

18. J. B. Manley, P. T. Anastas, and B. W. Cue Jr., "Frontiers in Green Chemistry: meeting the grand challenges for sustainability in R&D and manufacturing," Journal of Cleaner Production, vol. 16, no. 6, pp. 743–750, 2008.

19. W. Lilienblum, W. Dekant, H. Foth et al., "Alternative methods to safety studies in experimental animals: role in the risk assessment of chemicals under the new European Chemicals Legislation (REACH)," Archives of Toxicology, vol. 82, no. 4, pp. 211–236, 2008.

20. Regulation (EC) No 1272/2008 of the European parliament and of the council of 16 December 2008 on classification, labelling and packaging of substances and mixtures, amending and repealing Directives 67/548/EEC and 1999/45/EC, and amending Regulation (EC) No 1907/2006, http://eur-lex.europa.eu/LexUriServ/LexUriServ.do?uri=OJ:L:2008:353:0001:01:EN:HTML.

21. Regulation of the European parliament and of the council on classification, labelling and packaging of substances and mixtures, amending and repealing directives 67/548/EEC and 1999/45/EC, and amending regulation (EC) No 1907/2006, http://echa.europa.eu/fi/addressing-chemicals-of-concern/harmonised-classification-and-labelling/annex-vi-to-clp.

22. C&L Inventory database, 2014, http://echa.europa.eu/fi/information-on-chemicals/cl-inventory-database;jsessionid=F950E8759BC3897F24C96 2972266BCD9.live1.

23. D. Delledonne, F. Rivetti, and U. Romano, "Developments in the production and application of dimethylcarbonate," Applied Catalysis A: General, vol. 221, no. 1-2, pp. 241–251, 2001.

24. H. M. Wang, H. Wang, N. Zhao, W. Wei, and Y. Sun, "High-yield synthesis of dimethyl carbonate from urea and methanol using a catalytic distillation process," Industrial and Engineering Chemistry Research, vol. 46, no. 9, pp. 2683–2687, 2007.

25. M. A. Pacheco and C. L. Marshall, "Review of dimethyl carbonate (DMC) manufacture and its characteristics as a fuel additive," Energy & Fuels, vol. 11, no. 1, pp. 2–29, 1997.

26. F. Rivetti, U. Romano, and D. Delledone, "Dimethyl carbonate and its production technology," in Green Chemistry, P. T. Anastas and T. C. Williamson, Eds., vol. 626 of ACS Symposium Series, pp. 70–80, 1996.

27. Y. Katrib, G. Deiber, P. Mirabel et al., "Atmospheric loss processes of dimethyl and diethyl carbonate,"Journal of Atmospheric Chemistry, vol. 43, no. 3, pp. 151–174, 2002.

28. Y. Yu, X. Liu, W. Zhang et al., "Electrosynthesis of dimethyl carbonate from methanol and carbon monoxide under mild conditions," Industrial and Engineering Chemistry Research, vol. 52, no. 21, pp. 6901–6907, 2013.

29. M. Wang, N. Zhao, W. Wei, and Y. Sun, "Synthesis of dimethyl carbonate from urea and methanol over ZnO," Industrial and Engineering Chemistry Research, vol. 44, no. 19, pp. 7596–7599, 2005.

30. W. Zhao, F. Wang, W. Peng et al., "Synthesis of dimethyl carbonate from methyl carbamate and methanol with zinc compounds as catalysts," Industrial and Engineering Chemistry Research, vol. 47, no. 16, pp. 5913–5917, 2008.

31. P. Unnikrishnan and D. Srinivas, "Calcined, rare earth modified hydrotalcite as a solid, reusable catalyst for dimethyl carbonate synthesis," Industrial and Engineering Chemistry Research, vol. 51, no. 18, pp. 6356–6363, 2012.

32. E. Pongrácz, E. Turpeinen, R. Raudaskoski, D. Ballivet-Tkatchenko, and R. L. Keiski, "CO$_2$: from waste to resource for methanol-based processes," Proceedings of Institution of Civil Engineers: Waste and Resource Management, vol. 162, no. 4, pp. 215–220, 2009.

33. F. Cavani, G. Centi, S. Perathoner, and F. Trifiro, Sustainable Industrial Process, Wiley-VCH, New York, NY, USA, 2009.

34. D. Ballivet-Tkatchenko and S. Sorokina, "Open chain organic carbonates," in Carbon Dioxide Recovery and Utilization, M. Aresta, Ed., pp. 261–277, Kluwer Academic Publishers, Dordrecht, The Netherlands, 2003.

35. S. K. Kabra, E. Turpeinen, G. D. Yadav, and R. Keiski, "Direct synthesis of dimethyl carbonate from methanol and carbon dioxide: a thermodynamic and experimental study," The Journal of Supercritical Fluids. To be submitted.

36. P. Adhuri and G. D. Yadav, Insight into catalytic green chemistry and technology of industrial relevance [M.S. thesis], University of Mumbai, 2007.

37. H. Kawanami, A. Sasaki, K. Matsui, and Y. Ikushima, "A rapid and effective synthesis of propylene carbonate using a supercritical CO$_2$-ionic liquid system," Chemical Communications, vol. 9, no. 7, pp. 896–897, 2003.

38. M. Nicola Di, C. Fusi, F. Rivetti, and G. Sasselli, "Patent: process for producing dimethyl carbonate," Tech. Rep. EP 0460732, EniChem Synthesis S.p.A., 1991.

39. Y. Du, F. Cai, D.-L. Kong, and L.-N. He, "Organic solvent-free process for the synthesis of propylene carbonate from supercritical carbon dioxide and propylene oxide catalyzed by insoluble ion exchange resins," Green Chemistry, vol. 7, no. 7, pp. 518–523, 2005.

40. M. Cinelli, S. R. Coles, and K. Kirwan, "Analysis of the potentials of multi criteria decision analysis methods to conduct sustainability assessment," Ecological Indicators, vol. 46, pp. 138–148, 2014.

41. V. Belton and T. J. Stewart, Multiple Criterial Decision Analysis. An Integrated Approach, Kluwer Academic, Boston, Mass, USA, 2002.

42. T. Buchholz, E. Rametsteiner, T. A. Volk, and V. A. Luzadis, "Multi Criteria Analysis for bioenergy systems assessments," Energy Policy, vol. 37, no. 2, pp. 484–495, 2009.

43. L. C. Dias and A. R. Domingues, "On multi-criteria sustainability assessment: spider-gram surface and dependence biases," Applied Energy, vol. 113, pp. 159–163, 2014.

44. V. Eta, P. Mäki-Arvela, A.-R. Leino et al., "Synthesis of dimethyl carbonate from methanol and carbon dioxide: circumventing thermodynamic limitations," Industrial and Engineering Chemistry Research, vol. 49, no. 20, pp. 9609–9617, 2010.

Chapter 3

BIOLOGICAL AND CHEMICAL WASTEWATER TREATMENT PROCESSES

Mohamed Samer

Cairo University, Faculty of Agriculture, Department of Agricultural Engineering, Egypt

ABSTRACT

This chapter elucidates the technologies of biological and chemical wastewater treatment processes. The presented biological wastewater treatment processes include:

- bioremediation of wastewater that includes aerobic treatment (oxidation ponds, aeration lagoons, aerobic bioreactors, activated sludge, percolating or trickling filters, biological filters, rotating biological contactors, biological removal of nutrients) and anaerobic treatment (anaerobic bioreactors, anaerobic lagoons);
- phytoremediation of wastewater that includes constructed wetlands, rhizofiltration, rhizodegradation, phytodegradation, phytoaccumulation, phytotransformation, and hyperaccumulators; and
- mycoremediation of wastewater.

The discussed chemical wastewater treatment processes include chemical precipitation (coagulation, flocculation), ion exchange, neutralization, adsorption, and disinfection (chlorination/dechlorination, ozone, UV light). Additionally, this chapter elucidates and illustrates the wastewater treatment plants in terms of plant sizing, plant layout, plant design, and plant location.

INTRODUCTION

The chapter concerns with wastewater treatment engineering, with focus on the biological and chemical treatment processes. It aims at providing a brief and obvious description of the treatment methods, designs, schematics, and specifications. The chapter also answers an important question on how the different processes are interrelated and the correct order of these processes in relation to each other. The main objective of this work was to summarize the work of the eminent scientists in this field in order to provide a clear but concise chapter that can be used as a quick reference for environmental engineers and researchers, and to be effectively implemented in higher education teaching undergraduate and graduate students, as well as extension and outreach.

CHAPTER DESCRIPTION AND CONTENTS OVERVIEW

The chapter describes the biological and chemical wastewater treatment processes that include:

- Bioremediation of wastewater using oxidation ponds, aeration lagoons, anaerobic lagoons, aerobic and anaerobic bioreactors, activated sludge, percolating or trickling filters, biological filters, rotating biological contactors, and biological removal of nutrients;
- Mycoremediation of wastewater using bioreactors;
- Phytoremediation of wastewater that includes: constructed wetlands, rhizofiltration, rhizodegradation, phytodegradation, phytoaccumulation, phytotransformation, and hyperaccumulators;
- Vermifiltration and vermicomposting;
- Microbial fuel cells for electricity production from wastewater;
- Chemical wastewater treatment processes that include: chemical precipitation, ion exchange, neutralization, adsorption and disinfection (chlorination/dechlorination, ozone, ultraviolet radiation);
- Wastewater treatment plants. The chapter elucidates and illustrates the plant sizing, plant layout, plant design, and plant location.

OVERVIEW

Wastewater Treatment Techniques

Wastewater, or sewage, originates from human and home wastewaters, industrial wastes, animal wastes, rain runoff, and groundwater infiltration. Generally, wastewater is the flow of used water from a neighborhood. The wastewater consists of 99.9% water by weight, where the remaining 0.1% is suspended or dissolved material. This solid material is a mixture of excrements, detergents, food leftovers, grease, oils, salts, plastics, heavy metals, sands, and grits [1, 2]. Types of wastewaters include: municipal wastewater, industrial wastewaters, mixtures of industrial/domestic wastewaters, and agricultural wastewaters. Typical agricultural industries include: dairy processing industries, meat processing factories, juice and beverage industries, slaughterhouses, vegetable processing facilities, rendering plants, and drainage water of irrigation systems.

Subsequent to primary treatment of wastewater, i.e., physical treatment of wastewater, it still contains large amounts of dissolved and colloidal material that must be removed before discharge. The issue is how to transform the dissolved materials or particulate matters that are too little for sedimentation into larger particles to allow the separation processes to eliminate them. This can be accomplished by secondary treatment, i.e., biological treatment. The treatment of wastewater subsequent to the removal of suspended solids by microorganisms such as algae, fungi, or bacteria under aerobic or anaerobic conditions during which organic matter in wastewater is oxidized or incorporated into cells that can be eliminated by removal process or sedimentation is termed biological treatment. Biological treatment is termed secondary treatment. Chemical treatment, or tertiary treatment, using chemical materials will react with a portion of the undesired chemicals and heavy metals, but a portion of the polluting material will remain unaffected. Additionally, the cost of chemical additives and the environmental problem of disposing large amounts of chemical sludge make this treatment process deficient [1]. Alternatively, the biological treatment must be implemented. This treatment process implements naturally occurring microorganisms to transform the dissolved organic matter into a dense biomass that can be separated from the treated wastewater by the sedimentation process. In fact, the microorganisms utilize the dissolved organic matter as food for themselves, where the generated sludge will be far less for chemical treatment. In practice, therefore, secondary treatment tends to be a biological process with chemical treatment implemented for the removal of toxic compounds.

Aims of Wastewater Treatment

The goals of treating the wastewaters are:

- Transforming the materials available in the wastewater into secure end products that are able to be safely disposed off into domestic water devoid of any negative environmental effects;
- Protecting public health;
- Ensuring that wastewaters are efficiently handled on a trustworthy basis without annoyance or offense;
- Recycling and recovering the valuable components available in wastewaters;
- Affording feasible treatment processes and disposal techniques;

- Complying with the legislations, acts and legal standards, and approval conditions of discharge and disposal.

Biological Treatment Processes

The secondary treatment can be defined as "treatment of wastewater by a process involving biological treatment with a secondary sedimentation". In other words, the secondary treatment is a biological process. The settled wastewater is introduced into a specially designed bioreactor where under aerobic or anaerobic conditions the organic matter is utilized by microorganisms such as bacteria (aerobically or anaerobically), algae, and fungi (aerobically). The bioreactor affords appropriate bioenvironmental conditions for the microorganisms to reproduce and use the dissolved organic matter as energy for themselves. Provided that oxygen and food, in the form of settled wastewater, are supplied to the microorganisms, the biological oxidation process of dissolved organic matter will be maintained. The biological process is mostly carried out bacteria that form the basic trophic level (the level of an organism is the position it occupies in a food chain) of the food chain inside the bioreactor. The bioconversion of dissolved organic matter into thick bacterial biomass can fundamentally purify the wastewater. Subsequently, it is crucial to separate the microbial biomass from the treated wastewater though sedimentation. This secondary sedimentation is basically similar to primary sedimentation except that the sludge contains bacterial cells rather than fecal solids. The biological removal of organic matter from settled wastewater is conducted by microorganisms, mainly heterotrophic bacteria but also occasionally fungi. The microorganisms are able to decompose the organic matter through two different biological processes: biological oxidation and biosynthesis [1]. The biological oxidation forms some end-products, such as minerals, that remain in the solution and are discharged with the effluent (Eq. 1). The biosynthesis transforms the colloidal and dissolved organic matter into new cells that form in turn the dense biomass that can be then removed by sedimentation (Eq. 2). Figure 1summarizes these processes. On the other hand, algal photosynthesis plays an important role in some cases (Figure 2).

Oxidation:

$$COHNS + O_2 + Bacteria \rightarrow CO_2 + NH_3 + Other\ end\ products$$
(Organic matter) $\qquad$ +Energy

$$(1)$$

Biosynthesis:

$$COHNS + O_2 + Bacteria \rightarrow C_5H_7NO_2$$
(Organic matter) $\qquad\qquad$ (New cells)

$$(2)$$

Useful Terms

The following terms are the most used in biological treatment processes [2]:

a. DO: Dissolved Oxygen (mg L^{-1})

b. BOD: Biochemical Oxygen Demand (mg L^{-1})

c. BOD$_5$: BOD (mg L^{-1}), incubation at 15°C for 5 days

d. COD: Chemical Oxygen Demand (mg L^{-1})

e. CBOD: Carbonaceous BOD (mg L^{-1})

f. NBOD: Nitrogenous (mg L^{-1})

g. SOD: Sediment Oxygen Demand (mg L^{-1})

h. TBOD: Total BOD (mg L^{-1})

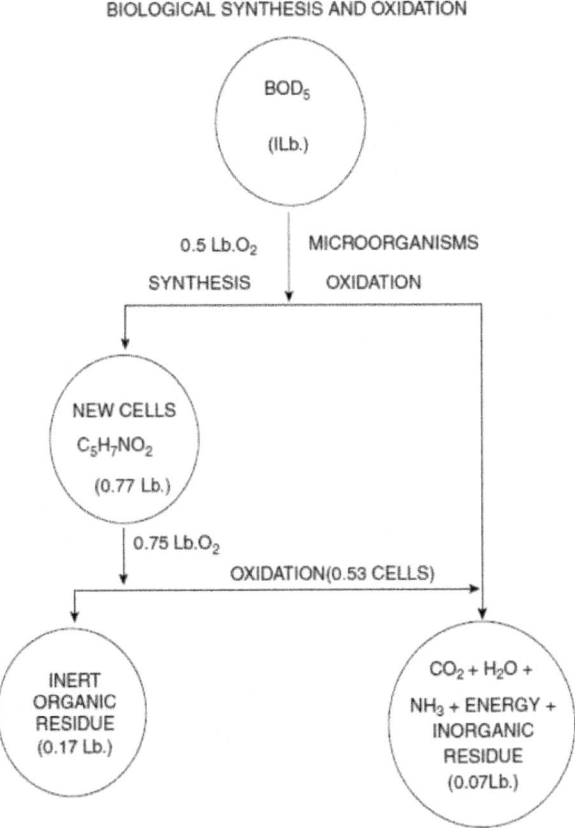

Figure 1. Biological synthesis and oxidation [3].

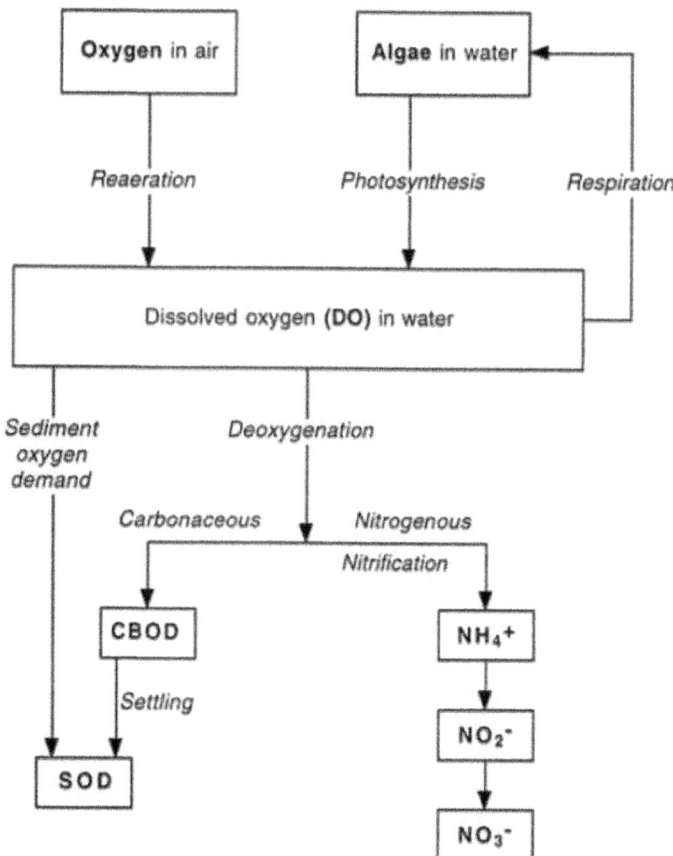

Figure 2. Photosynthesis and oxidation [2].

Chemical Treatment Processes

In early wastewater treatment technologies, chemical treatment has preceded biological treatment. Recently, the biological treatment precedes chemical treatment in the treatment process. Chemical treatment is now considered as a tertiary treatment that can be more broadly defined as "treatment of wastewater by a process involving chemical treatment". The mostly implemented chemical treatment processes are: chemical precipitation, neutralization, adsorption, disinfection (chlorine, ozone, ultraviolet light), and ion exchange.

BIOLOGICAL TREATMENT OF WASTEWATER

Biological Growth Equation

The biological growth can be described according to the Monod equation:

$$\mu = (\lambda S) / (K_S + S)$$

Where, μ is the specific growth rate coefficient; is the maximum growth rate coefficient that occurs at $0.5\,\mu_{max}$; S is the concentration of limiting nutrient, that is BOD and COD; and KS is the Monod coefficient [3].

Generally, the bacterial growth can be explained by the following simplified figure:

Organics+Bacteria+Nutrients+Oxygen→NewBacteria+CO$_2$+H$_2$O+Residual Organics+InorganicsOrganics+Bacteria+Nutrients+Oxygen→NewBacteria+CO$_2$+H$_2$O+ResidualOrganics+Inorganics

Several bioenvironmental factors affect the activity of bacteria and the rate of biochemical reactions. The most important factors are: temperature, pH, dissolved oxygen, nutrient concentration, and toxic materials. All these factors can be controlled within a biological treatment system and/or a bioreactor in order to ensure that the microbial growth is maintained under optimum bioenvironmental conditions. The majority of biological treatment systems operate in the mesophilic temperature range, where the optimal temperature ranges from 20°C to 40°C. Aeration tanks and percolating filters operate at the temperature of the wastewater that ranges from 12°C to 25°C; although in percolating filters, the air temperature and the ventilation rate may have a significant effect on heat loss. The higher temperatures increase the biological activity and metabolism, which result in increasing the substrate removal rate. However, the increased metabolism at the higher temperatures may lead to problems of oxygen limitations.

Bacterial Kinetics

The bacterial kinetics can be shown in Figures 3 and 4. The microbial growth curve that shows bacterial density and specific growth rate at the different growth phases is shown in Figure 3. The microbial growth curves that compare the total biomass and the variable biomass are shown in Figure 4.

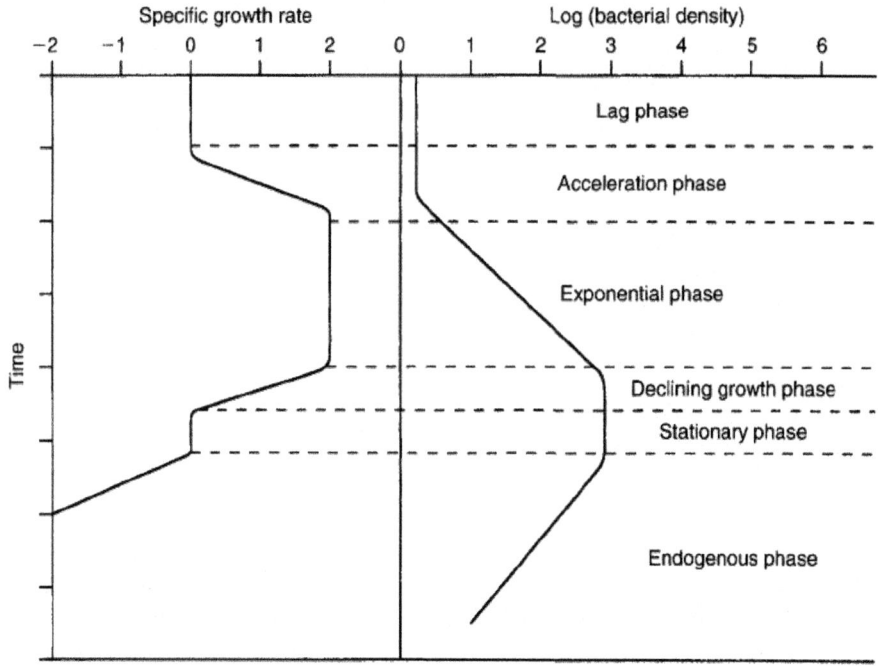

Figure 3. Microbial growth curve [1].

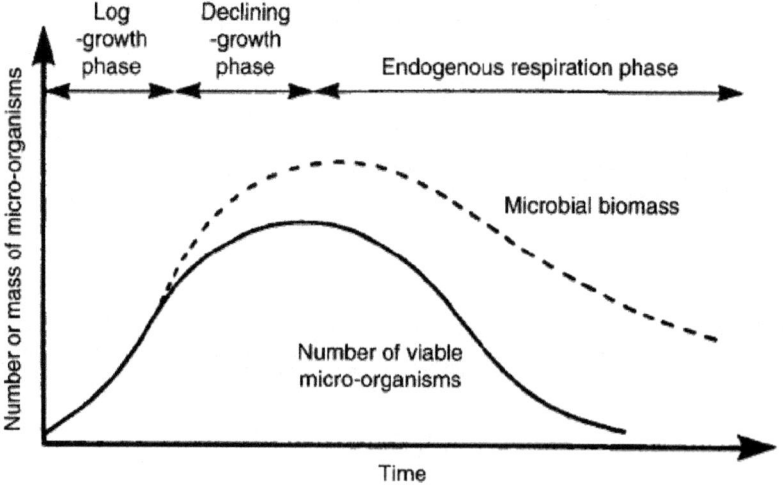

Figure 4. Microbial growth curves [1].

Principles of Biological Treatment

The principles of biological treatment of wastewater were stated by [3]. The following is a summary of the principles:

- The biological systems are very sensitive for extreme variations in hydraulic loads. Diurnal variations of greater than 250% are problematic because they will create biomass loss in the clarifiers.

- The growth rate of microorganisms is highly dependent on temperature. A 10°C reduction in wastewater temperature dramatically decreases the biological reaction rates to half.

- BOD is efficiently treated in the range of 60 to 500 mg L^{-1}. Wastewaters in excess of 500 mg L^{-1}BODs have been treated successfully if sufficient dilution is applied in the treatment process, or if an anaerobic process was implemented as a pretreatment process.

- The biological treatment is effective in removing up to 95% of the BOD. Large tanks are required in order to eliminate the entire BOD, which is not feasible.

- The biological treatment systems are unable to handle "shock loads" efficiently. Equalization is necessary if the variation in strength of the wastewater is more than 150% or if that wastewater at its peak concentration is in excess of 1,000 mg L^{-1} BOD.

- The carbon:nitrogen:phosphorus (C:N:P) ratio of wastewater is usually ideal. The C:N:P ratio of industrial wastewaters should range from 100:20:1 to 100:5:1 for a most advantageous biological process.

- If the C:N:P ratio of the wastewater is strong in an element in comparison to the other elements, then poor treatment will result. This is especially true if the wastewater is very strong in carbon. The wastewater should also be neither very weak nor very strong in an element; although very weak is acceptable, it is difficult to treat.

- Oils and solids cannot be handled in a biological treatment system because they negatively affect the treatment process. These wastes should be pretreated to remove solids and oils.

- Toxic and biological-resistant materials require special consideration and may require pretreatment before being introduced into a biological treatment system.

- Although the capacity of the wastewater to utilize oxygen is unlimited, the capacity of any aeration system is limited in terms of oxygen transfer.

Bioremediation of Wastewater

Bioremediation is a treatment process that involves the implementation of microorganisms to remove pollutants from a contaminated setting. Bioremediation can be defined as "treatment that implements natural organisms to decompose hazardous materials into less toxic or nontoxic materials". Some examples of bioremediation-related technologies are phytoremediation, bioaugmentation, rhizofiltration, and biostimulation. The microorganisms implemented to carry out the bioremediation are called bioremediators. However, some pollutants are not easily removed or decomposed by bioremediation. For example, heavy metals such as lead and cadmium are not eagerly captured by bioremediators. Example of bioremediation: fish bone char has been shown to bioremediate small amounts of cadmium, copper, and zinc.

The bioremediation of wastewater can be achieved by autotrophs or heterotrophs. A heterotroph is an organism that is unable to fix carbon and utilizes organic carbon for its growth. Heterotrophs are divided based on their source of energy. If the heterotroph utilizes light as its source of energy, then it is considered a photoheterotroph. If the heterotroph utilizes organic and/or inorganic compounds as energy sources, it is then considered a chemoheterotroph. Autotrophs, such as plants and algae, that are able to utilize energy from sunlight are called photoautotrophs. Autotrophs that utilize inorganic compounds to produce organic compounds such as carbohydrates, fats, and proteins from inorganic carbon dioxide are called lithoautotrophs. These reduced carbon compounds can be utilized as energy sources by autotrophs and provide the energy in food consumed by heterotrophs. Over 95% of all organisms are heterotrophic.

Aerobic Treatment

Aeration has been used to remove trace organic volatile compounds (VOCs) in water. It has also been employed to transfer a substance, such as oxygen, from air or a gas phase into water in a process called "gas adsorption" or "oxidation", i.e., to oxidize iron and/or manganese. Aeration also provides the escape of dissolved gases, such as CO_2 and H_2S. Air stripping has been also utilized effectively to remove NH_3 from wastewater and to remove volatile tastes and other such substances in water [2]. Samer [4] and Samer et al. [5] mentioned that aerobic treatment with biowastes is effective in reducing harmful gaseous emissions as greenhouse gases (CH_4 and N_2O) and ammonia.

Oxidation Ponds

Oxidation ponds (Figure 5) are aerobic systems where the oxygen required by the heterotrophic bacteria (a heterotroph is an organism that cannot fix carbon and uses organic carbon for growth) is provided not only by transfer from the atmosphere but also by photosynthetic algae. The algae are restricted to the euphotic zone (sunlight zone), which is often only a few centimeters deep. Ponds are constructed to a depth of between 1.2 and 1.8 m to ensure maximum penetration of sunlight, and appear dark green in color due to dense algal development. Samer [6] and Samer et al. [7] illustrated the structures and constructions of the aerobic treatment tanks and the used building materials.

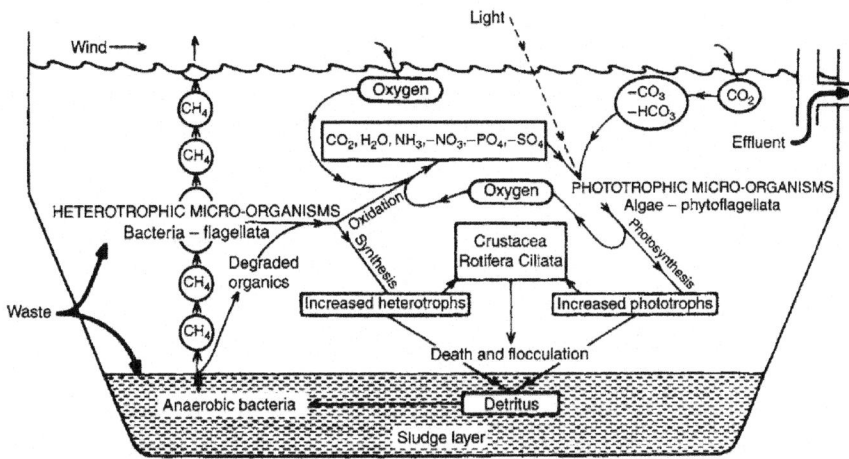

Figure 5. Aerobic system/oxidation pond [1].

In oxidation ponds, the algae use the inorganic compounds (N, P, CO_2) released by aerobic bacteria for growth using sunlight for energy. They release oxygen into the solution that in turn is utilized by the bacteria, completing the symbiotic cycle. There are two distinct zones in facultative ponds: the upper aerobic zone where bacterial (facultative) activity occurs and a lower anaerobic zone where solids settle out of suspension to form a sludge that is degraded anaerobically.

Aeration Lagoons

Aeration lagoons are profound (3–4 m) compared to oxidation ponds, where oxygen is provided by aerators but not by the photosynthetic activity of algae as in the oxidation ponds. The aerators keep the microbial biomass suspended and provide sufficient dissolved oxygen that allows maximal aerobic activity.

On the other hand, bubble aeration is commonly used where the bubbles are generated by compressed air pumped through plastic tubing laid through the base of the lagoon. A predominately bacterial biomass develops and, whereas there is neither sedimentation nor sludge return, this procedure counts on adequate mixed liquor formed in the tank/lagoon. Therefore, the aeration lagoons are suitable for strong but degradable wastewater such as wastewaters of food industries. The hydraulic retention time (HRT) ranges from 3 to 8 days based on treatment level, strength, and temperature of the influent. Generally, HRT of about 5 days at 20°C achieves 85% removal of BOD in household wastewater. However, if the temperature falls by 10°C, then the BOD removal will decrease to 65% [1].

Anaerobic Treatment

The anaerobic treatments are implemented to treat wastewaters rich in biodegradable organic matter (BOD >500 mg L^{-1}) and for further treatment of sedimentation sludges. Strong organic wastewaters containing large amounts of biodegradable materials are discharged mainly by agricultural and food processing industries. These wastewaters are difficult to be treated aerobically due to the troubles and expenses of fulfillment of the elevated oxygen demand to preserve the aerobic conditions [1]. In contrast, anaerobic degradation occurs in the absence of oxygen. Although the anaerobic treatment is time-consuming, it has a multitude of advantages in treating strong organic wastewaters. These advantages include elevated levels of purification, aptitude to handle high organic loads, generating small amounts of sludges that are usually very stable, and production of methane (inert combustible gas) as end-product.

Anaerobic digestion is a complex multistep process in terms of chemistry and microbiology. Organic materials are degraded into basic constituents, finally to methane gas under the absence of an electron acceptor such as oxygen [8]. The basic metabolic pathway of anaerobic digestion is shown in Figures 6and 7. To achieve this pathway, the presence of very different and closely dependent microbial population is required.

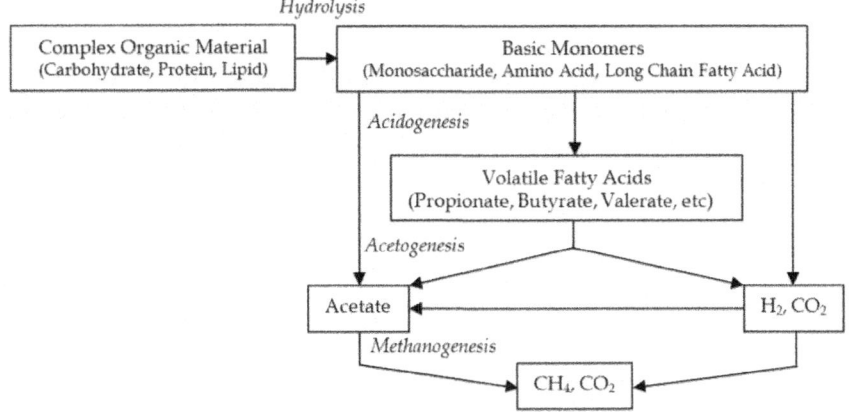

Figure 6. Steps of the anaerobic digestion process [8].

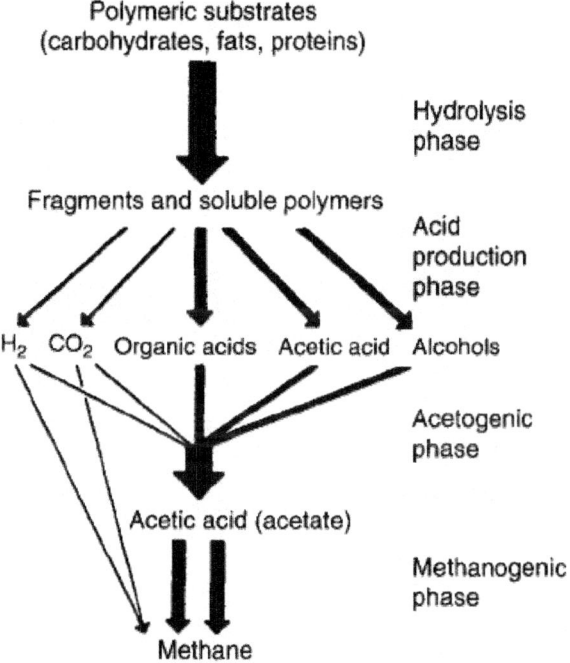

Figure 7. Major steps in anaerobic decomposition [1].

Suitable wastewaters include livestock manure, food processing effluents, petroleum wastes (if the toxicity is controlled), and canning and dyestuff wastes where soluble organic matters are implemented in the treatment. Most anaerobic processes (solids fermentation) occur in two predetermined temperature ranges: mesophilic or thermophilic. The temperature ranges are 30–38°C and 38–50°C, respectively [3]. In contrast to aerobic systems, absolute stabilization of organic matter is not achievable under anaerobic conditions. Therefore, subsequent aerobic treatment of the anaerobic effluents is usually essential. The final waste matter discharged by the anaerobic treatment includes solubilized organic matter that is acquiescent to aerobic treatment demonstrating the possibility of installing collective anaerobic and aerobic units in series [1].

Anaerobic Digesters

Samer [9] elucidated and illustrated the structures and constructions of the anaerobic digesters and the used building materials. Samer [10] developed an expert system for planning and designing biogas plants. Figures 8 to 13 show different types of anaerobic digesters. While Figures 14 and 15 show some industrial applications. Table 1 shows the advantages and disadvantages of anaerobic treatment compared to aerobic treatment.

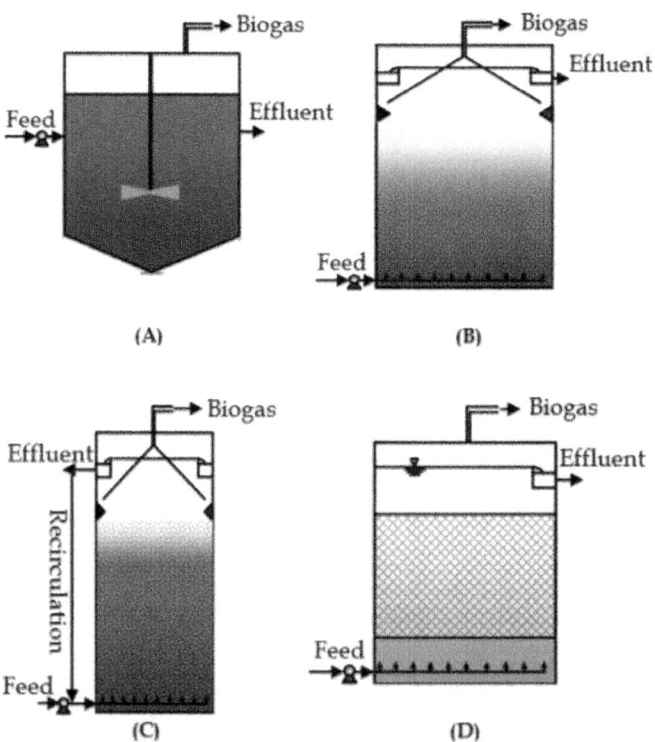

Figure 8. Most commonly used anaerobic reactor types: (A) Completely mixed anaerobic digester, (B) UASB reactor, (C) AFB or EGSB reactor, and (D) Upflow AF [8].

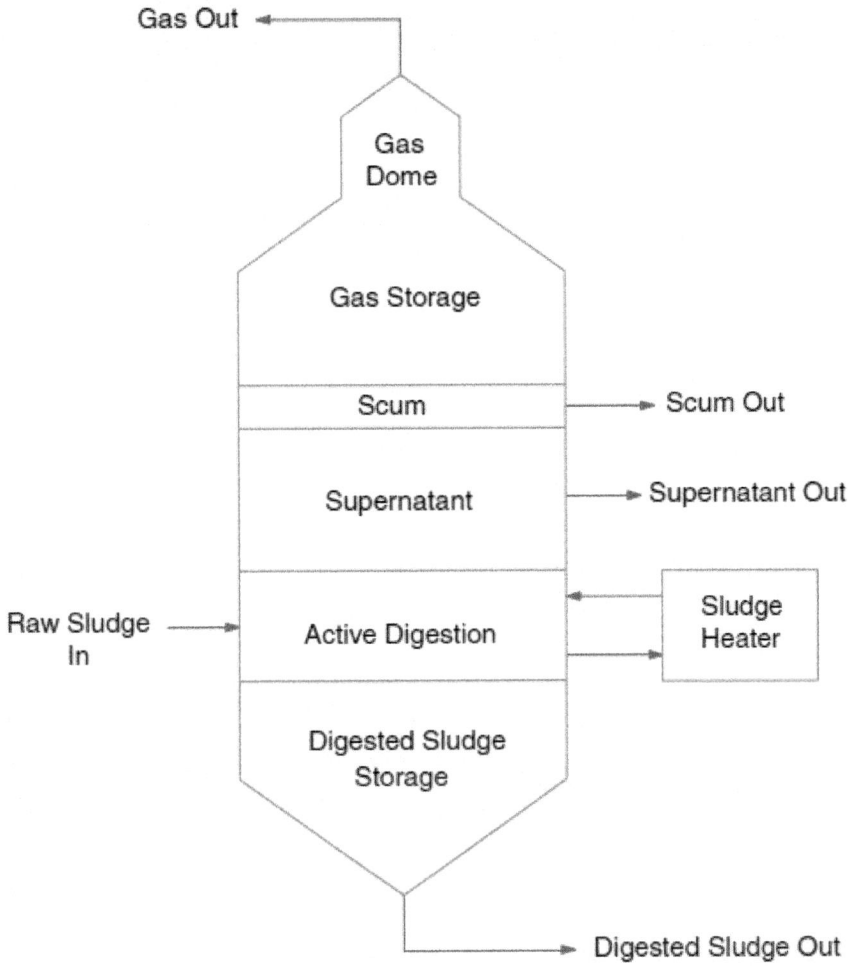

Figure 9. Single-stage conventional anaerobic digester [3].

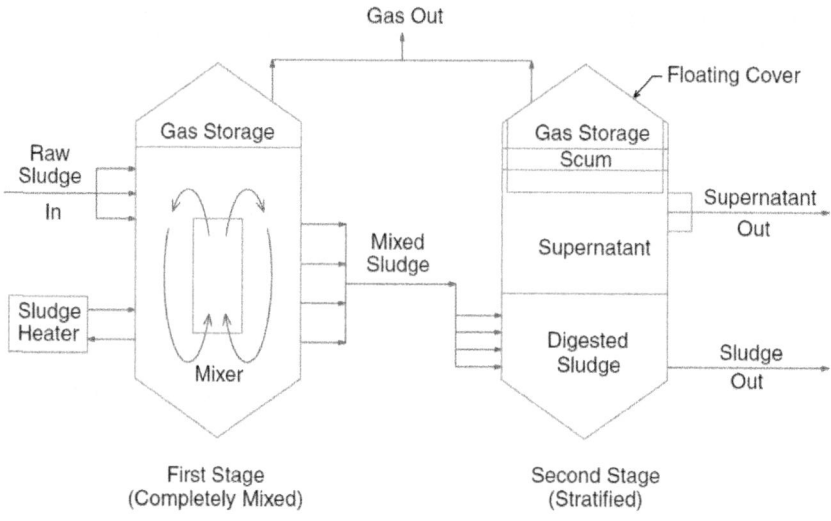

Figure 10. Dual-stage high rate digester [3].

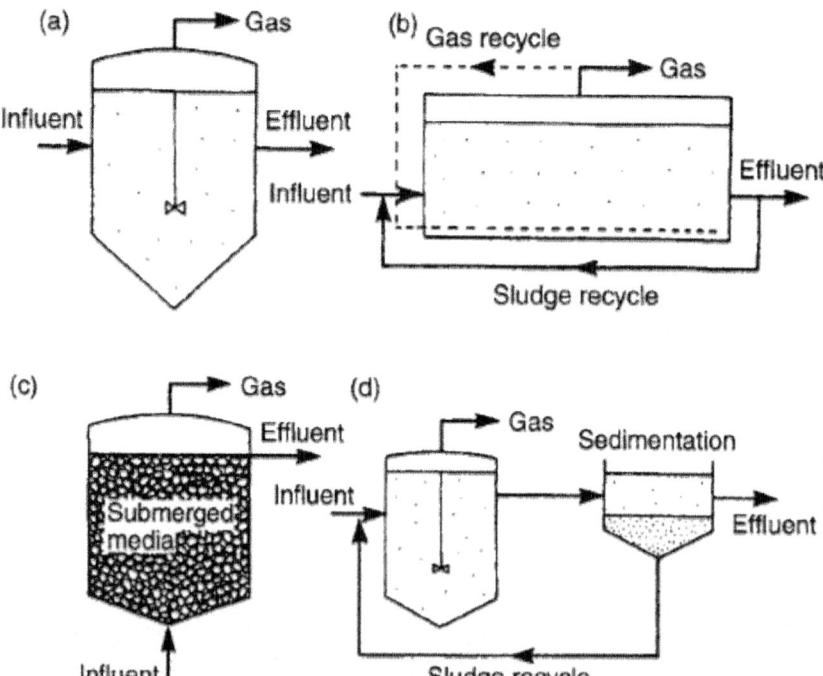

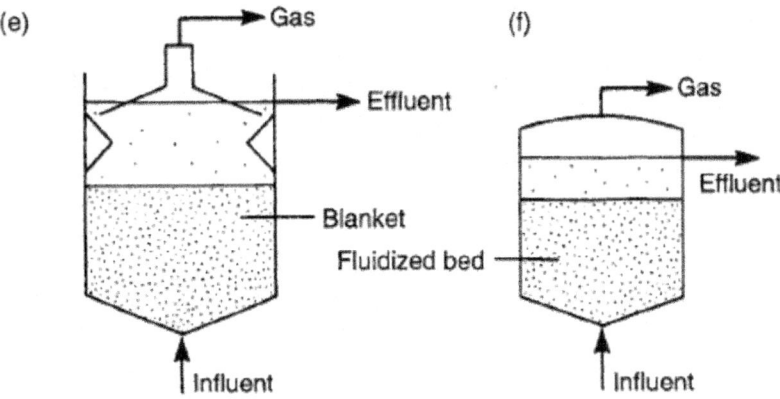

Figure 11. Schematic representation of digester types. Flow-through (A–B) and contact systems (C–F) [1].

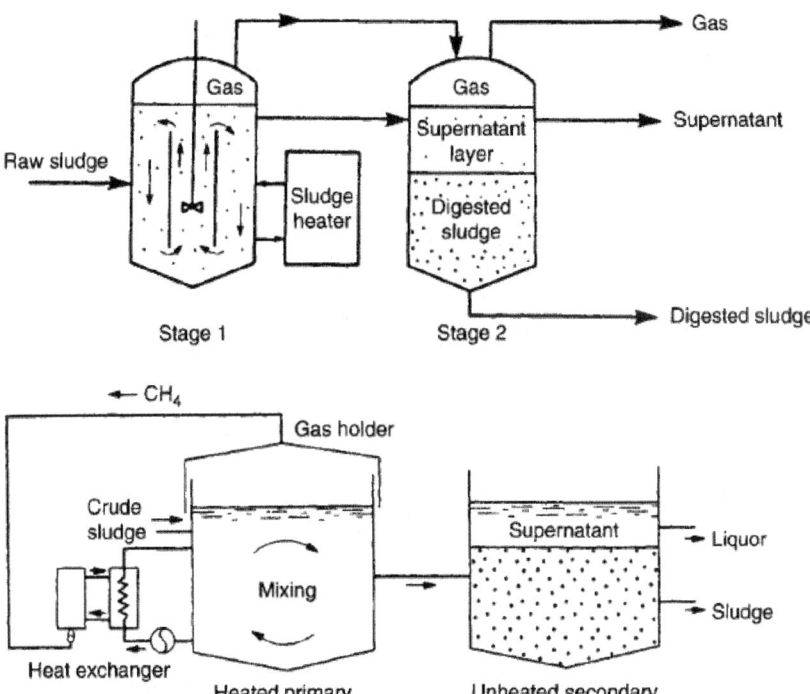

Figure 12. The upper scheme shows a two-stage anaerobic sludge digester, while the lower scheme shows the conventional sludge digestion plant [1].

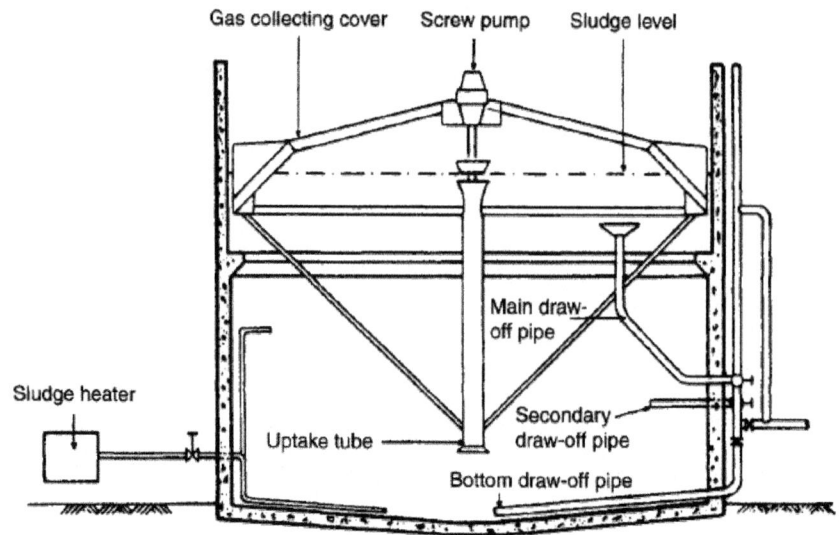

Figure 13. Primary digestion tank with screw mixing pump and external heater [1].

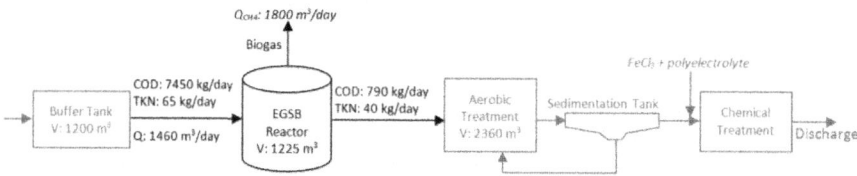

Figure 14. Wastewater treatment plant for corn processing industry [8].

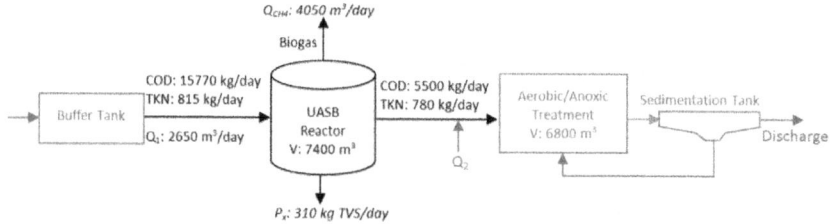

Figure 15. Mass balance study for a wastewater treatment plant of the baker's yeast industry [8].

By definition, the anaerobic treatment is conducted without oxygen. It is different from an anoxic process, which is a reduced environment in contrast to an environment without oxygen. Both processes are anoxic, but anaerobic is an environment beyond anoxic where the oxidation reduction potential

(ORP) values are highly negative. In the anaerobic process, nitrate is reduced to ammonia and nitrogen gas, and sulfate (SO_3^{2-}) is reduced to hydrogen sulfide (H_2S). Phosphate is also reduced because it is often transformed through the ADP–ATP chain [3].

Table 1. The advantages and disadvantages of anaerobic treatment compared to aerobic treatment [1]

The advantages and disadvantages of anaerobic treatment compared to aerobic treatment

Advantages	*Disadvantages*
Low operational costs	High capital costs
	Generally require heating
Low sludge production	Low retention times required (>24 h)
Reactors sealed giving no odour or aerosols	Corrosive and malodorous compounds produced during anaerobiosis
Sludge is highly stabilized	Not as effective as aerobic stabilization for pathogen destruction
Methane gas produced as end product	Hydrogen sulphide also produced
Low nutrient requirement due to lower growth rate of anaerobes	Reactor may require additional alkalinity
Can be operated seasonally	Slow growth rate of anaerobes can result in long initial start-up of reactors and recovery periods
Rapid start-up possible after acclimation	Only used as pre-treatment for liquid wastes

Anaerobic Lagoons

An anaerobic lagoon is a deep lagoon, fundamentally without dissolved oxygen, that enforces anaerobic conditions. The anaerobic process occurs in deep ground ponds, and such basins are implemented for anaerobic pretreatment. The anaerobic lagoons are not aerated, heated, or mixed. The depth of an anaerobic lagoon should be typically deeper than 2.5 m, where deeper lagoons are more efficient. Such depths diminish the amount of oxygen diffused from the surface, allowing anaerobic conditions to prevail (U.S. EPA, 2002). Figures 16 to 18 show different types of anaerobic lagoons.

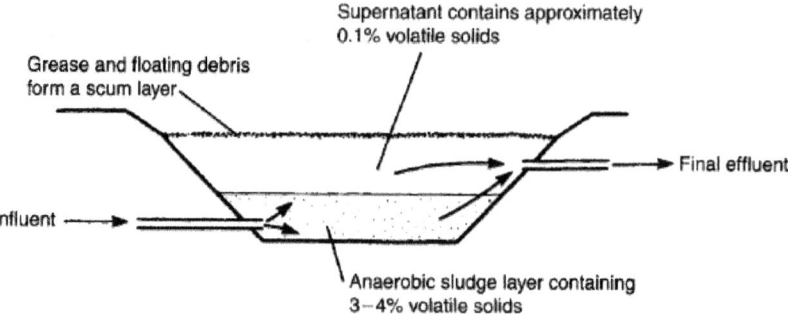

Figure 16. Anaerobic lagoon for strong wastewater treatment, such as meat processing wastewater [1].

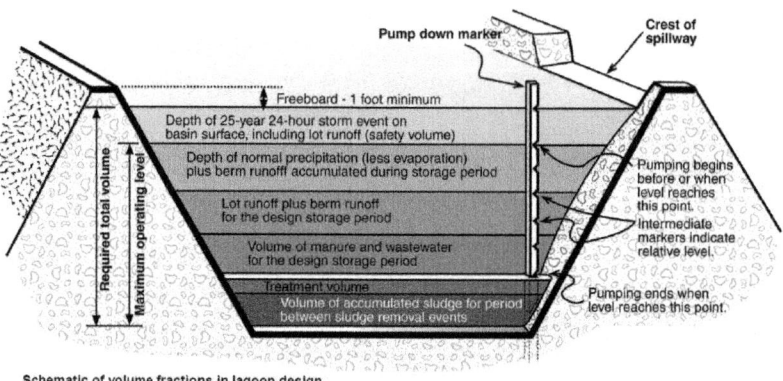

Figure 17. Schematic of volume fractions in anaerobic lagoon design [11].

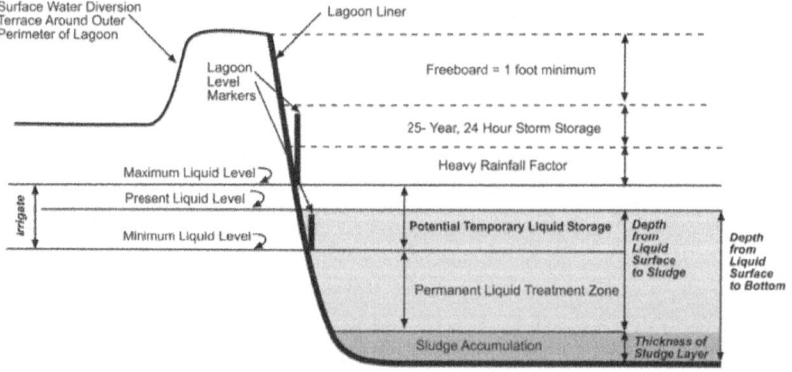

Figure 18. Anaerobic wastewater treatment lagoon [12].

Bioreactors

A bioreactor can be defined as "engineered or manufactured apparatus or system that controls the embraced or encompassed bioenvironment". Precisely, the bioreactor is a vessel in which a biochemical process is conducted, where it involves microorganisms (e.g., bacteria, algae, fungi) or biochemical substances (e.g., enzymes) derived from such microorganisms. The treatment can be conducted under either aerobic or anaerobic conditions. The bioreactors are commonly made of stainless steel, usually cylindrical in shape and range in size from liters to cubic meters. The bioreactors are classified as batch, plug, or continuous flow reactors (e.g., continuous stirred-tank bioreactor).

Mycoremediation is a type of bioremediation where fungi are implemented to break down the contaminants. The term "mycoremediation" refers particularly to the implementation of fungal "mycelia" in bioremediation. The principal role of fungi in the ecological system is the breakdown of pollutants, which is performed by the mycelium. The mycelium, the vegetative part of a fungus, secretes enzymes and acids that biodegrade lignin and cellulose that are the main components of vegetative fibers. Lignin and cellulose are organic compounds composed of long chains of carbon and hydrogen, and therefore they are structurally similar to several organic pollutants. One key issue is specifying the right fungus to break down a determined pollutant. Similarly, mycofiltration is a process that uses fungal mycelia to filter toxic compounds from wastewater. In an experiment, wastewater contaminated with diesel oil was inoculated with mycelia of oyster mushrooms. One month later, more than 93% of many of the polycyclic aromatic hydrocarbons (PAH) had been reduced to non-toxic components in the mycelial-inoculated samples. The natural microbial community participates with the fungi to break down contaminants, eventually into CO_2 and H_2O. Wood-degrading fungi are particularly effective in breaking down aromatic pollutants (toxic components of petroleum), as well as chlorinated compounds (certain persistent pesticides). Figures 19 to 22 show different types and designs of bioreactors.

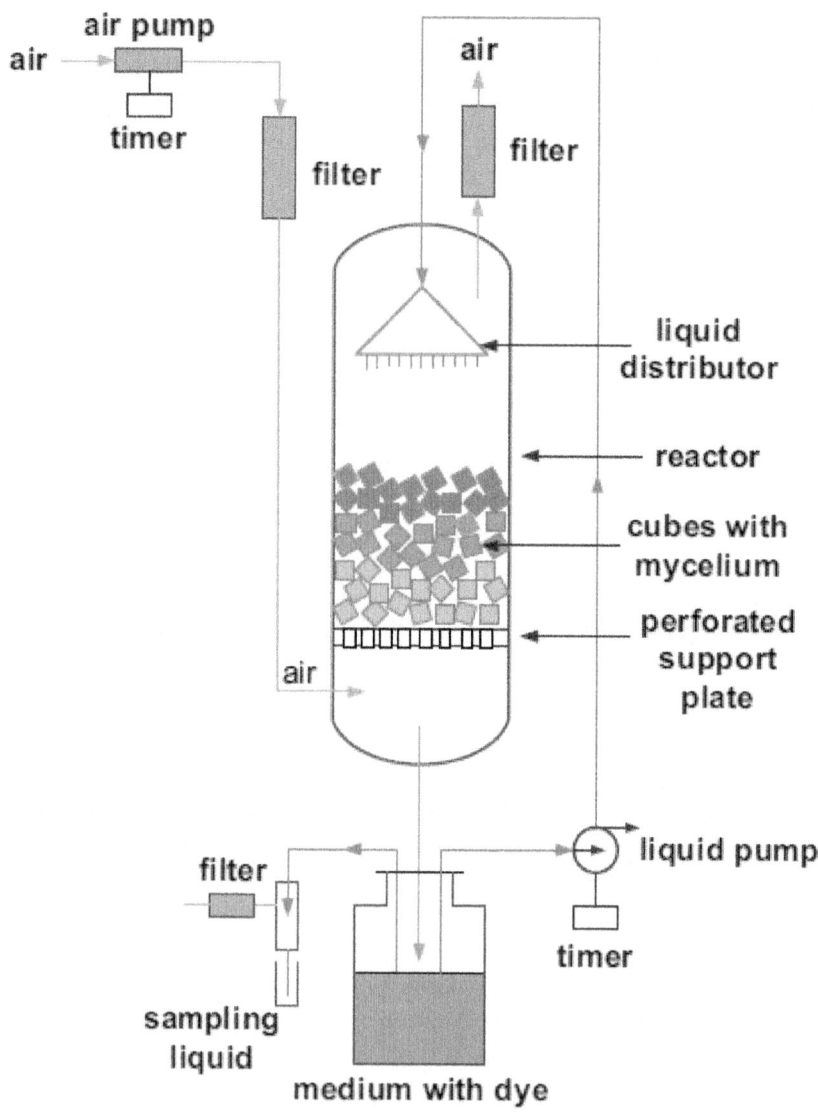

Figure 19. A bioreactor for fungal degradation: trickle bed bioreactor [13].

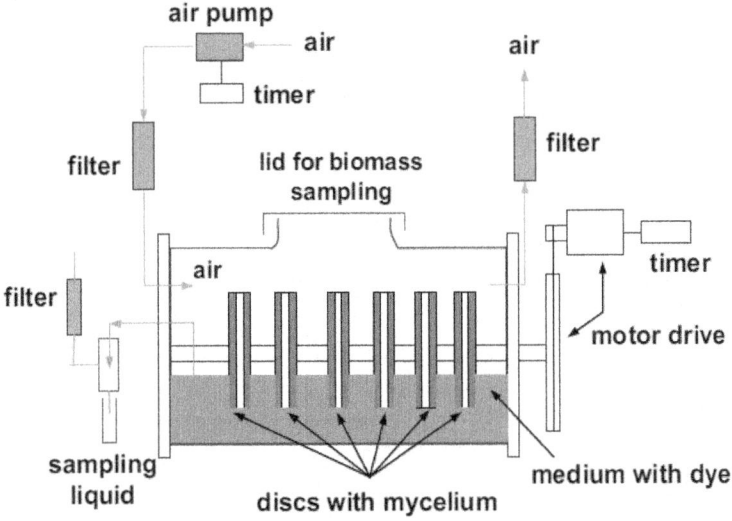

Figure 20. A bioreactor for fungal degradation: rotating disc bioreactor [13].

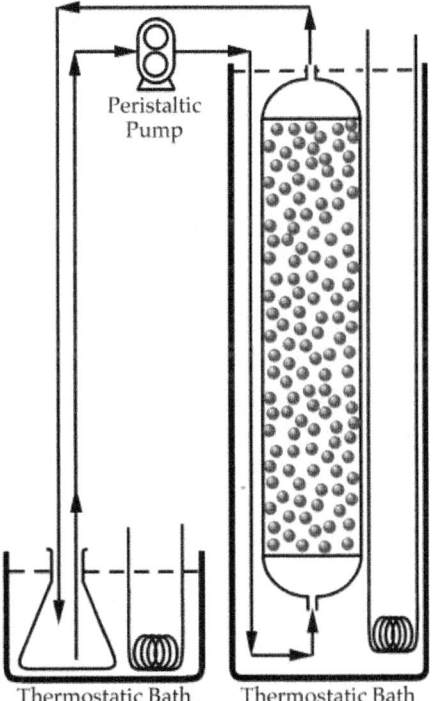

Figure 21. Fluidized bed bioreactor [14].

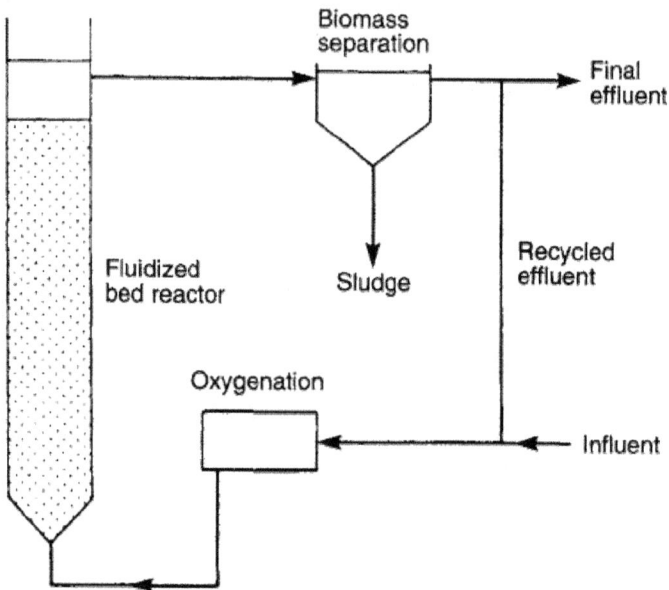

Figure 22. Typical design of fluidized bed reactor system [1].

Activated Sludge

The activated sludge process is based on a mixture of thick bacterial population suspended in the wastewater under aerobic conditions. With unlimited nutrients and oxygen, high rates of bacterial growth and respiration can be attained, which results in the consumption of the available organic matter to either oxidized end-products (e.g., CO_2, NO_3^-, SO_4^{2-}, and PO_4^{3-}) or biosynthesis of new microorganisms. The activated sludge process is based on five interdependent elements, which are: bioreactor, activated sludge, aeration and mixing system, sedimentation tank, and returned sludge [1]. The biological process using activated sludge is a commonly used method for the treatment of wastewater, where the running costs are inexpensive (Figure 23). However, a huge quantity of surplus sludge is produced in wastewater treatment plants (WWTPs) which is an enormous burden in both economical and environmental aspects. The excess sludge contains a lot of moisture and is not easy to treat. The byproducts of WWTPs are dewatered, dried, and finally burnt into ashes. Some are used in farm lands as compost fertilizer [15]. However, it is suggested that the dried byproducts of WWTPs are fed into the pyrolysis process rather than the burning process.

The sludge volume index (SVI) is an estimation that specifies the tendency of aerated solids, i.e., activated sludge solids, to become dense or concentrated through the thickening process. SVI can be computed as follows: (a) allowing a mixed liquor sample from the aeration tank to sediment in 30 min; (b) determining the concentration of the suspended solids for a sample of the same mixed liquor; (c) SVI is then computed as ratio of the measured wet volume (mL/L) of the settled sludge to the dry weight concentration of MLSS in g/L (Source: Office of Water Programs, Sacramento State, USA).

During the treatment of wastewater in aeration tanks through the activated sludge process (Table 2) there are suspended solids, where the concentration of the suspended solids is termed as mixed liquor suspended solids (MLSS), which is measured in milligrams per liter (mg L^{-1}). Mixed liquor is a mixture of raw wastewater and activated sludge in an aeration tank. MLSS consists mainly of microorganisms and non-biodegradable suspended solids. MLSS is the effective and active portion of the activated sludge process that ensures that there is adequate quantity of viable biomass available to degrade the supplied quantity of organic pollutants at any time. This is termed as Food to Microorganism Ratio (F/M Ratio) or food to mass ratio. If this ratio is kept at the suitable level, then the biomass will be able to consume high quantities of the food, which reduces the loss of residual food in the discharge. In other words, the more the biomass consumes food the lower the BOD will be in the treated effluent. It is important that MLSS eliminates BOD in order to purify the wastewater for further usage and hygiene. Raw sewage is introduced into the wastewater treatment process with concentration of several hundred mg L^{-1} of BOD. The concentration of BOD in wastewater is reduced to less than 2 mg L^{-1} after being treated with MLSS and other treatment methods, which is considered to be safe water to use.

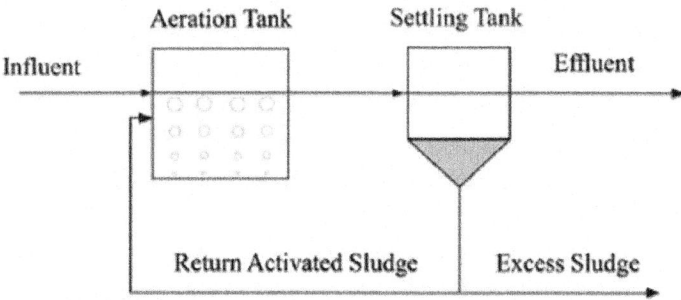

Figure 23. Activated sludge [15].

Table 2. Conventional activated sludge [15]

Specification	Value	Unit
BOD-Sludge Loading	0.40	mg L^{-1}
BOD-Volume Loading	0.20	mg L^{-1}
MLSS	2000	mg L^{-1}
COD of Influent	300	mg L^{-1}
Amount of Influent	4.48	L^{d-1}
Aeration Rate	3.00	L mi^{n-1}

The biological treatment process is the most commonly implemented method for the treatment of domestic sewage. This method implements bacterial populations that possess superior sedimentation characteristics. The living microorganisms break down the organic matter in the wastewater and consequently purify the wastewater from biological waste [15].

According to [1], the main components of all activated sludge systems are:

- The bioreactor: it can be a lagoon, tank, or ditch. The main characteristic of a bioreactor is that it contains sufficiently aerated and mixed contents. The bioreactor is also known as the aeration tank.

- Activated sludge: it is the bacterial biomass inside the bioreactor that consists mostly of bacteria and other flora and microfauna. The sludge is a flocculent suspension of these microorganisms and is usually termed as the mixed liquor suspended solids (MLSS) that ranges between 2,000 and 5,000 mg L^{-1}.

- Aeration and mixing system: the aeration and mixing of the activated sludge and the raw influent are necessary. While these processes can be accomplished separately, they are usually conducted using a single system of either surface aeration or diffused air.

- Sedimentation tank: clarification or settlement of the activated sludge discharged from the aeration tank is essential. This separates the bacterial biomass from the treated wastewater.

- Returned sludge: the settled activated sludge in the sedimentation tank is returned to the bioreactor to maintain the microbial population at a required concentration to guarantee persistence of treatment process.

Several parameters should be considered while operating activated sludge plants. The most important parameters are: (1) biomass control, (2) plant loading, (3) sludge settleability, and (4) sludge activity. The main operational variable is the aeration, where its major functions are: (1) ensuring a sufficient and continuous supply of dissolved oxygen (DO) for the bacterial population, (2) keeping the bacteria and the biomass suspended, and (3) mixing the influent wastewater with the biomass and removing from the solution the excessive CO_2 resulting from oxidation of organic matter [1].

There are several types of activated sludge processes, e.g., conventional activated sludge plant (Figure 24), complete mix plant (Figure 25), contact stabilization plant (Figure 26), and step aeration plant (Figure 27). Figure 28 shows the food pyramid that represents the feeding relationships within the activated sludge process.

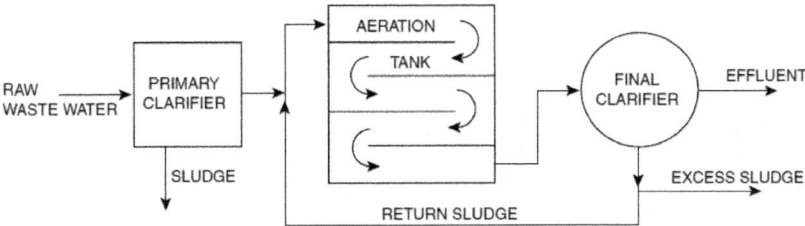

CONVENTIONAL ACTIVATED SLUDGE PLANT

Figure 24. Conventional activated sludge plant [3].

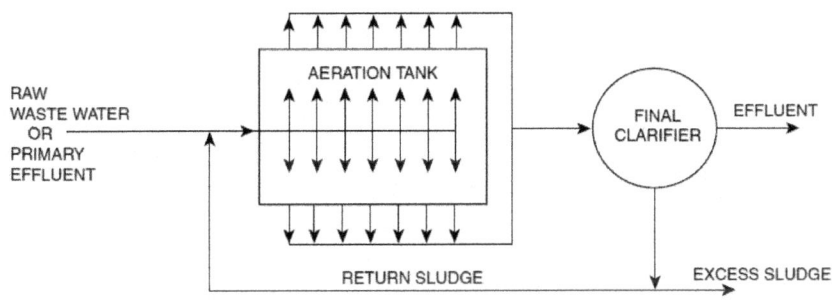

COMPLETE MIX PLANT

Figure 25. Complete mix plant [3].

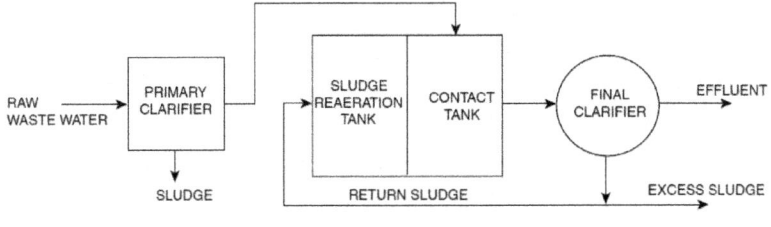

CONTACT STABILIZATION PLANT

Figure 26. Contact stabilization plant [3].

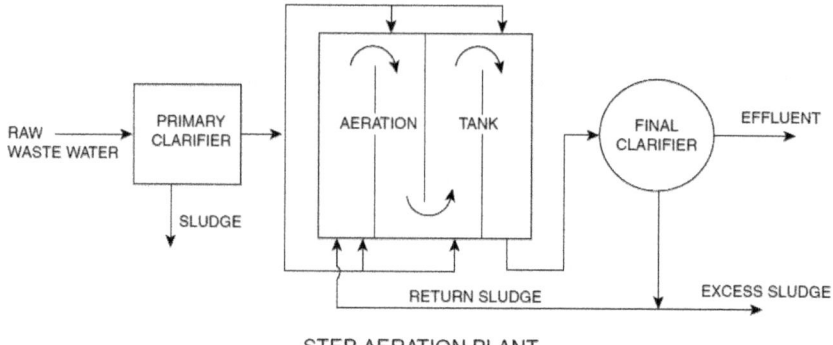

STEP AERATION PLANT

Figure 27. Step aeration plant [3].

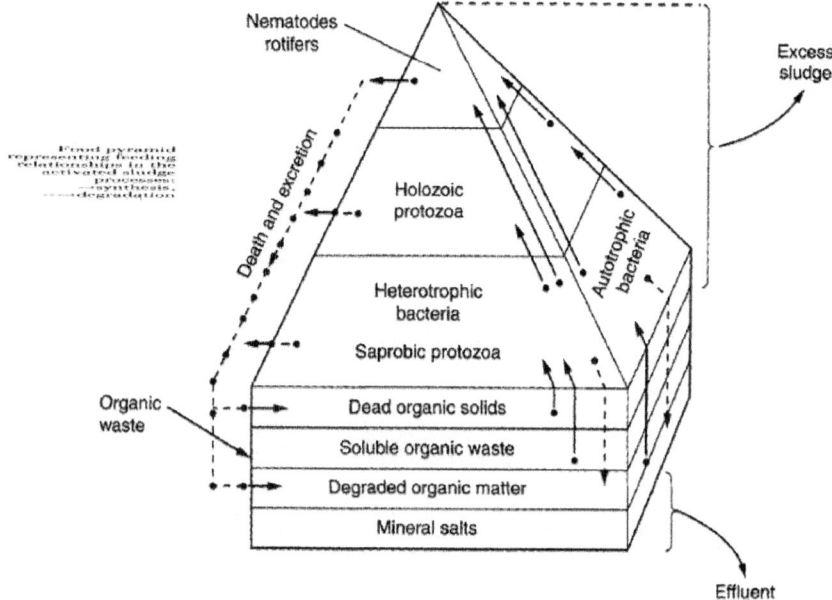

Figure 28. Food pyramid illustrating the feeding relationships within the activated sludge process [1].

Biological Filters

The main systems of operation of biological filters are:

- single filtration,
- recirculation,
- ADF, and

- two-stage filtration with high-rate primary biotower (Figure 29).

There are several types of biological filters, for example, submerged aerated filters that are widely known as biological aerated filters (BAFs) and are the commonly implemented design (Figure 30), and the percolating (trickling) filters (Figure 31). The BAFs implement either the sunken granular media with upward (Figure 30a) or download (Figure 30b) flows, or floating granular media with upward flow (Figure 30c), which is the most common design of BAFs. In order to compare the biological filters and the activated sludge systems (Figures 31 and 32), the comparison is based on the oxidation that can be accomplished by three processes:

- Spreading the wastewater into a thin film of liquid with a large surface area, consequently the required oxygen can be supplied by gaseous diffusion, which is the case of the percolating filters.
- Aerating the wastewater by pumping air in the form of bubbles or stirring forcefully, which is the case of the activated sludge process.
- Implementing algae to produce oxygen by photosynthesis, which is the case of the stabilization ponds.

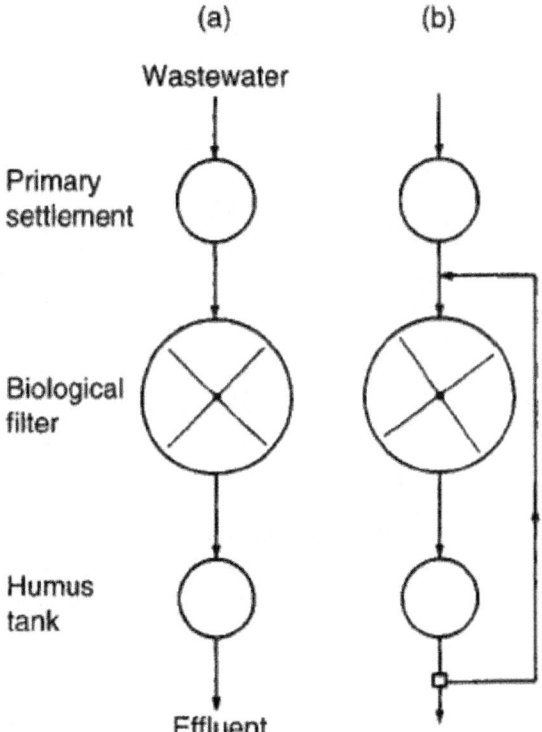

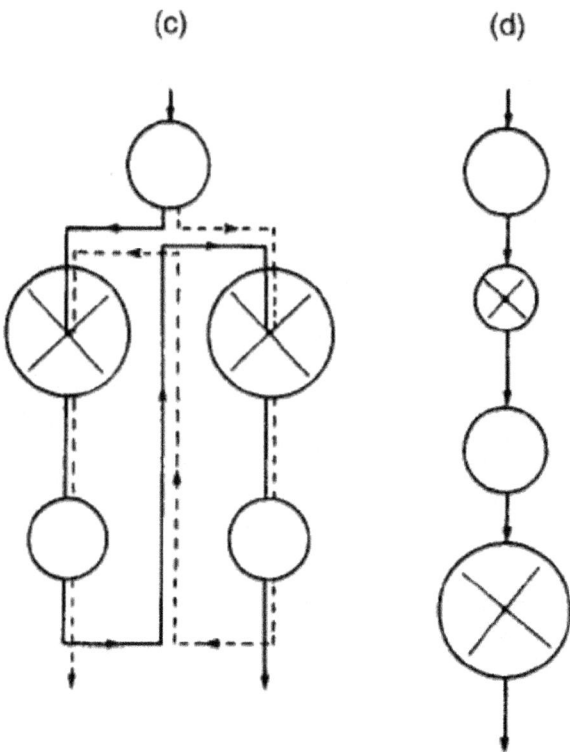

Figure 29. The main systems of operation of biological filters [1].

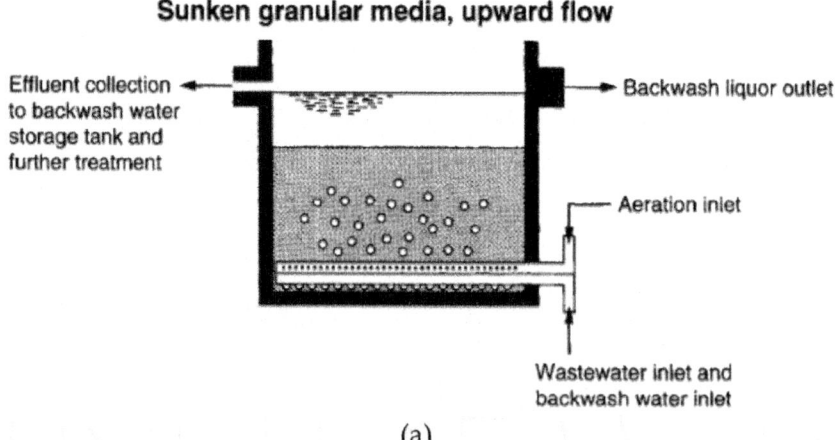

Sunken granular media, downward flow

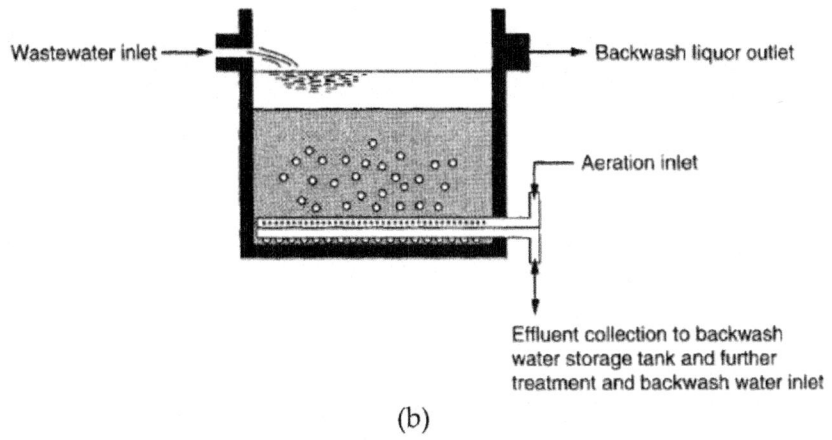

(b)

Floating granular media, upward flow

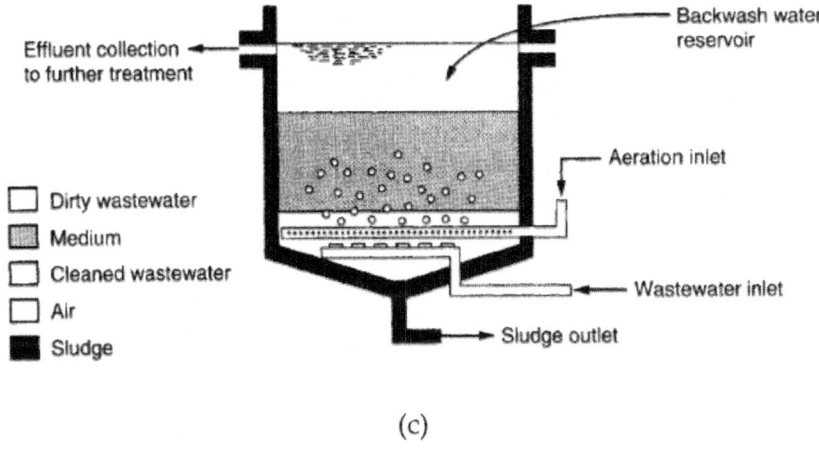

(c)

Figure 30. Biological aerated filters [1].

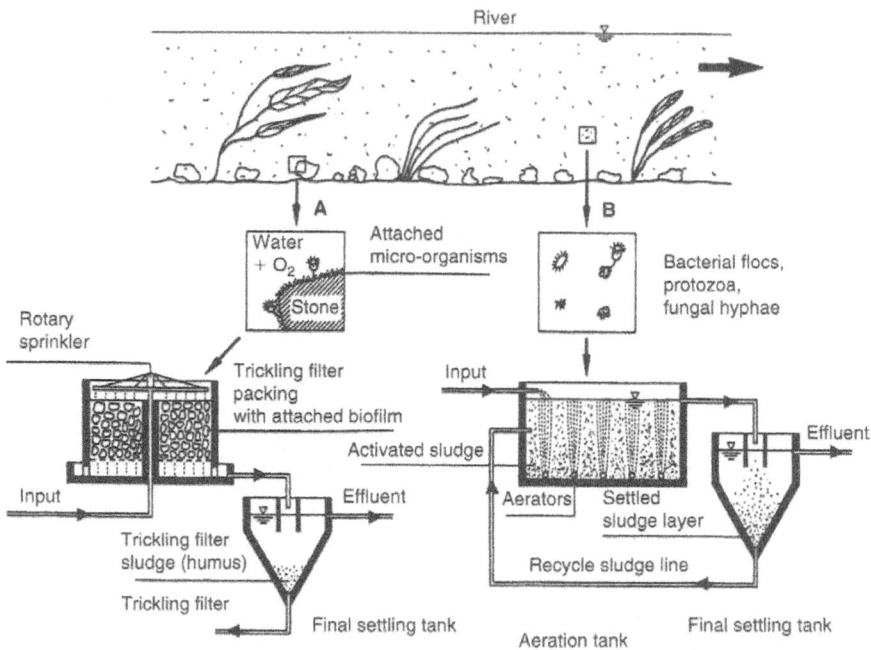

Figure 31. Relationship between the natural bacterial populations in rivers and the development of (A) trickling (percolating) filter and (B) activated sludge system [1].

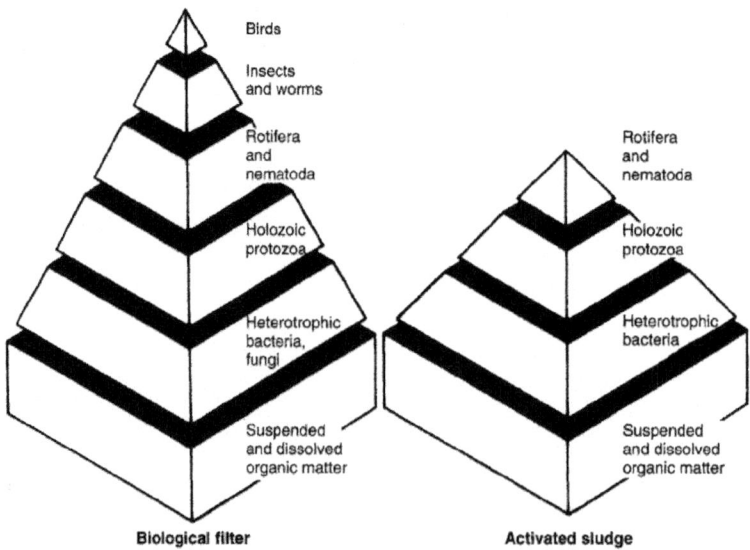

Figure 32. Comparison of the food chain pyramids for biological filters and activated sludge systems [1].

Rotating Biological Contactors

The rotating biological contactors (RBC) system (Figure 33) can be implemented to amend and improve the available treatment processes as the secondary or tertiary treatment processes. The RBC is successfully implemented in all three steps of the biological treatment, which are BOD_5 removal, nitrification, and denitrification. The process is a fixed-biofilm of either aerobic or anaerobic biological treatment system for removal of nitrogenous and carbonaceous compounds from wastewater (Figure 34). The RBC installations (Figure 35) were designed for removal of BOD_5 or ammonia nitrogen (NH_3-N), or both, from wastewater [1, 2].

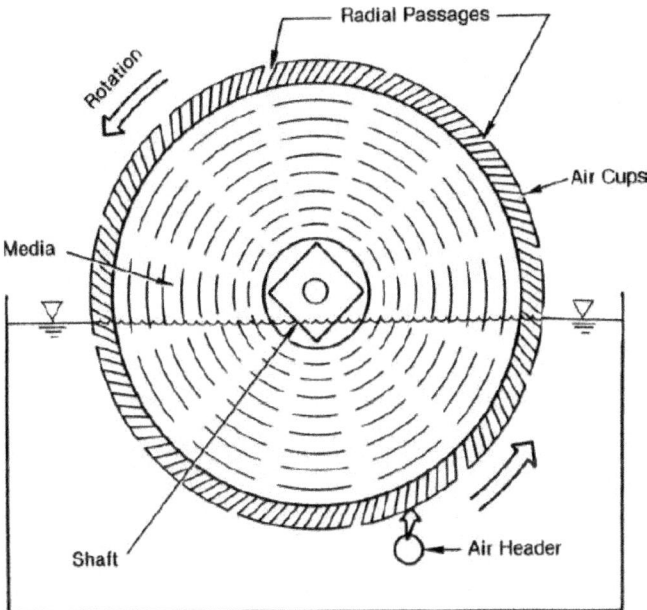

Figure 33. Schematic diagram of air-drive RBC [2].

The RBC consists of media, shaft, drive, bearings, and cover (Figure 34). The RBC hardware consists of a large diameter and closely spaced circular plastic media that is mounted on a horizontal shaft supported by bearings and is slowly rotated by an electric motor. The plastic media are made of corrugated polystyrene or polyethylene material with different designs, dimensions, and densities. The model designs are based on increasing surface area and firmness, allowing a winding wastewater flow path and stimulating air turbulence [1, 2].

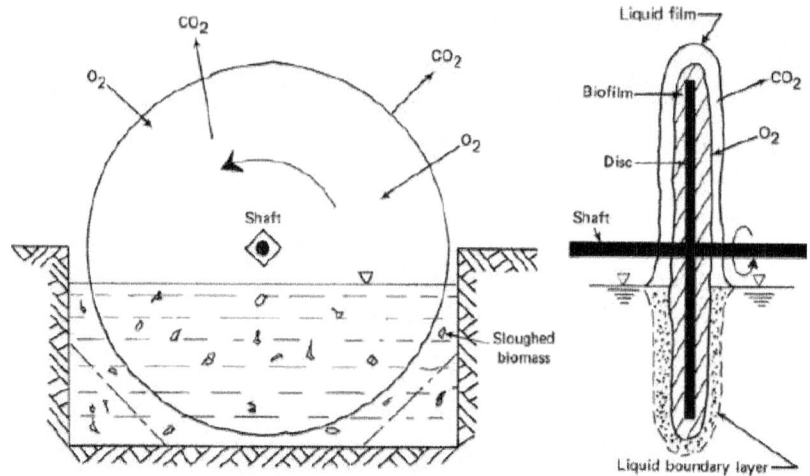

Figure 34. Mechanism of attached growth media in an RBC system [2].

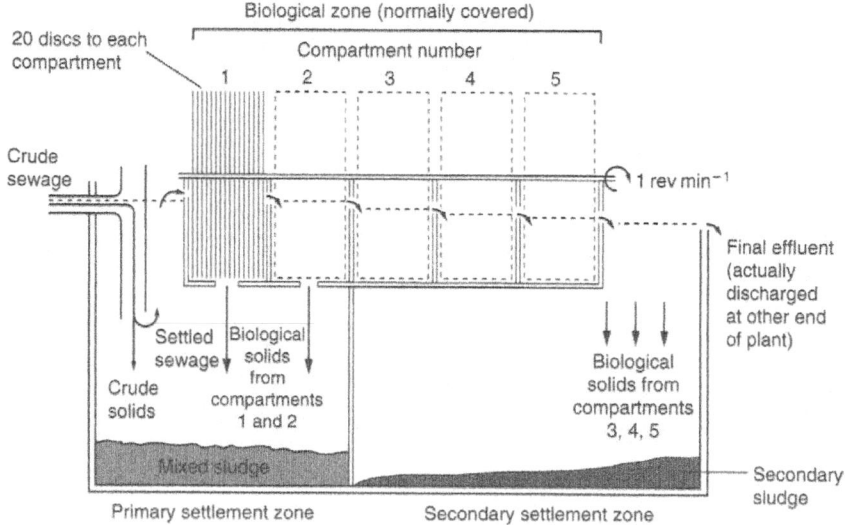

Figure 35. RBC system [1].

Biological Removal of Nutrients

Biological Phosphorous Removal

It is widely agreed that microorganisms utilize acetate and fatty acids to accumulate polyphosphates as poly-β-hydroxybutyrate, which is an acid polymer. The precise mechanism is based on the production and regeneration of adenosine diphosphate (ADP) within the bacteria, and it involves the adenosine triphosphate (ATP). Phosphate removal requires true anaerobic conditions, which occur only when there is no other oxygen donor [3]. Figure 36 shows a phosphate removal process. This process needs long narrow tanks for maintenance of plug flow.

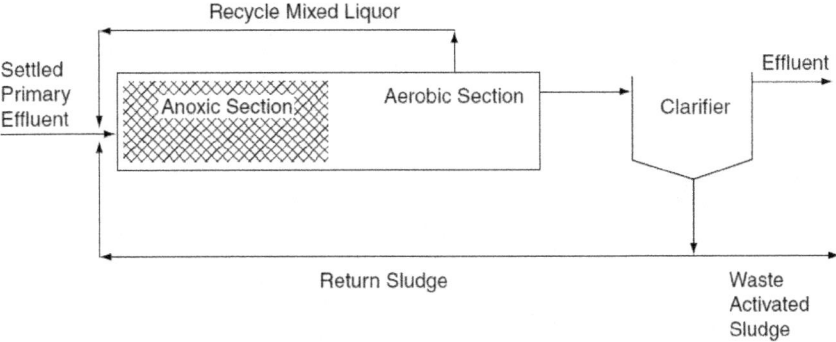

Figure 36. Phosphate removal process [3].

Biological Removal of Nitrogen

The nitrification and denitrification processes are responsible for N_2O production (Figure 37). Figure 38 shows a nitrification/denitrification system for biological removal of nitrogen.

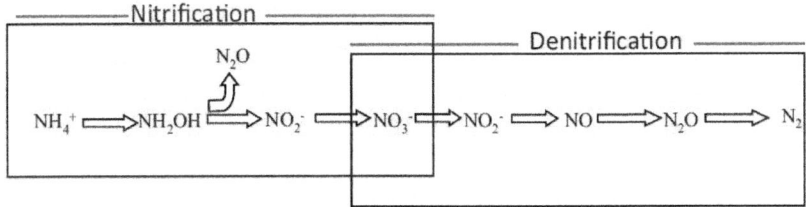

Schematic chemical representation of two processes responsible for N₂O production

Figure 37. Schematic illustration of nitrification and denitrification processes that are responsible for N_2O release [16].

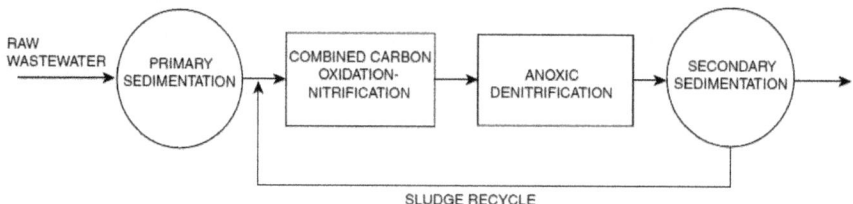

Figure 38. Nitrification/denitrification system for biological removal of nitrogen [3].

Phytoremediation

Phytoremediation is a treatment process that solves environmental problems by implementing plants that abate environmental pollution without excavating the pollutants and disposing them elsewhere. Phytoremediation is the abatement of pollutant concentrations in contaminated soils or water using plants that are able to accumulate, degrade, or eliminate heavy metals, pesticides, solvents, explosives, crude oils and its derivatives, and a multitude of other contaminants and pollutants from water and soils. Figures 39 through 44 show the designs of constructed wetlands where the phytoremediation takes place.

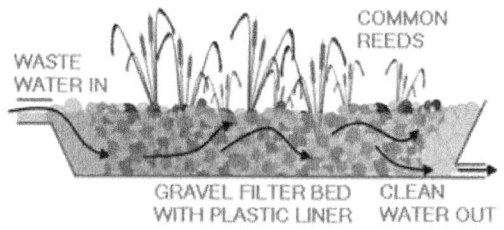

Figure 39. Cross-sectional view of a typical subsurface flow constructed wetland [17].

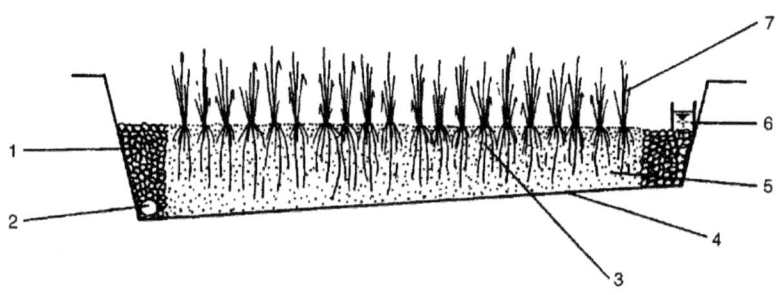

Figure 40. Components of a horizontal flow reed bed: (1) drainage zone consisting of large rocks, (2) drainage tube of treated effluent, (3) root zone, (4) impermeable liner, (5) soil or gravel, (6) wastewater distribution system, and (7) reeds [1].

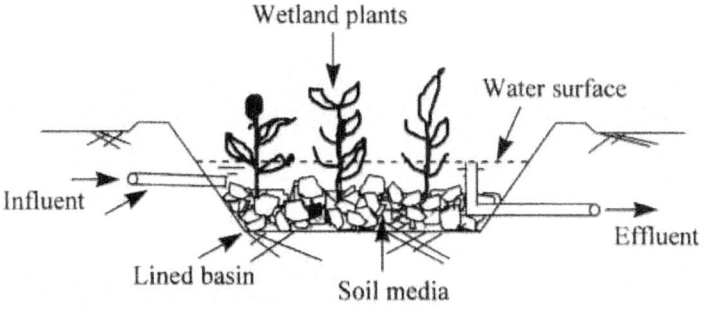

Figure 41. Free water surface system [18].

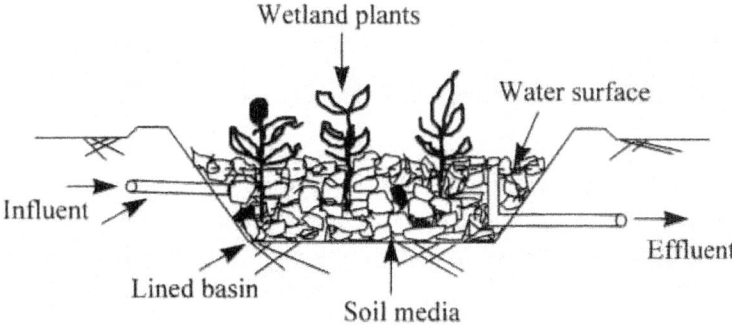

Figure 42. Sub-surface flow system [18].

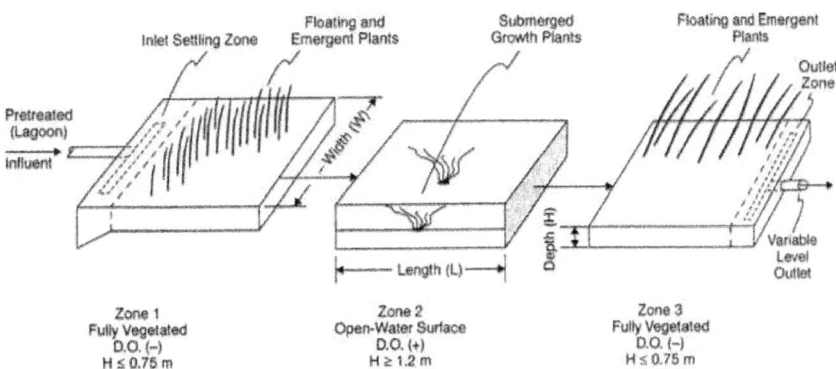

Figure 43. Components of a free water surface constructed wetland [2].

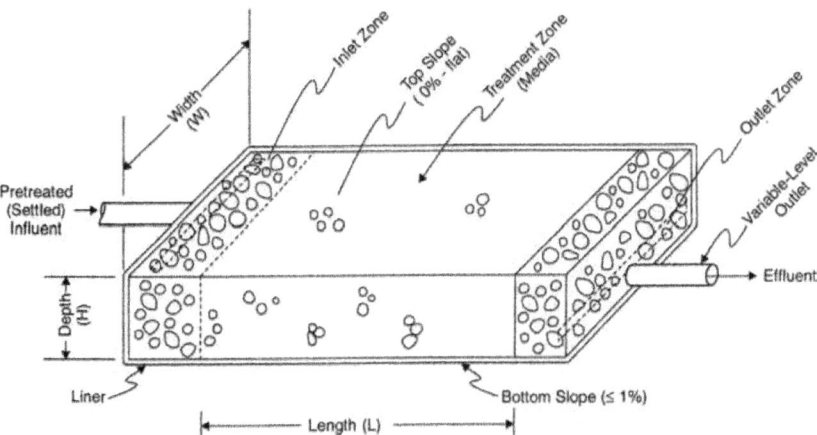

Figure 44. Components of a vegetated submerged bed system [2].

The incorporation of heavy metals, such as mercury, into the food chain may be a deteriorating matter. Phytoremediation is useful in these situations, where natural plants or transgenic plants are able to phytodegrade and phytoaccumulate these toxic contaminants in their above-ground parts, which will be then harvested for extraction. The heavy metals in the harvested biomass can be further concentrated by incineration and recycled for industrial implementation. Rhizofiltration is a sort of phytoremediation that involves filtering wastewater through a mass of roots to remove toxic substances or excess nutrients. Phytoaccumulation or phytoextraction implements plants or algae to remove pollutants and contaminants from wastewater into plant biomass that can be harvested. Organisms that accumulate over than usual amounts of pollutants from soils are termed hyperaccumulators, where a multitude of tables that show the different hyperaccumulators are available and should be referred to. In the case of organic pollutants, such as pesticides, explosives, solvents, industrial chemicals, and other xenobiotic substances, certain plants render these substances non-toxic by their metabolism and this process is called phytotransformation. In other cases, microorganisms that live in symbiosis with plant roots are able to metabolize these pollutants in wastewater. Figure 45 shows the tissues where the rhizofiltration, phytodegradation, and phytoaccumulation take place.

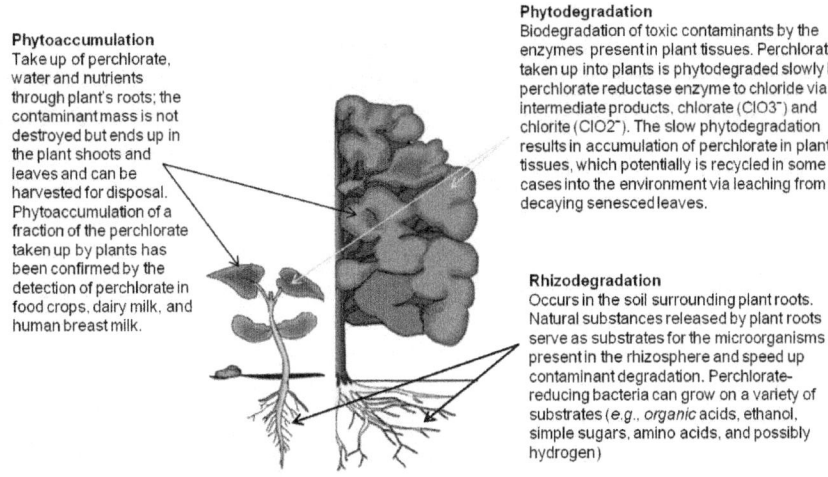

Phytoaccumulation
Take up of perchlorate, water and nutrients through plant's roots; the contaminant mass is not destroyed but ends up in the plant shoots and leaves and can be harvested for disposal. Phytoaccumulation of a fraction of the perchlorate taken up by plants has been confirmed by the detection of perchlorate in food crops, dairy milk, and human breast milk.

Phytodegradation
Biodegradation of toxic contaminants by the enzymes present in plant tissues. Perchlorate taken up into plants is phytodegraded slowly by perchlorate reductase enzyme to chloride via intermediate products, chlorate ($ClO3^-$) and chlorite ($ClO2^-$). The slow phytodegradation results in accumulation of perchlorate in plant tissues, which potentially is recycled in some cases into the environment via leaching from decaying senesced leaves.

Rhizodegradation
Occurs in the soil surrounding plant roots. Natural substances released by plant roots serve as substrates for the microorganisms present in the rhizosphere and speed up contaminant degradation. Perchlorate-reducing bacteria can grow on a variety of substrates (*e.g., organic* acids, ethanol, simple sugars, amino acids, and possibly hydrogen)

Figure 45. Rhizofiltration, phytodegradation, and phytoaccumulation [19].

Vermifiltration

Vermiculture, or worm farming, is the implementation of some species of earthworm, such as *Eisenia fetida* (known as red wiggler, brandling, or manure worm) and *Lumbricus rubellus*, to make vermicompost, also known as worm compost, vermicast, worm castings, worm humus, or worm manure, which is the end-product of the breakdown of organic matter and considered to be a nutrient-rich biofertilizer and soil conditioner. Vermiculture can be implemented to transform livestock manure, food leftovers, and organic matters into a nutrient-rich biofertilizer.

The potential use of earthworms to break down and manage sewage sludge began in the late 1970s [20] and was termed vermicomposting. The introduction of earthworms to the filtration systems, termed vermifiltration systems, was advocated by José Toha in 1992 [21]. Vermifilter is widely used to treat wastewater, and appeared to have high treatment efficiency, including synchronous stabilization of wastewater and sludge [22, 23, 24]. Vermifiltration is a feasible treatment method to reduce and stabilize liquid-state sewage sludge under optimal conditions [24, 25, 26]. Vermicomposting involves the joint action of earthworms and microorganisms [24, 27, 28], and significantly enhances the breakdown of sludge. Earthworms operate as mechanical blenders and by comminuting the organic matter they modify its physical and chemical composition, steadily decreasing the C:N ratio, increasing the surface area exposed to microorganisms, and making it much more suitable for bacterial activity and further breakdown. Throughout the passageway is the earthworm gut, they move fragments and bacteria-rich excrements, consequently homogenizing the organic matter [29]. An intensified bacterial diversity was found in vermifilter, compared with conventional biofilter without earthworms [25]. The principle of using earthworms to treat sewage sludge is based on the perception that there is a net loss of biomass and energy when the food chain is extended [25]. Compared to other technologies of liquid-state sludge stabilization, such as anaerobic digestion and aerobic digestion [30], vermifiltration is a low-cost and an ecologically sound technique, and more suitable for sewage sludge treatment of small or developing-countries' WWTPs [23, 24, 25, 26, 31]. Figure 46 illustrates schematic diagram of a vermifilter, where the earthworms are in the filter bed.

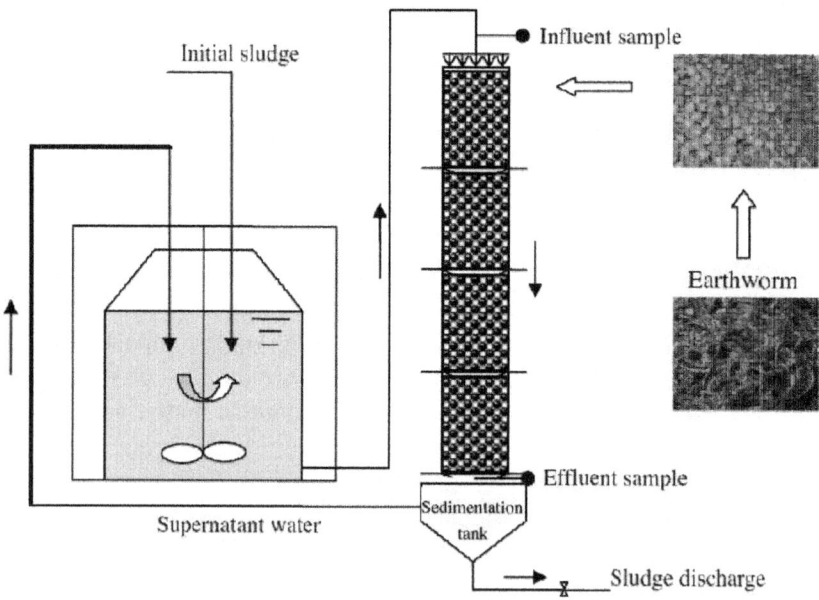

Figure 46. Schematic diagram of a vermifilter [24].

An important application is in livestock manure treatment as shown in Figure 47, where manure is flushed out from the livestock building to a raw effluent tank then the raw effluent is screened to separate the solid waste from manure. The screened effluent is then introduced to the vermifilter to produce the vermicompost. The vermifiltered effluent is then stored in a sedimentation tank. Afterwards, the vermifiltered effluent is introduced to constructed wetlands where the phytoremediation process takes place. The purified water can be then used to flush the water from the livestock building.

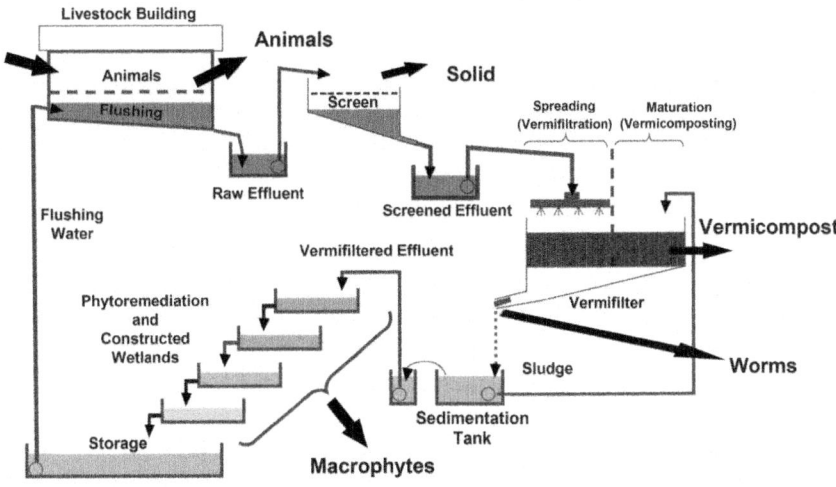

Figure 47. Schematic diagram of a manure treatment system containing vermifiltration and phytoremediation processes (Amended and redrawn from Morand et al. [32]).

Microbial Fuel Cells

The microbial fuel cells (MFCs) allow bacteria to grow on the anode by oxidizing the organic matter that result in releasing electrons. The cathode is sparked with air to provide dissolved oxygen for the reaction of electrons, protons, and oxygen on the cathode, which result in completing the electrical circuit and producing electrical energy (Figure 48).

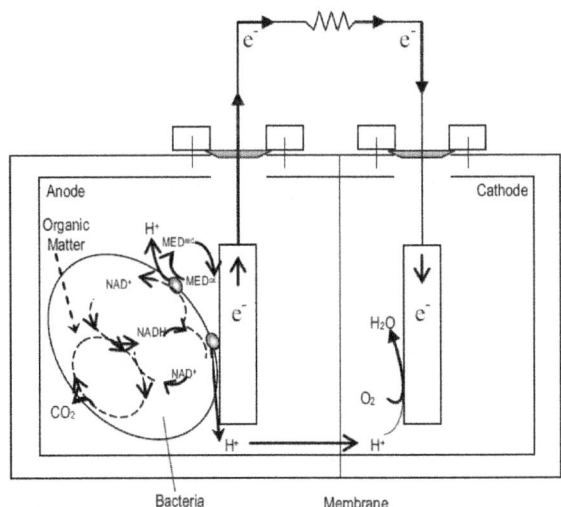

Figure 48. Schematic diagram of the essential components of an MFC [33].

CHEMICAL TREATMENT OF WASTEWATER

Chemical Precipitation

The dissolved inorganic components can be removed by adding an acid or alkali, by changing the temperature, or by precipitation as a solid. The precipitate can be removed by sedimentation, flotation, or other solid removal processes [1]. Although chemical precipitation (coagulation, flocculation) is still implemented, it is highly recommended to substitute the chemical precipitation process by phytoremediation (see previous section), where the trend is to ramp up the implementation of bioremediation and phytoremediation to reduce the use of chemicals, which is in line with the "Green Development".

Neutralization

Neutralization is controlling the pH of the wastewater whether it is acidic or alkaline to keep the pH around 7. The lack of sufficient alkalinity will require the addition of a base (Table 3) to adjust the pH to the acceptable range. Lime (CaO), calcium hydroxide ($Ca(OH)_2$), sodium hydroxide (NaOH), and sodium carbonate (Na_2CO_3), also known as soda ash, are the most common chemicals used to adjust the pH [34]. The lack of sufficient acidity will require the addition of an acid to adjust the pH to the acceptable range. Sulfuric acid (H_2SO_4) and carbonic acid (H_2CO_3) are the most common chemicals used to adjust the pH.

Table 3. Neutralization: Case of acidic wastewater [34]

Neutralization reactions
To neutralize sulfuric acid with
Lime: $\quad H_2SO_4 + CaO \rightleftharpoons CaSO_4 + H_2O$
Calcium hydroxide: $H_2SO_4 + Ca(OH)_2 \rightleftharpoons CaSO_4 + H_2O$
Sodium hydroxide: $H_2SO_4 + NaOH \rightleftharpoons NaSO_4 + 2H_2O$
Soda ash: $\quad H_2SO_4 + Na_2CO_3 \rightleftharpoons Na_2SO_4 + H_2O + CO_2$
To neutralize hydrochloric acid with
Lime: $\quad 2HCl + CaO \rightleftharpoons CaCl_2 + H_2O$
Calcium hydroxide: $HCl + Ca(OH)_2 \rightleftharpoons CaCl_2 + H_2O$
Sodium hydroxide: $HCl + NaOH \rightleftharpoons NaCl + H_2O$
Soda ash: $\quad HCl + Na_2CO_3 \rightleftharpoons NaCl + H_2O + CO_2$

Note: a stoichiometric reaction will yield a pH of 7.0.

[i] - Note: a stoichiometric reaction will yield a pH of 7.0

Adsorption

Adsorption is a physical process where soluble molecules (adsorbate) are removed by attachment to the surface of a solid substrate (adsorbent). Adsorbents should have an extremely high specific surface area. Examples of adsorbents include activated alumina, clay colloids, hydroxides, resins, and activated carbon. The surface of the adsorbent should be free of adsorbate. Therefore, the adsorbent should be activated before use. A wide range of organic materials can be removed by adsorption, including detergents and toxic compounds. The most widely used adsorbent is activated carbon, which can be produced by pyrolytic carbonization of biomass [1]. Figure 49 illustrates the difference between absorption and adsorption. Activated carbon is the most implemented adsorbent and is a sort of carbon processed to be riddled with small, low-volume pores that enlarge the surface area available for adsorption. Owing to its high level of microporosity, 1 g of activated carbon has a surface area larger than 500 m^2, which was determined by gas adsorption. Figure 50 shows a bed carbon adsorption unit. Note that the carbon can be regenerated by thermal oxidation or steam oxidation and reused. The adsorption capacity, one of the most important characteristics of an adsorbent, can be calculated as follows:

$$\underset{mg/g}{Adsorption\ Capacity} = \underset{mg}{Adsorbate} / \underset{g}{Adsorbent}$$

The factors that affect adsorption are [3]:

- Particle diameter: the adsorption is inversely proportional to the particle size of the adsorbent, and directly proportional to surface area.

- Adsorbate concentration: the adsorption is directly proportional to adsorbate concentration.

- Temperature: the adsorption is directly proportional to temperature.

- Molecular weight: generally, the adsorption is inversely proportional to molecular weight depending upon the compound weight and configuration of pores diffusion control.

- pH: the adsorption is inversely proportional to pH due to surface charge.

- Individual properties of adsorbate and adsorbent are difficult to compare.

- Iodine number: is the mass of iodine (g) that is consumed by 100 g of a substance.

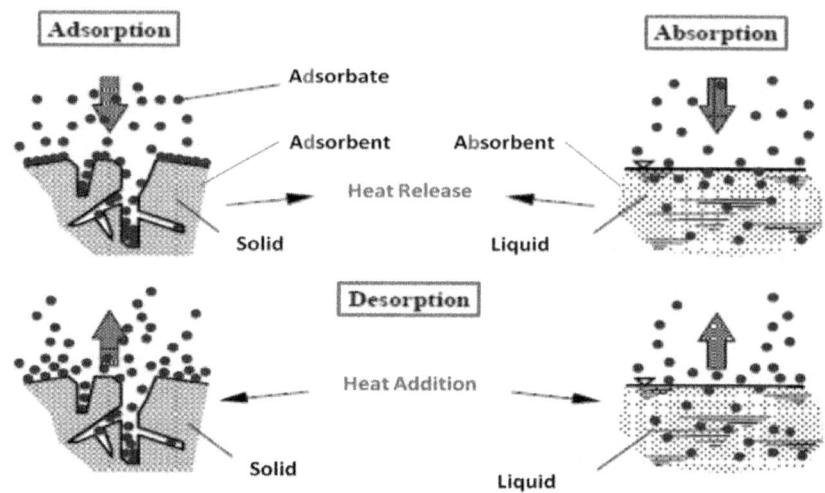

Figure 49. A comparison between absorption and adsorption.

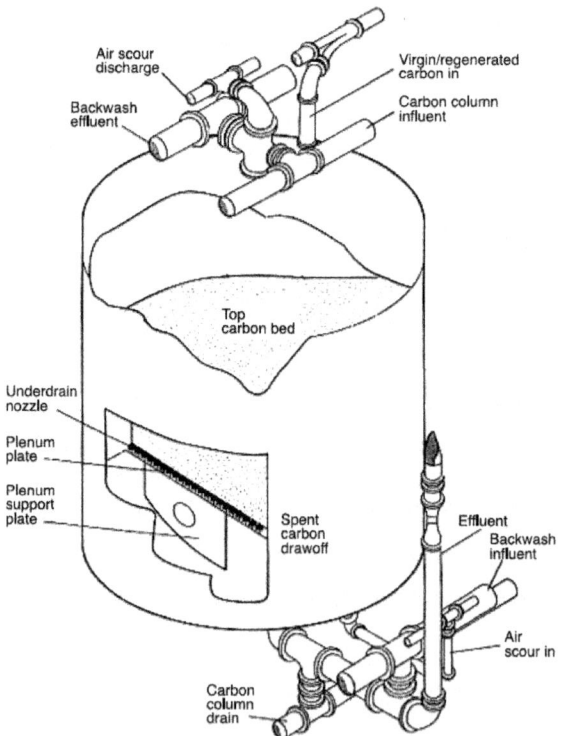

Figure 50. A bed carbon adsorption unit [35].

Disinfection

The disinfection of wastewater is the last treatment step of the tertiary treatment process. Disinfection is a chemical treatment process conducted by treating the effluent with the selected disinfectant to exterminate or at least inactivate the pathogens. The rationales behind effluent disinfection are to protect public health by exterminating or inactivating the pathogens such as microbes, viruses, and protozoan, and to meet the wastewater discharge standards. The purpose of disinfection is the protection of the microbial wastewater quality. The ideal disinfectant should have bacterial toxicity, is inexpensive, not dangerous to handle, and should have reliable means of detecting the presence of a residual. The chemical disinfection agents include chlorine, ozone, ultraviolet radiation, chlorine dioxide, and bromine [3].

Chlorine

Chlorine is one of the oldest disinfection agents used, which is one of the safest and most reliable. It has extremely good properties, which conform to the aspects of the ideal disinfectant. Effective chlorine disinfection depends upon its chemical form in wastewater. The influencing factors are pH, temperature, and organic content in the wastewater [3]. When chlorine gas is dissolved in wastewater, it rapidly hydrolyzes to hydrochloric acid (HCl) and hypochlorous acid (HOCl) as shown in the following chemical equation:

$$Cl_2 + H_2O \leftrightarrow H^+ + Cl^- + HOCl$$

Free ammonia combines with the HOCl form of chlorine to form chloramines in a three-step reaction, as follows:

$$NH_3 + HOCl \rightarrow NH_2Cl + H_2O$$
$$NH_2Cl + HOCl \rightarrow NHCl_2 + H_2O$$
$$NHCL_2 + HOCl \rightarrow NCl_3 + H_2O$$

Figure 51 illustrates the chlorination curve, where the formation of chloramines occurs at the breakpoint. The free chlorine residual first rises then falls until the reaction with ammonia has been completed. As additional chlorine is applied and ammonia is consumed, the chlorine residual rises again.

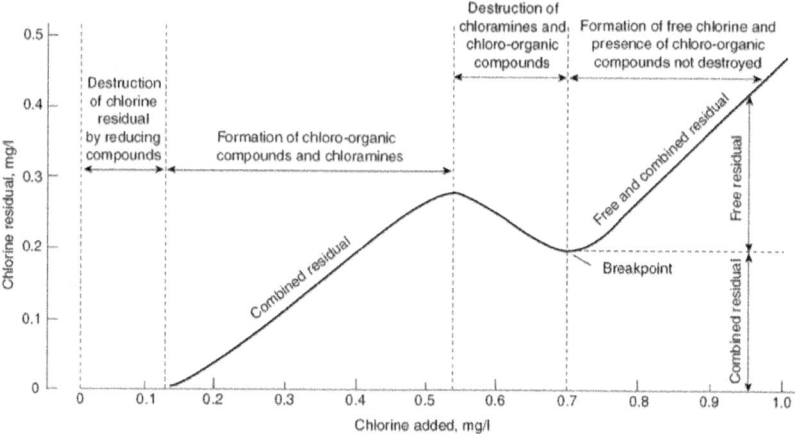

Break point chlorination by the formation of chloramines. The free chlorine residuals first rises then falls until the reaction with ammonia have been completed. As additional chlorine is applied and ammonia is consumed, the chlorine residual again rises.

Figure 51. Chlorination curve [3].

Dechlorination is a very important process, where activated carbon, sulfur compounds, hydrogen sulfide, and ammonia can be implemented to minimize the residual chlorine in a disinfected effluent prior to discharge. Activated carbon and sulfur compounds are the most widely used [3]. The commonly used sulfur compounds are sulfur dioxide (SO_2), sodium metabisulfite (NaS_2O_5), sodium bisulfate ($NaHSO_3$), and sodium sulfite (Na_2SO_3). The dechlorination reactions with the abovementioned compounds are described in the following equations:

$$SO_2 + 2H_2O + Cl_2 \rightarrow H_2SO_4 + 2HCl$$
$$SO_2 + H_2O + HOCl \rightarrow 3H^+ + Cl^- + SO_4^{2-}$$
$$Na_2S_2O_5 + 2Cl_2 + 3H_2O \rightarrow 2NaHSO_4 + 4HCl$$
$$NaHSO_3 + H_2O + Cl_2 \rightarrow NaHSO_4 + 2HCl$$

Ozone

Ozone (O_3) is a very strong oxidant typically used in wastewater treatment. Ozone is able to oxidize a multitude of organic and inorganic compounds in wastewater. These reactions cause an ozone demand in the treated wastewater, which should be fulfilled throughout wastewater ozonation prior to developing an assessable residual. Ozone should be generated at the point of application for use in wastewater treatment as ozone is an unstable molecule [3]. Figure 52 illustrates the corona discharge method for making ozone. Ozone is generally formed by combining an oxygen atom with an oxygen molecule (O_2) as follows:

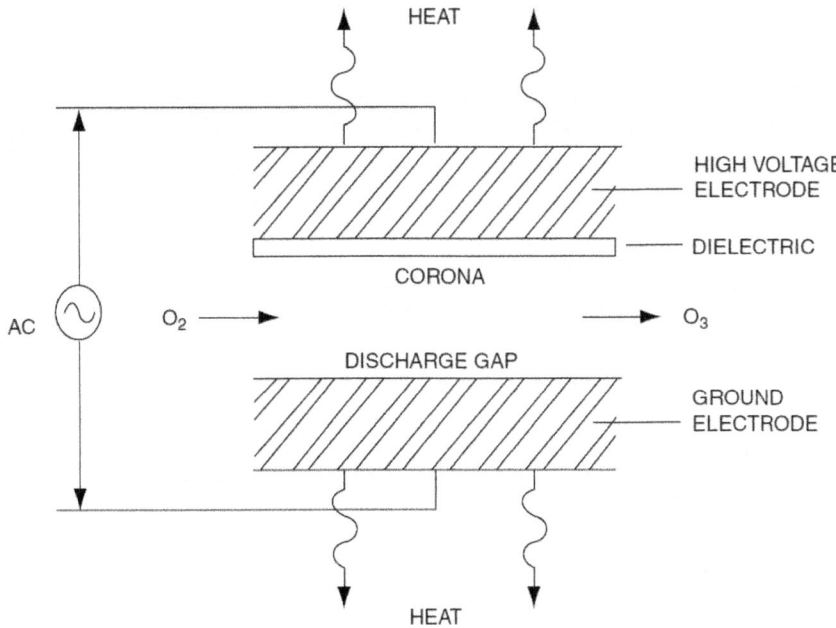

Figure 52. Schematic drawing of corona discharge method for making ozone [3].

Ultraviolet Light

Ultraviolet (UV) radiation is a microbial disinfectant that leaves no residual. It requires clear, un-turbid, and non-colored water for its implementation. The commercial UV disinfection systems use low- to medium-powered UV lamps with a wavelength of 354 nm [3]. The UV dosage can be calculated as follows:

$D = I \cdot t$

where, D is the UV dose (mW. s/cm^2); I is the intensity (mW/cm^2); and t is the exposure time (s).

The advantages of UV radiation are: (1) directly effective against the DNA of many microorganisms, (2) not reactive with other forms of carbonaceous demand, and (3) provides superior bactericidal kill values while not leaving any residues. The advantage is often the disadvantage, because power fluctuations, variations in hydraulic flow rates, and color or turbidity can cause the treatment to be ineffective [3]. Additionally, cell recovery and re-growth of the damaged organisms because of the inactivation of their predators and competitors has come to light.

Ion Exchange

Ion exchange (IX) is a reversible reaction in which a charged ion in a solution is exchanged with a similarly charged ion which is electrostatically attached to an immobile solid particle. The most common implementation of ion exchange method in wastewater treatment is for softening, where polyvalent cations (e.g., calcium and magnesium) are exchanged with sodium [36]. Practically, wastewater is introduced into a bed of resin. The resin is manufactured by converting a polymerization of organic compounds into a porous matrix. Typically, sodium is exchanged with cations in the solution [34]. The bed is shut down when it becomes saturated with the exchanged ions, where it should be regenerated by passing a concentrated solution of sodium back through the bed. Figure 53shows the schematic illustration of organic cation-exchange bead. Figure 54 shows a typical ion exchange resin column. Table 4 shows the ion preference and affinity for some selected compounds.

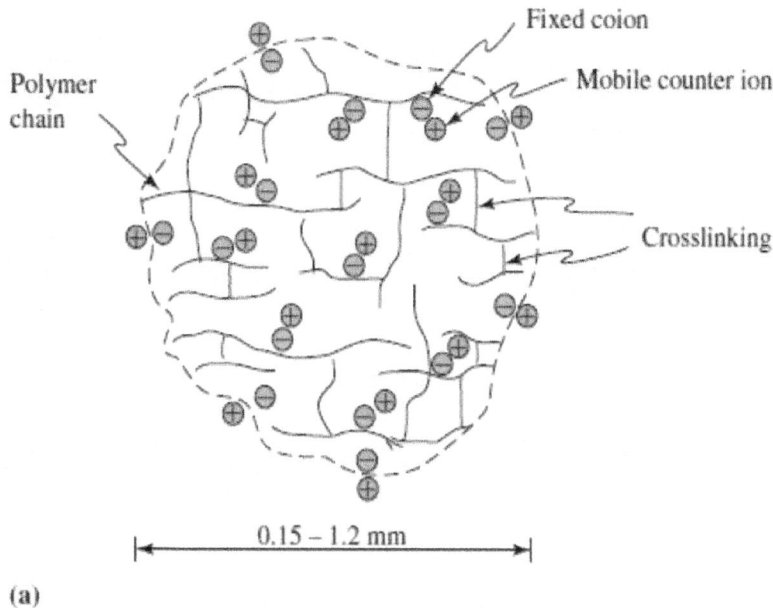

(a)

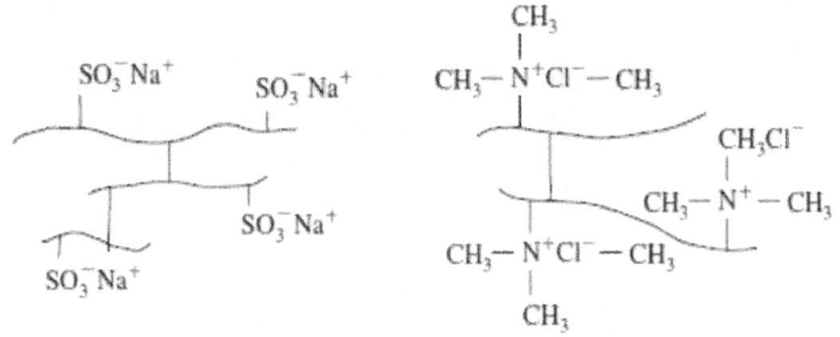

$SO_3^-Na^+$ $SO_3^-Na^+$

$SO_3^-Na^+$

$SO_3^-Na^+$

CH_3

$CH_3-N^+Cl^--CH_3$

CH_3Cl^-

$CH_3-N^+-CH_3$

$CH_3-N^+Cl^--CH_3$

CH_3

(b)

Schematic of organic cation-exchange bead:

(a) The bead is shown as a polystyrene polymer cross-linked with divinyl benzene with fixed coions (minus charge) balanced by mobile positively charged counterions (positive charge).

(b) strong cation exchange resin on left (Na⁻ form) and strong-base on right (Cl⁻ form).

Figure 53. Schematic illustration of organic cation-exchange bead [34].

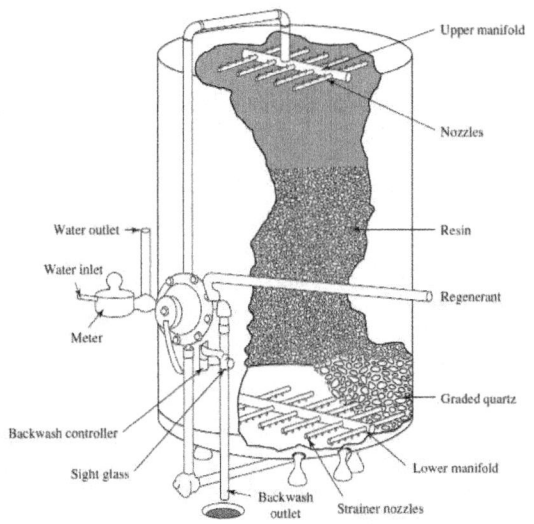

Typical ion exchange resin column.
(*Source:* U.S. EPA, 1981.)

Figure 54. Typical ion exchange resin column [37].

Table 4. Ion preference and affinity for some selected compounds [3]

Strong Acid Cation Exchanger	Strong Base Anion Exchanger	Weak Acid Cation Exchanger	Weak Base Anion Exchanger	Weak Acid Chelate Exchanger
Barium (2+)	Iodide (1−)	Hydrogen (1+)	Hydroxide (1−)	Copper (2+)
Lead (2+)	Nitrate (1−)	Copper (2+)	Sulfate (2−)	Iron (2+)
Mercury (2+)	Bisulfite (1−)	Cobalt (2+)	Chromate (2−)	Nickel (2+)
Copper (1+)	Chloride (1−)	Nickel (2+)	Phosphate (2−)	Lead (2+)
Calcium (2+)	Cyanide (1−)	Calcium (2+)	Chloride (1−)	Manganese (2+)
Nickel (2+)	Bicarbonate (1−)	Magnesium (2+)		Calcium (2+)
Cadmium (2+)	Hydroxide (1−)	Sodium (1+)		Magnesium (2+)
Copper (2+)	Fluoride (1−)			Sodium (1+)
Cobalt (2+)	Sulfate (2−)			
Zinc (2+)				
Cesium (1+)				
Iron (2+)				
Magnesium (2+)				
Potassium (1+)				
Manganese (2+)				
Ammonia (1+)				
Sodium (1+)				
Hydrogen (1+)				
Lithium (1+)				

Physicochemical Treatment Processes

The principal advanced physicochemical wastewater treatment processes are elucidated in Table 5.

Table 5. Principal advanced physicochemical wastewater treatment processes [1]

Principal advanced physico-chemical wastewater treatment processes

Process	*Removal function*
Filtration	Suspended solids
Air stripping	Ammonia
Breakpoint chlorination	Ammonia
Ion exchange	Nitrate, dissolved inorganic solids
Chemical precipitation	Phosphorus, dissolved inorganic solids
Carbon adsorption	Toxic compounds, refractory organics
Chemical oxidation	Toxic compounds, refractory organics
Ultrafiltration	Dissolved inorganic solids
Reverse osmosis	Dissolved inorganic solids
Electrodialysis	Dissolved inorganic solids
Volatilization and gas stripping	Volatile organic compounds

WASTEWATER TREATMENT PLANTS

This section shows some examples of WWTPs as shown in Figure 55 (a, b, and c) and Figure 56. On the other hand, there are some computer programs for planning and designing WWTPs (Figures 57, 58, and 59).

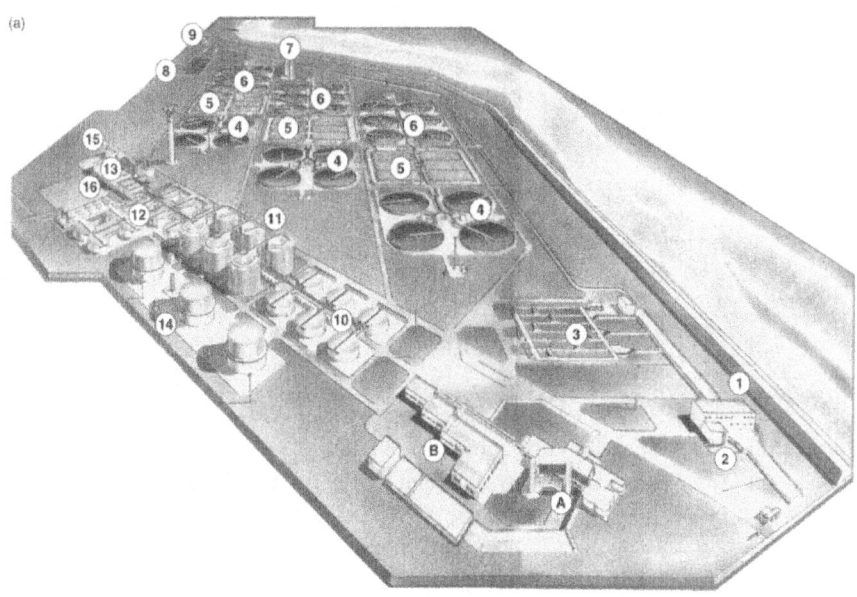

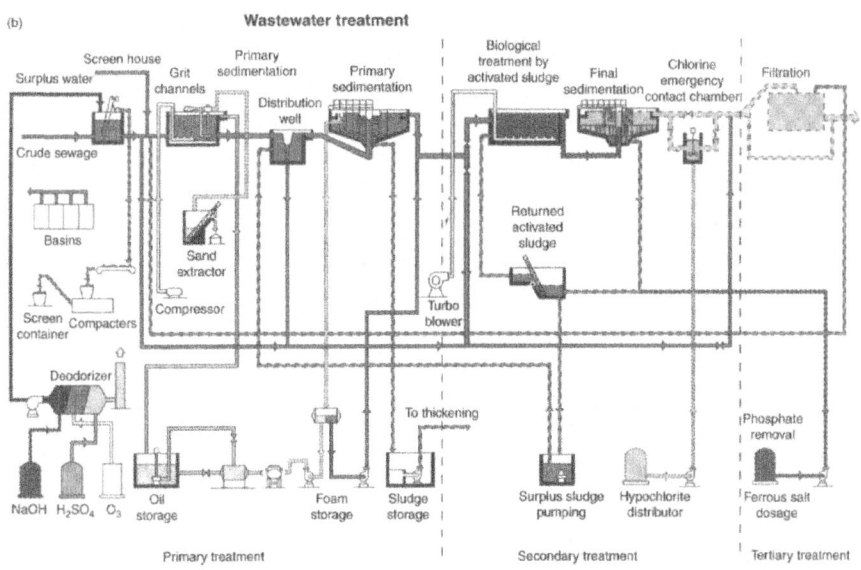

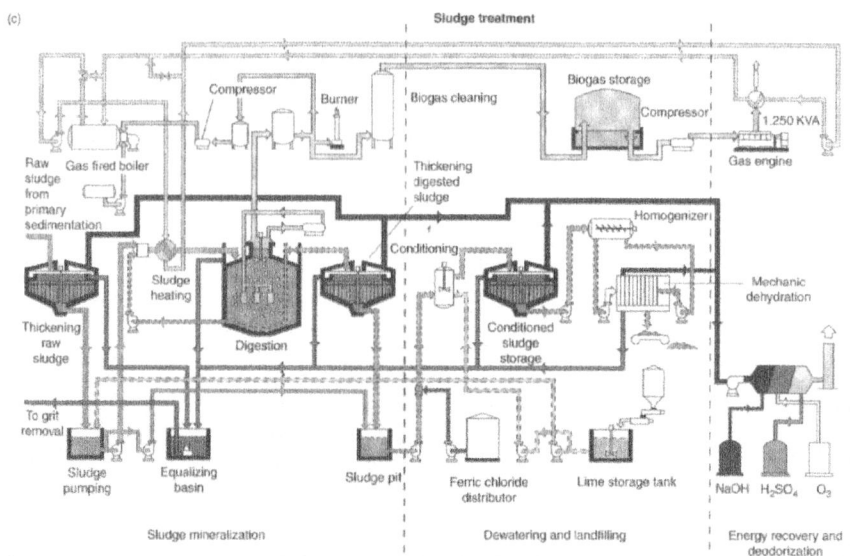

Figure 55. WWTP showing: (a) layout of the plant, (b) wastewater process flow diagrams, and (c) sludge process flow diagram. **Wastewater treatment**: 1. Storm water overflow; 2. screening; 3. grit removal; 4. primary sedimentation; 5. aeration tanks; 6. Secondary sedimentation; 7. emergency chlorination; 8. filtration; 9. effluent outfall. **Sludge treatment**: 10. raw sludge thickeners; 11. digestion tanks; 12. digested sludge thickeners; 13. power house; 14. biogas storage; 15. filter press house; 16. transformer station. A and B are administrative areas [1].

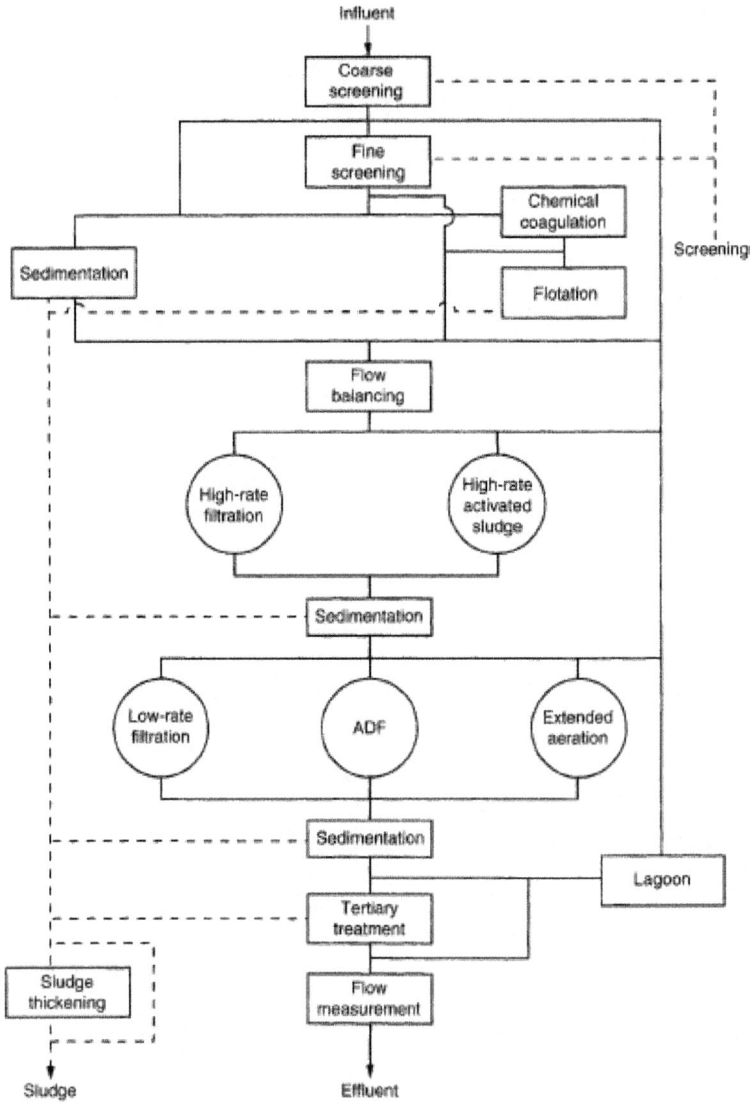

Figure 56. Summary of the main process options commonly employed at both domestic and industrial WWTPs. Not all of these unit processes may be selected, but the order of their use remains the same [1].

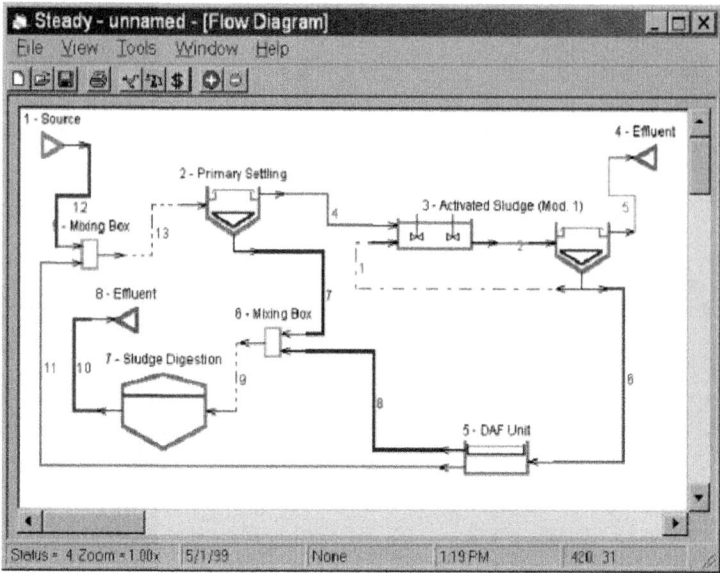

Figure 57. Screenshot of the STEADY program [3].

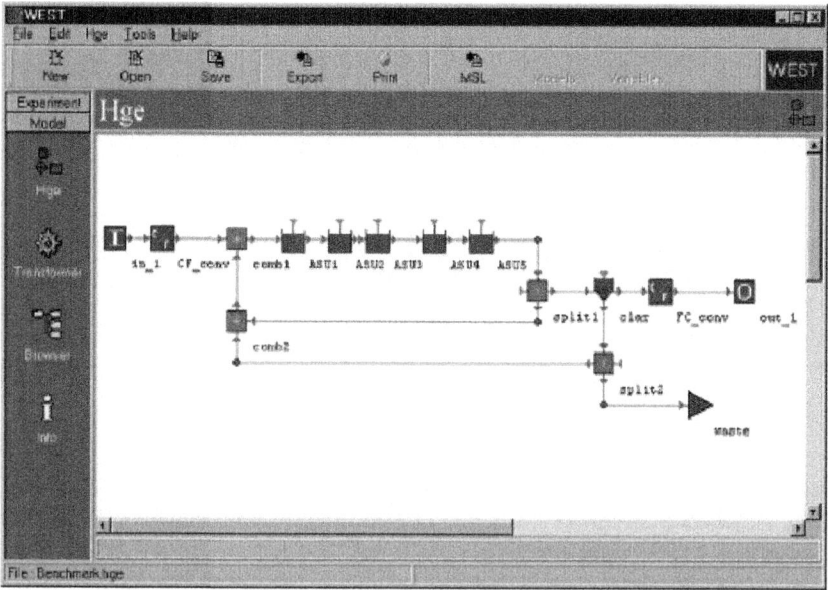

Figure 58. WEST software typical plant configuration [3].

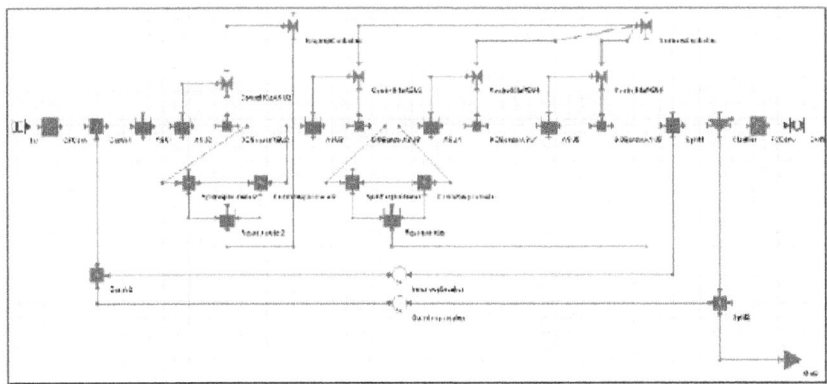

Figure 59. WEST configuration for multitank system [3].

CONCLUSIONS

According to this study, it can be conclude that:

- The trend is to ramp up the implementation of bioremediation, phytoremediation, and mycoremediation to reduce the use of chemicals, which is in line with the "Green Development".

- The recent developments elucidate that subsequent to the physical treatment processes (the primary treatment) the biological treatment processes come in turn as secondary treatment and precede the chemical treatment processes, which constitute the tertiary treatment.

- Microbial fuel cells, phytoremediation, and mycoremediation are the focus of the future development in this field.

REFERENCES

1. Gray N. F. (2005). *Water Technology: An Introduction for Environmental Scientists and Engineers (2*nd *Edition)*, Elsevier Science & Technology Books, ISBN 0750666331, Amsterdam, The Netherlands.

2. Lin, S. D. (2007). *Water and Wastewater Calculations Manual (2*nd *Edition)*, McGraw-Hill Companies, Inc., ISBN 0-07-154266-3, New York, USA.

3. Russell D. L. (2006). *Practical Wastewater Treatment*, John Wiley & Sons, Inc., ISBN-13: 978-0-471-78044-1, Hoboken, New Jersey, USA.

4. Samer M. (2015). GHG Emission from Livestock Manure and its Mitigation Strategies, In: *Climate Change Impact on Livestock: Adaptation and Mitigation*, V. Sejian, J. Gaughan, L. Baumgard & C.

Prasad (Eds.), In Press, Springer International, ISBN 978-81-322-2264-4, Germany.

5. Samer M., Mostafa E. & Hassan A. M. (2014). Slurry Treatment with Food Industry Wastes for Reducing Methane, Nitrous Oxide and Ammonia Emissions. *Misr Journal of Agricultural Engineering*, 31 (4): 1523–1548.

6. Samer M. (2011). How to Construct Manure Storages and Handling Systems? *IST Transactions of Biosystems and Agricultural Engineering*, Vol. 1, No. 1, pp. 1–7, ISSN 1913-8741.

7. Samer M., Grimm H., Hatem M., Doluschitz R. & Jungbluth T. (2008). Mathematical Modeling and Spark Mapping for Construction of Aerobic Treatment Systems and their Manure Handling System, *Proceedings of International Conference on Agricultural Engineering*, Book of Abstracts p. 28, EurAgEng, Crete, Greece, June 23-25, 2008.

8. Ersahin M. E., Ozgun H., Dereli R. K. & Ozturk I. (2011). Anaerobic Treatment of Industrial Effluents: An Overview of Applications, In: *Waste Water - Treatment and Reutilization,* F. Einschlag (Ed.), pp. 3–28, InTech, ISBN 978-978-953-307-249-4, Rijeka, Croatia.

9. Samer M. (2012). Biogas Plant Constructions, In: *Biogas*, S. Kumar (Ed.), pp. 343–368, InTech, ISBN 978-953-51-0204-5, Rijeka, Croatia.

10. Samer M. (2010). A Software Program for Planning and Designing Biogas Plants. *Transactions of the ASABE*, Vol. 53, No. 4, pp. 1277–1285, ISSN 2151-0032.

11. Pfost D. L. & Fulhage C. D. (2000). Anaerobic Lagoons for Storage/ Treatment of Livestock Manure, EQ 387, MU Extension University of Missouri-Columbia, USA.

12. Westerman P. W., Shaffer K. A. & Rice J. M. (2008). Sludge Survey Methods for Anaerobic Lagoons, AG-639W, North Carolina Cooperative Extension Service, College of Agriculture & Life Sciences, North Carolina State University, USA.

13. Pavko A. (2011). Fungal Decolourization and Degradation of Synthetic Dyes Some Chemical Engineering Aspects, In: *Waste Water - Treatment and Reutilization*, F. Einschlag (Ed.), pp. 65–88, InTech, ISBN 978-978-953-307-249-4, Rijeka, Croatia.

14. Diano N. & Mita D.G. (2011). Removal of Endocrine Disruptors in Waste Waters by Means of Bioreactors, In: *Waste Water - Treatment and Reutilization*, F. Einschlag (Ed.), pp. 29-48, InTech, ISBN 978-978-953-307-249-4, Rijeka, Croatia.

15. Kabir M., Suzuki M. & Yoshimura N. (2011). Excess Sludge Reduction in Waste Water Treatment Plants, In: *Waste Water - Treatment and Reutilization*, F. Einschlag (Ed.), pp. 133–150, InTech, ISBN 978-978-953-307-249-4, Rijeka, Croatia.

16. Chadwick D., Sommer S., Thorman R., Fangueiro D., Cardenas L., Amon B. & Misselbrook T. (2011). Manure Management: Implications for Greenhouse Gas Emissions. *Animal Feed Science and Technology*, 166–167(2011): 514– 531, ISSN 0377-8401.

17. Morris R. H. & Knowles P. (2011). Measurement Techniques for Wastewater Filtration Systems, In: *Waste Water - Treatment and Reutilization*, F. Einschlag (Ed.), pp. 109–132, InTech, ISBN 978-978-953-307-249-4, Rijeka, Croatia

18. ESCWA (2003). Economic and Social Commission for Western Asia.

19. Kucharzyk K. H., Soule T., Paszczynski A. J. & Hess T. F. (2011). Perchlorate: Status and Overview of New Remedial Technologies, In: *Waste Water - Treatment and Reutilization*, F. Einschlag (Ed.), pp. 171–194, InTech, ISBN 978-978-953-307-249-4, Rijeka, Croatia.

20. Li, X., Xing M., Yang J. & Huang Z. (2011). Compositional and functional features of humic acid-like fractions from vermicomposting of sewage sludge and cow dung, *J. Hazard. Mater.* 185: 740–748.

21. Li Y., Robin P., Cluzeau D., Bouché M., Qiu J., Laplanche A., Hassouna M., Morand P., Dappelo C. & Callarec J. (2008). Vermifiltration as a stage in reuse of swine wastewater: monitoring methodology on an experimental farm, *Ecol. Eng.* 32: 301–309.

22. Li Y., Xiao Y., Qiu J., Dai Y. & Robin P. (2009). Continuous village sewage treatment by vermifiltration and activated sludge process, *Water Sci. Technol.* 60: 3001–3010.

23. Xing M., Yang J., Wang Y., Liu J. & Yu F. (2011). A comparative study of synchronous treatment of sewage and sludge by two vermifiltration using epigeic earthworm Eisenia fetida, *J. Hazard. Mater.* 185 (2–3): 881–888.

24. Xing M., Li X., Yang J., Lv B. & Lu Y. (2012). Performance and mechanism of vermifiltration system for liquid-state sewage sludge treatment using molecular and stable isotopic techniques. *Chemical Engineering Journal* 197: 143–150.

25. Zhao L., Wang Y., Yang J., Xing M., Li X., Yi D. & Deng D. (2010). Earthworm-microorganism interactions: a strategy to stabilize domestic wastewater sludge, *Water Res.* 44: 2572–2582.

26. Xing M., Zhao L., Yang J., Huang Z. & Xu Z. (2011). Distribution and transformation of organic matter during liquid-state vermiconversion of activated sludge using elemental analysis and spectroscopic evaluation, *Environ. Eng. Sci.* 28 (9): 619–626.

27. Suthar S. (2008). Development of a novel epigeic–anecic-based polyculture vermireactor for efficient treatment of municipal sewage water sludge,*Int. J. Environ. Waster Manage.* 2: 84–101.

28. Xing M., Li X., Yang J., Huang Z. & Lu Y. (2011). Changes in the chemical characteristics of water-extracted organic matter from vermicomposting of sewage sludge and cow dung, *J. Hazard. Mater.* http://dx.doi.org/10.1016/j.jhazmat.2011.11.070.

29. Domínguez J. & Edwards C.A. (2004). Vermicomposting organic wastes, in: S.H. Shakir Hanna, W.Z.A. Mikhaïl (Eds.), *Soil Zoology for Sustainable Development in the 21st Century*, Cairo, pp. 369–395.

30. Fytili D. & Zabaniotou A. (2008). Utilization of sewage sludge in EU application of old and new methods – a review, *Renew. Sust. Energ. Rev.* 12: 116–140.

31. Sinha R.K., Bharambe G. & Chaudhari U. (2008). Sewage treatment by vermifiltration with synchronous treatment of sludge by earthworms: A low-cost sustainable technology over conventional systems with potential for decentralization, *Environmentalist.* 28: 409–420.

32. Morand P., Robin P., Qiu J.P., Li Y., Cluzeau D., Hamon G., Amblard C., Fievet S., Oudart D., Pain le Quere C., Pourcher A.-M., Escande A., Picot B. & Landrain B. (2009). Biomass production and water purification from fresh liquid manure by vermiculture, macrophytes ponds and constructed wetlands to recover nutrients and recycle water for flushing in pig housing. Proceedings of the International Congress "Ecological engineering: From concepts to applications Foreword (EECA)", 2-4 December 2009, Paris, France.

33. Logan B. (2008). *Microbial Fuel Cells*, Wiley. Hoboken, NJ, EEUU. 200 pp.

34. Davis M. L. (2010). *Water and Wastewater Engineering: Design Principles and Practice*, McGraw-Hill Companies, Inc., ISBN 978-0-07-171385-6, New York, USA.

35. Sincero, A. P. & G. A. Sincero (2003). *Physical-Chemical Treatment of Water and Wastewater,* CRC Press LLC, ISBN 1-84339-028-0, Boca Raton, Florida, USA.

36. Clifford D. A. (1999). Ion Exchange and Inorganic Adsorption, In: *Water Quality and Treatment (5th Edition)*, R. D. Letterman (Ed.), pp. 9.1–9.91., American Water Works Association, McGraw-Hill, New York, USA.

37. U.S. EPA (1981) Development Document for Effluent Limitations: Guideline and Standards for the Metal Finishing Point Source Category, U.S. Environmental Protection Agency, EPA/440/1-83-091, Washington, D.C., USA.

38. U.S. EPA (2002). Wastewater Technology Fact Sheet: Anaerobic Lagoons, U.S. Environmental Protection Agency, EPA 832-F-02-009, Washington, D.C., USA.

Chapter 4

RECENT DEVELOPMENTS IN CHEMICAL SYNTHESIS WITH BIOCATALYSTS IN IONIC LIQUIDS

Mahesh K. Potdar, Geoffrey F. Kelso, Lachlan Schwarz, Chunfang Zhang and Milton T. W. Hearn

Centre for Green Chemistry, School of Chemistry, Monash University, Melbourne, Victoria 3800, Australia

ABSTRACT

Over the past decade, a variety of ionic liquids have emerged as greener solvents for use in the chemical manufacturing industries. Their unique properties have attracted the interest of chemists worldwide to employ them as replacement for conventional solvents in a diverse range of chemical transformations including biotransformations. Biocatalysts are often regarded as green catalysts compared to conventional chemical catalysts in organic synthesis owing to their properties of low toxicity, biodegradability, excellent selectivity and good catalytic performance under mild reaction conditions. Similarly, a selected number of specific ionic liquids can be considered as greener solvents superior to organic solvents owing to their negligible vapor pressure, low flammability, low toxicity and ability to dissolve a wide range of organic and biological substances, including proteins. A combination of biocatalysts and ionic liquids thus appears to be a logical and promising opportunity for industrial use as an alternative to conventional organic chemistry processes employing organic solvents. This article provides an overview of recent developments in this field with special emphasis on the application of more sustainable enzyme-catalyzed reactions and separation processes employing ionic liquids, driven by advances in fundamental knowledge, process optimization and industrial deployment.

INTRODUCTION

As a consequence of rapid developments in many fields of synthetic organic chemistry, researchers from both academia and industry now have to pay much greater attention to the detrimental effects of chemicals and chemistry processes on the environment. Against the backdrop of climate change and increasing public concern about global sustainability, the increasing needs for environmental protection and sustainable development are forcing chemists to rethink the way in which chemistry is conducted, resulting in new synthetic strategies that are more atom efficient, achieve better mass intensification factors, generate less waste and utilize less hazardous conditions through the deployment of the twelve principles of Green Chemistry [1].

Various approaches can be practiced in green organic synthesis, including ways to enhance atom utilization, replace stoichiometric reagents with catalysts, use benign solvents or solvent-free processes, design and generate safer chemical products that are biodegradable, and reduce or totally eliminate waste products. A variety of next-generation solvents, including liquid polymers, fluorous solvents, supercritical fluids and ionic liquids have been explored as greener alternatives to those conventional organic solvents that exhibit undesirable flammable, explosive or toxic properties.

Ionic liquids, as the name suggests, are liquids comprised of ions. In practical terms, it is desirable for an ionic liquid to exist in the liquid state at temperatures below 100 °C and, preferentially, at ambient temperature. Although ionic liquids have been known for at least 100 years, their application in various chemistry processes has only been studied extensively over the last three decades [2]. Research into ionic liquids has ranged from their application as solvents in electrochemistry [3], organic and inorganic transformations [2,4], biotechnology processes [5], as solar cell electrolytes [6], and as supported phases in organic chemistry [7,8]. For organic synthesis with transition metal catalysis, ionic liquids were initially employed as biphasic systems [9]. Since then, their application has expanded to include the development of task-specific ionic liquids, wherein the constituent ions of the ionic liquid are designed to fulfill the requirements of a specific reaction [10,11] or alternatively a biotransformation [12,13]. In addition, substrates have been covalently attached to an ionic liquid prior to the reaction, after which the desired product is isolated and the ionic liquid recycled [14,15].

Ionic liquids have long been considered as green alternatives to organic solvents, although extensive debate and some controversy have surrounded their "so called" green properties. For example, although a specific ionic liquid may have negligible vapor pressure and low flammability, the anions and/or the cations, which comprise the ionic liquid, may individually, both, or when in combination, be toxic. Therefore, like any other solvent, all the potential detrimental properties, e.g., toxicity and low biodegradability, of a specific ionic liquid needs to be assessed before it can be regarded as a green alternative to conventional organic solvents. In addition to these considerations, the production and life-cycle of an ionic liquid also needs to be evaluated in order to fully determine its "greenness" compared to conventional organic solvents. A recent study by Deetlefs and Seddon on the "greenness" of various ionic liquids made by laboratory-scale synthetic methods found that there is considerable room for improvement in terms of E-factor, energy consumption and purification [16]. Hence, for future industrial processes it will be desirable from the outset to design the ionic liquid to be easily and cleanly synthesized, to have *inter alia* the requisite green properties of low vapor pressure, low flammability and low toxicity, and possess practical working attributes such as low viscosity and easy recyclability, by taking advantage of the enormous combinatorial and structural diversity of ionic liquids in terms of their anion/ cation composition [17,18].

Enzymes are nature's catalysts and have been employed for millennia in the manufacture of food and beverage products. With the advent of genetic engineering and advanced biotechnology processes, their use has greatly increased over several decades in industry for the synthesis of important pharmaceuticals, agrochemicals and fine chemicals [19,20]. Due to their excellent catalytic, stereo-selective and chemoselective properties, low toxicity, high biodegradability and efficiency under mild reaction conditions, enzymes are regarded as excellent green catalysts. Whilst some enzymes need to be associated with biological membranes to elicit function, many enzymes exist in nature in an aqueous environment. Many enzymes have the specific ability to perform well in organic solvents. For example, in his pioneering work, Klibanov showed that certain enzymes are stable and can catalyze reactions in various organic solvents like hexane, toluene, acetonitrile or tetrahydrofuran [21,22].

This versatile aspect of enzymes has prompted chemists to explore their utility in ionic liquids as the combination of biocatalysts and ionic liquids have the potential to provide a more sustainable approach for organic transformations to be carried out, particularly when the substrates have poor solubility in water. Interest in the use of ionic liquids dates back to 1984, when Magnuson *et al.* [23] investigated the effect of ethylammonium nitrate on the activity and stability of the enzyme alkaline phosphatase. Subsequently, considerable research effort has been committed, particularly since 2000, to evaluate the impact of the physicochemical properties and the design of ionic liquids suitable for the stabilization or activation of enzymes in several different formats, such as immobilization onto solid carriers, as sol-gel encapsulation, as solid phase complexes or following chemical modification [24,25,26]. It is widely appreciated that many ionic liquids exhibit considerable toxicity and poor biodegradability, which preclude their use in many cases as "green" solvents, whilst the effects of ionic liquids on the functional status and solubility of enzymes are difficult to accurately predict with the current state of knowledge.

Nevertheless, it is well known that the effect of ions from low to relatively high concentrations, e.g., in excess of 4 M for ammonium sulphate, on enzyme activity and stability usually follows the Hofmeister series, whereby strongly hydrated ions (called kosmotropes) increase the water structure and thus the conformational stability of proteins, whilst weakly hydrated ions (called chaotropes) generally operate in the opposite way, e.g., destabilize protein conformation. However, as a generalization, the behavior of the anions of more hydrophilic ionic liquids tends to correlate well with the Hofmeister series. For example, Yang [27] and Zhao [28] demonstrated that an ionic liquid composed of a chaotropic cation and a kosmotropic anion can favorably influence an enzyme to retain good activity and stability. When both the anion and cation are chaotropic, then loss of activity and stability tend to follow. The more hydrophilic ionic liquids that combine powerful hydrogen bonding capacity tend to act as water-mimicking liquids with the ability to dissolve an enzyme for monophasic catalysis while retaining a high level of enzyme activity [29,30].

The first examples of the use of biocatalysts in the presence of an ionic liquid were reported by Erbeldinger et al. [31] and Lau et al. [32] in 2000. Since then, a large amount of research has been performed in this field as illustrated by Figure 1, which summarizes the number of publications per year from 2000. The number of issued or pending patent applications mirrors a similar trend. The importance of this rapidly growing field of research has been documented by various authors, including the different research groups of Sheldon et al., Kragl et al. and Goto et al., highlighting practical developments, as well as providing compilations of data for different ionic liquids/enzyme combinations [13,24,25,26,33,34,35,36,37,38,39,40,41]. Collectively, these works have summarized different facets of the use of biocatalysts in ionic liquids, wherein key issues related to the effects of ionic liquids on the structure, activity and stability of enzymes have been explored and the impact of the water content in the ionic liquid assessed, together with several biocatalytic reactions employing a range of different enzymes and ionic liquid combinations for the design of reaction systems, including biocatalyst recovery, product isolation and choice of biphasic systems.

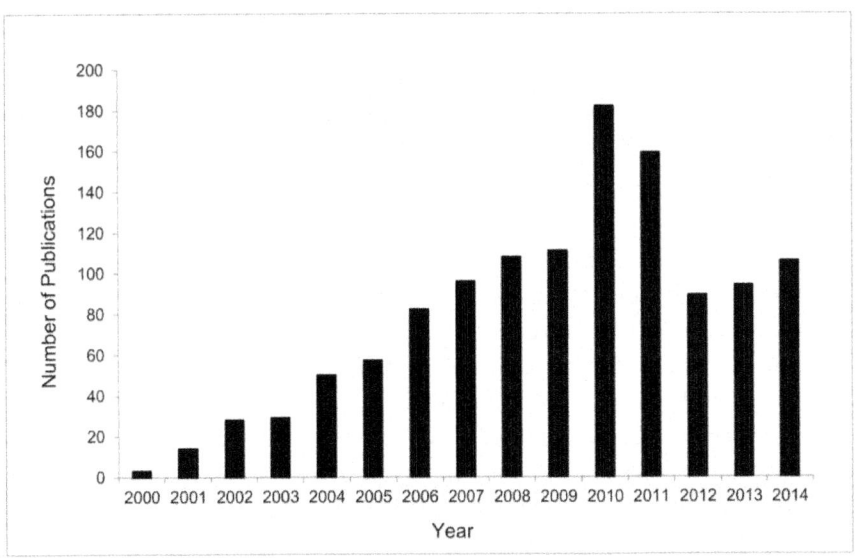

Figure 1. Year-wise reports published in the area of enzyme-based catalytic reactions in ionic liquids.

In this current article, an overview of innovations that have been developed in this field over the last decade is presented with a focus on the synthetic chemistry outcomes and benefits derived from carry out biocatalytic reactions in suitable ionic liquids. Key aspects behind different chemical synthetic processes that are destined upon scale-up to truly be more sustainable are described. Emphasis has thus been mainly placed on the synthetic processes, including approaches whereby the product can be efficiently separated and the ionic liquid and biocatalyst can be recycled or employed in continuous processes, rather than on the physicochemical properties of the ionic liquids *per se*, which have been eloquently summarized in earlier reviews, (e.g., [2,5,13,1 6,24,25,26,33,34,35,36,37,38,39,40]). Importantly, such uses of biocatalysts in appropriate ionic liquids offers important advantages for process intensification and enhancement of synthetic productivity for the manufacture of chiral molecules. An overview of the application of a special class of biodegradable ion-based liquids, known as deep eutectic solvents, in biocatalysis, and recent adaptations of the use of ionic liquids to whole-cell biocatalysis, which from a green chemical perspective have potential for new applications in chemical manufacturing, medicine and the environmental sciences are also provided.

CHEMICAL SYNTHESIS WITH ENZYMES IN IONIC LIQUIDS

Ionic Liquids as Solvents for Enzyme Catalysis

The first definitive example of enzyme-catalyzed reactions in a pure ionic liquid was reported by Sheldon's group. Thus, Lau *et al.* [32] reported lipase-catalyzed transesterification, amidation and epoxidation reactions in ionic liquids (Scheme 1). Reaction rates for the immobilized *Candida antarctica* lipase B (Novozym® 435)-catalyzed transesterification in anhydrous [bmim][PF$_6$] were found to be comparable, faster or slower than those in organic solvents depending on the type of ester and alcohol employed, with good to excellent conversions (56%–81%) obtained. The Novozym® 435-catalysed condensation of ammonia with octanoic acid in [bmim][BF$_4$] proceeded in quantitative conversion after 4 days. Epoxidation of cyclohexene by peroctanoic acid was also investigated using the Novozym® 435 to catalyze the production of the peroxy-acid *in situ* from octanoic acid and 60% aqueous H$_2$O$_2$ in [bmim][BF$_4$], documenting an example of a safer reaction process that did not require the use of a flammable solvent or the handling of a hazardous peroxy-acid. A yield of 83% was achieved after 24 h, comparable to that obtained when the same reaction was carried out in acetonitrile (93%). Overall, this pioneering study highlighted the potential of low vapor pressure, non-flammable ionic liquids as greener solvent alternatives to organic solvents for enzyme-catalyzed production of important industrial chemicals.

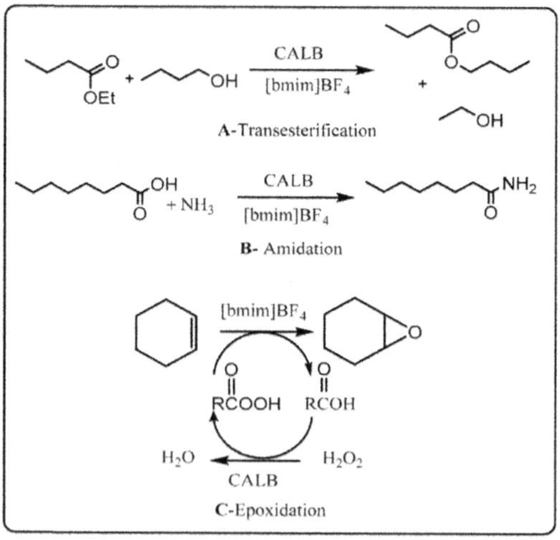

Scheme 1. *Candida antarctica* lipase (CALB)-catalyzed transesterification (**A**); amidation (**B**) and epoxidation (**C**) reactions in ionic liquids [32].

The important roles that enzymes can play as catalysts are particularly prominent in enantioselective transformations. Pioneering examples of enantioselective reactions using enzymes in ionic liquids were demonstrated independently in 2001 by Schöfer *et al.* [40] and Itoh *et al.* [41]. For example, Itoh *et al.* [41] carried out the enantioselective acylation of allylic alcohols in an ionic liquid solvent. The group successfully demonstrated that the lipase *Candida antarctica* lipase B (Novozym® 435) was anchored to the imidazolium based ionic liquid solvent. These experiments documented the power of a recyclable system of enzyme and ionic liquid for enantioselective transformation (Scheme 2). Overall, these demonstrations of the potential of lipase-catalyzed reactions in ionic liquids [32,40,41] represent landmark studies, since lipases are now the most widely used enzymes in organic synthesis.

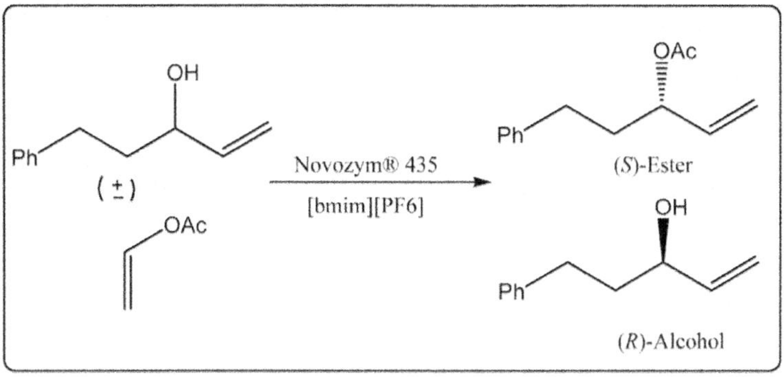

Scheme 2. *Candida antarctica* lipase (CALB)-catalyzed enantioselective transesterification of allylic alcohols in ionic liquid [41].

Subsequent to this pioneering study of lipase-catalyzed biocatalysis in ionic liquids, several other different reactions encompassing a wider array of substrates have been reported [13,42,43]. For example, Mohile *et al.* have demonstrated the use of ionic liquid as a co-solvent with aqueous buffer for the *Candida rugosa* lipase-catalyzed enantioselective hydrolysis of racemic butyl 2-(4-chlorophenoxy)propionate (Scheme 3) [44].

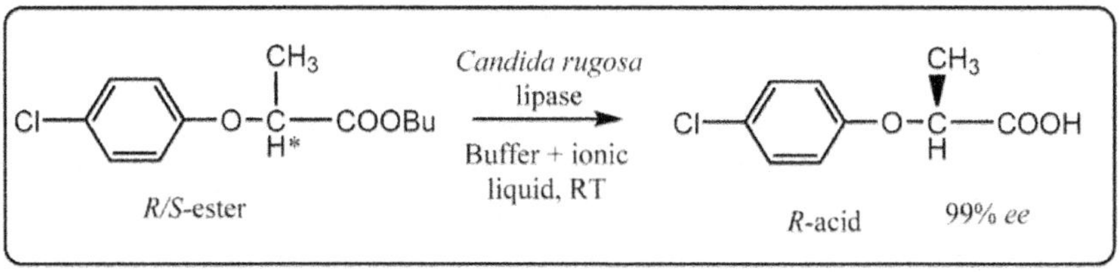

Scheme 3. Enantioselective hydrolysis of racemic butyl-2-(4-chlorophenoxy)propionate in an aq. buffer/ionic liquid system [44].

In biphasic systems employing the hydrophobic [bmim][PF$_6$] or 1-hexyl-3-methyl-imidazolium [hmim][BF$_4$] ionic liquids, reaction times increased by about 10-fold compared to when an aqueous buffer alone was employed, but a remarkably large enhancement in enantioselectivity up to 99% compared to 47%, respectively, was observed. Similar results have been obtained with a monophasic system when the more hydrophilic [bmim][BF$_4$] ionic liquid was used. In the case of the biphasic systems, the ionic liquids were able to be recycled with only marginally reduced conversion rates after four cycles and with no loss of enantioselectivity. In subsequent work [bmim][PF$_6$] was assessed as a solvent for the resolution of racemic alcohols by enantioselective acylation with succinic anhydride using *Pseudomonas cepacia* lipase supported on celite (lipase PS-C) as a catalyst (Scheme 4) [45].

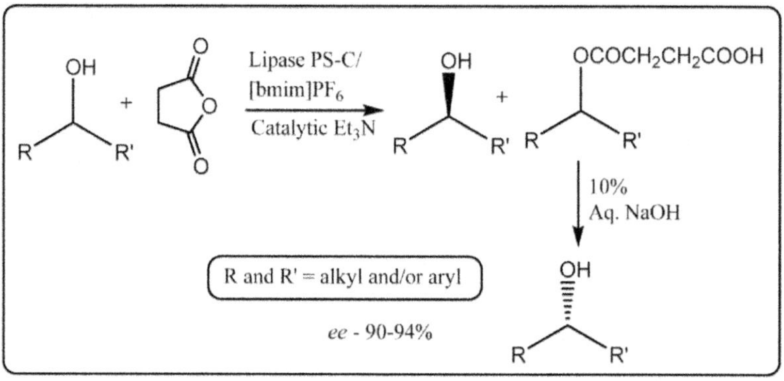

Scheme 4. Lipase PS-C catalyzed resolution of racemic alcohols by enantioselective acylation with succinic anhydride in [bmim][PF$_6$] [45].

A distinct advantage of using succinic anhydride as the acylating agent lies in the easy separation of the resultant ester by extraction from the ionic liquid into a mildly basic aqueous solution, followed by hydrolysis with a strong base to give back the enantiomerically enriched alcohol. The ionic liquid [bmim][PF$_6$] was found to be a suitable recyclable solvent for lipase PS-C-catalyzed enantioselective acylation, in up to 94% enantioselectivity. Including a catalytic amount of triethylamine as an additive enhanced the rate of the reaction by about 1.5-fold.

As an example of pharmaceutical synthesis, Lourenco *et al.* [46] developed an efficient lipase-mediated resolution of the HIV protease inhibitor Indinavir precursor (±)-*cis*-benzyl *N*-(1-hydroxyindan-2-yl) carbamate in [aliq][NCN$_2$] ionic liquid (Scheme 5). Novozym® 435 was again used to selectively acylate the 1*S*,2*R*-*cis*-benzyl enantiomer with vinyl acetate in up to 97% enantioselectivity. The catalyst was removed by filtration and the product and unreacted 1*R*,2*S*-enantiomer were then isolated from the non-volatile ionic liquid by sublimation, and then separated by column chromatography. This approach allowed reuse of the [aliq][NCN$_2$] and enzyme with comparable yield and enantioselectivity.

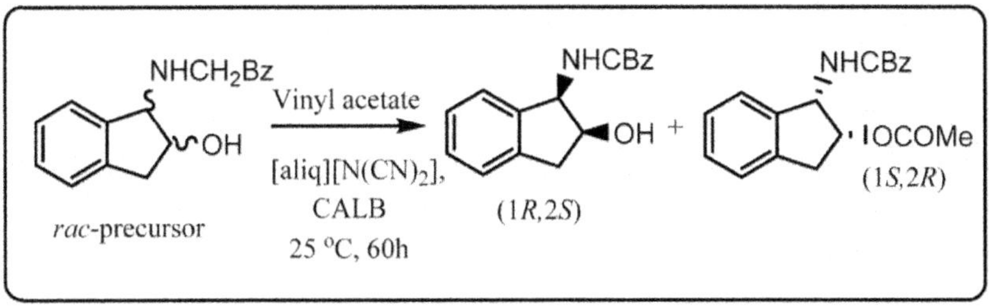

Scheme 5. Resolution of *cis*-benzyl *N*-(1-hydroxyindan-2-yl)carbamate using CALB (Novozym 435®) in ionic liquid [46].

Mai *et al.* have demonstrated the enzymatic synthesis of sugar fatty acids in ionic liquids [47]. Sugar fatty acid esters are environment friendly bio-surfactants known for their non-toxic, non-ionic, and high biodegradability. Transesterification of vinyl laurate with glucose was catalyzed by Novozym® 435 in a mixture of [Bmim][TfO]/[Bmim][Tf$_2$N] ionic liquid (1:1 *v/v*) in 96% conversion (Scheme 6). The mixture of [Bmim][TfO]/[Bmim][Tf$_2$N] permitted good solubility of glucose (12 g/L) and enzyme stability critical for its reuse. The reaction was scaled up to 2.5 L with comparable conversion efficiency. After 10 cycles, the activity of Novozym® 435 was 75% of the initial activity.

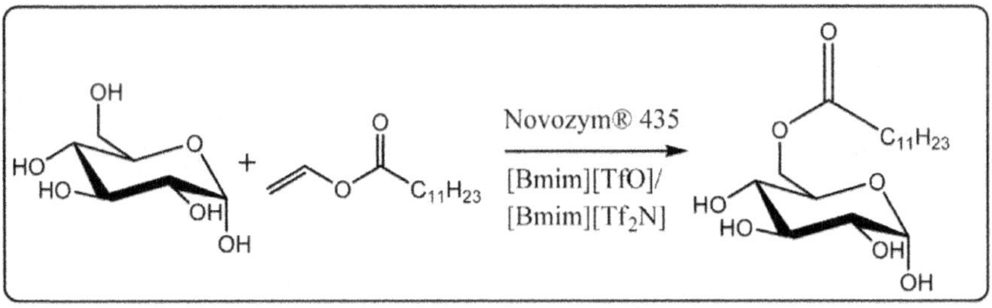

Scheme 6. Transesterification of vinyl laurate with glucose in ionic liquids catalyzed by Novozym® 435 [47].

Developments in Sustainable Biocatalysis Employing Ionic Liquids

Ionic Liquid-Coated Enzymes as Heterogeneous Catalysts

Initial applications of ionic liquids in biocatalysis concentrated on their use as a replacement for conventional solvents, with subsequent studies yielding greater understanding of how they interact with enzymes. An example of such an application was reported by Lee *et al.* [48], in which an ionic liquid-coated enzyme (ILCE) was prepared as a heterogeneous catalyst for use in the enantioselective transesterification of vinyl acetate with racemic alcohols in toluene. In this work, a lipase from *Pseudomonas cepacia* (PS) was first dissolved in the ionic liquid 1-(3'-phenylpropyl)-3-methylimid-azolium hexafluorophosphate [ppmim][PF$_6$], which has a melting point of 53 °C. Upon cooling the solution to room temperature, a solid ILCE formed, which was then used as a heterogeneous catalyst for the conversion of the substrate to product in toluene. Compared to the native enzyme in toluene, the ILCE enhanced the enantioselectivity of the products by 1.5–2 folds. Furthermore, the ILCE was easily removed from the reaction mixture by filtration and reused as an ionic liquid coated biocatalyst for up to 5 times with comparable activity.

Similarly, Itoh *et al.* [49] have prepared a 1-butyl-2,3-dimethylimidazolium [bdmim][cetyl-PEG10-sulfate] coated lipase by extracting a commercial lipase immobilized onto a ceramic matrix (lipase PS-C) into aqueous buffer containing the ionic liquid followed by freeze-drying. The ILCE was used as a heterogeneous catalyst for transesterification of vinyl acetate with racemic alcohols in diisopropyl ether (Scheme 7) and gave rate enhancements of up to 1000-fold compared to lipase PS-C while maintaining excellent enantioselectivity (>99%). Reusing the ILCE up to five times gave comparable product yield and enantioselectivity. To demonstrate the versatility of the protocol, similar activation effects were also demonstrated for [bdmim][cetyl-PEG10-sulfate]-coated *Candida Rugosa* lipase.

Electron microscopy of the ILCE showed it possessed a higher surface area than the commercial lipase PS-C immobilized on ceramic, whilst MALDI-TOF mass spectrometry experiments indicated the enzyme in the ILCE forms a discrete non-covalent complex with the ionic liquids. Together, these results suggested that the enhanced activity of the ILCE may be due to easier access of the substrate to the more porous ILCE and favorable enzyme flexibility and conformational changes imparted by its non-covalent interactions with the ionic liquid.

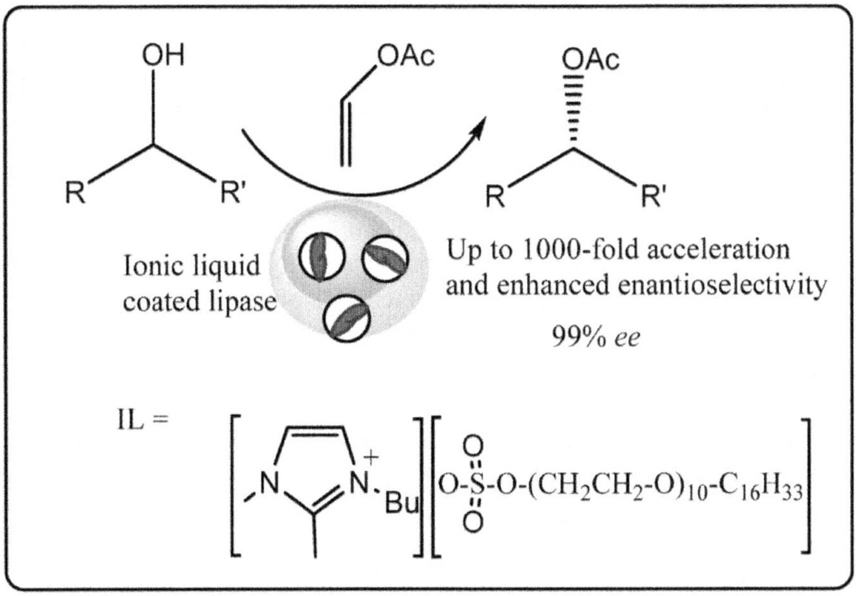

Scheme 7. ILCE-catalyzed enantioselective transesterification of vinyl acetate with racemic alcohols [48].

Encapsulation of enzymes in polymers prepared from ionic liquid monomers has also been investigated [50]. A useful feature in designing ionic liquids is that pendant functional groups can be incorporated into the cation or anion for polymerization. Thus, 1-vinyl-3-ethylimidazolium bromide was polymerized with the cross-linker N,N'-methylenebis(acrylamide) in the presence of native horse-radish peroxidase (HRP) or HRP modified with a polyethylene comb polymer (PM$_{13}$-HRP) to protect it from conformational deactivation by the ionic liquid bromide anion (Figure 2). This method led to the formation of enzyme-encapsulated polymer micro particles, which functioned as heterogeneous catalysts for the oxidation of guaiacol by H$_2$O$_2$ in water. The activity of the encapsulated PM$_{13}$-HRP was found to be 3 times higher than that of encapsulated native HRP and 2 times higher than PM$_{13}$-HRP encapsulated in non-ionic polyacrylamide micro particles. Furthermore, the encapsulated PM$_{13}$-HRP was easily recycled by centrifugation, with reduction in activity occurring only at the first recycling stage, which was attributed to loss of PM$_{13}$-HRP loosely immobilized on the surface of the micro particles.

Cross-linked enzyme aggregates (CLEAs) of mung bean epoxide hydrolases (*mb*EH) has been used to catalyze the asymmetric hydrolysis of styrene oxide in a water-ionic liquid biphasic system (Scheme 8) [51]. Styrene oxide remains dispersed in the ionic liquid [bmim][PF$_6$] and the hydrolysis product, (*R*)-phenylethanediol ((*R*)-PED), was obtained in the aqueous phase, facilitating its easy separation from starting material and avoiding non-enzymatic hydrolysis. On a 500 mL preparative scale the yield of (*R*)-PED was 49% in 94.6% enantioselectivity.

Rehmann and co-workers have recently demonstrated that a variety of ionic liquids can stabilize the activity of the laccase from *Trametes versicolour* [52]. Laccase, an oxidative mediator system, is important owing to its potential for numerous green oxidation processes but its activity is inhibited by oxidation mediators like TEMPO (2,2,6,6-tetramethyl-1-piperidinyl)oxidanyl, ABTS (2,2'-azino-bis-(3-ethylbenzo-thiazoline-6-sulfonic acid) or 4-hydroxybenzyl alcohol. Such inactivation was avoided by using a biphasic ionic liquid-water system, wherein the mediator predominantly partitions into the ionic liquid, restricting its contact with the laccase, which predominantly resided in the aqueous phase.

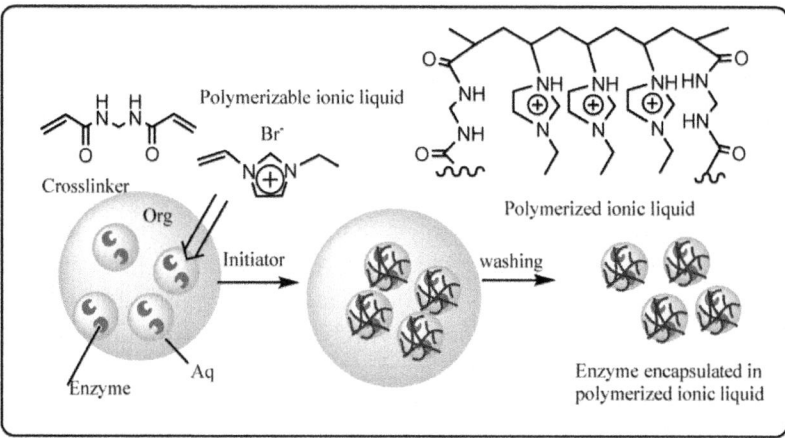

Figure 2. Enzyme encapsulation in an ionic liquid-based polymer [50].

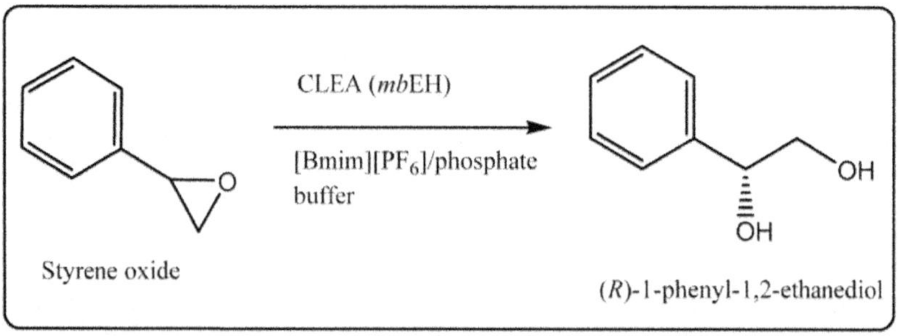

Scheme 8. Cross linked enzyme aggregate of *mb*EH catalyzed asymmetric hydrolysis of styrene oxide [51].

Role of Enzymes and Ionic Liquid System in Biomass Conversion and Biofuel Production

To address the expected increase in demand for energy and transport fuel from a rapidly increasing global population and developing economies, the sustainable conversion of renewable resources to liquid transport fuel and commodity chemicals has become an important area of research. Lignocellulosic (woody) biomass is the key sustainable biomass resource for transformation into biofuels and bio-based chemical products. However, the chemical composition of lignocellulosic biomass differs significantly from that of fossil fuels and efficient catalytic processes need to be developed in order to convert it to fuel precursors and platform chemicals, and the use of ionic liquids and enzymes in addressing this challenge is an active area of research.

Over the past decade, there has been increased interest in using cellulose from renewable biomass resources as a more sustainable alternative raw material to petrochemical feedstock for producing chemical products. A barrier to chemically transform cellulose to other products is its poor solubility in many different solvents. Although ionic liquids containing anions, which are strong hydrogen-bond acceptors, e.g., halides, have been found to dissolve cellulose, these types of ionic liquids tend to denature enzymes, hampering biocatalytic processing of the dissolved cellulose [53,54]. The recent observations of Zhao *et al.* [55] with polyethylene glycol-based ionic liquids that are able to dissolve carbohydrates and cellulose, and retain enzyme activity has provided one option to overcome these limitations. Moreover, these systems are also relevant to the transesterification of methylmethacrylate with glucose or cellulose using Novozym® 435 with the biocatalytic reaction proceeded in up to 89% conversion and 66% yield [55].

Currently, research efforts are also being focused on the use of ionic liquid and deep eutectic solvents for lignocellulosic pre-treatment [56,57] in conjunction with enzymatic hydrolysis of cellulose to fermentable sugars [58,59]. Lignocellulosic biomass is not amenable to facile microbial or enzymatic industrial biotransformation, which limits its economic conversion to fuel precursors and platform chemicals by a biotechnological process. Moniruzzaman and Ono have investigated ways to overcome this barrier by developing an enzymatic biomass process employing an ionic liquid pre-treatment to enable efficient access for enzyme laccase to reactive sites [60]. This enabled an efficient delignification in an ionic liquid/aqueous buffer system, giving cellulose fibers in improved yields and shorter reaction times (Figure 3).

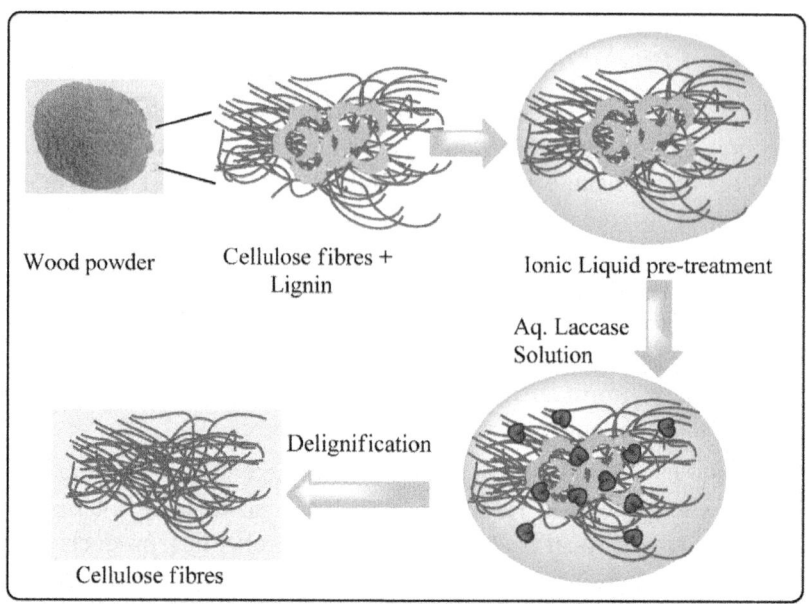

Wood powder   Cellulose fibres +
                      Lignin

Ionic Liquid pre-treatment

Aq. Laccase
Solution

Delignification

Cellulose fibres

Figure 3. Delignification of woody biomass using ionic liquid and enzymes [60].

Biodiesel is a transport fuel that consists largely of fatty acid methyl esters (FAMEs) prepared by transesterification of natural triglycerides in plant oils or animal fats with methanol. These fuels have generated interest as renewable, low-carbon impact alternatives to fossil fuels. Current biodiesel production uses homogenous alkaline catalysts for transesterification and although this process is efficient, it complicates downstream product separation and increases energy costs [61]. Biocatalytic production of biodiesel in ionic liquids merits consideration as a green and potentially environmentally friendlier alternative to eliminate issues associated with product separation, homogenous catalyst recycling and energy consumption. The first demonstration of the potential of biocatalytic biodiesel production in ionic liquids was reported by Ha *et al.* [62], who used the Novozym® 435 lipase to catalyze the transesterification of soya bean oil with methanol. 1-Ethyl-3-methylimidazolium trifluoromethanesulfonate ([Emim][TfO]) was found to be the best solvent, giving an 80% yield of fatty acid methyl esters in 12 h compared to 65% in *tert*-butanol and 10% under solvent-free reaction conditions. An additional advantage of the ionic liquid system was phase separation of the biodiesel as it was formed. Gamba *et al.* subsequently developed an efficient biphasic system for biodiesel production comprised of a [bmim]bis(trifluoro-methylsulfonyl)-imide ([NTf$_2$]) ionic liquid phase containing methanol and lipase, and a soya bean oil phase (Figure 4) [63].

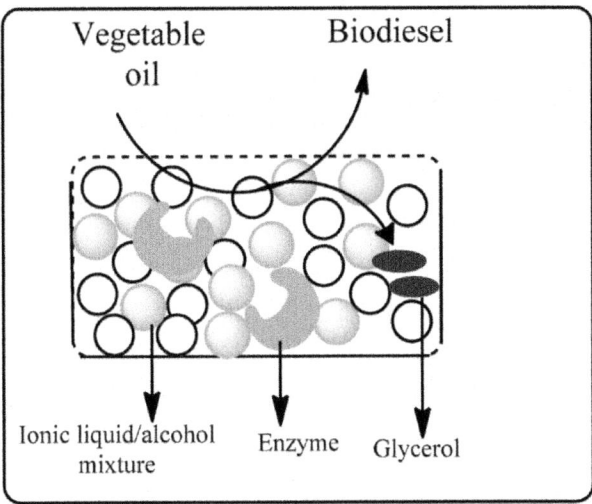

Figure 4. Lipase-catalyzed production of biodiesel in a biphasic ionic liquid-soybean oil system with concomitant separation of the glycerol by-product [63].

This transesterification proceeded at room temperature to give biodiesel in 95% yield in 24 h and was enhanced by the addition of water. Furthermore, the ionic liquid phase served as an extracting agent for removing glycerol from the reaction mixture and the biodiesel was separated by simple decantation. The ionic liquid phase containing the enzyme was recycled four times without significant loss of catalytic activity. After these four cycles the enzyme was filtered from the ionic liquid and the glycerol by-product extracted with water in 98% purity.

Further, Zhao *et al.* have reported the synthesis and application of novel triethyleneglycol-functionalised ionic liquids (Figure 5) in biodiesel production [64]. These ionic liquids dissolved both triglycerides and lipases, and a high level of catalytic activity was observed with different commercially available lipases; quantitative conversions of Miglyol® oil 812 prepared from coconut and palm kernel oils was achieved at 96 h in a 70/30 (*v*/*v*) ionic liquid/ MeOH mixture. Overall, these preliminary studies on the transesterification of vegetable oils in ionic liquids indicate they are promising solvent systems for industrial biocatalytic biodiesel production.

The enzyme catalysed esterification of soybean oil in [Emim][PF$_6$] with microwave heating has been demonstrated by Yu *et al.* to give faster conversion with a 1.8-7.8-fold increase in enzyme activity compared to *t*-butanol and a solvent-free system using conventional heating [65]. This is the first example of using synergistic effect of microwave irradiation and ionic liquids for enzyme catalyzed biofuel production. The study also demonstrated recycling of the ionic liquid/enzyme phase for five cycles with only a slight decrease (8%) in enzyme activity.

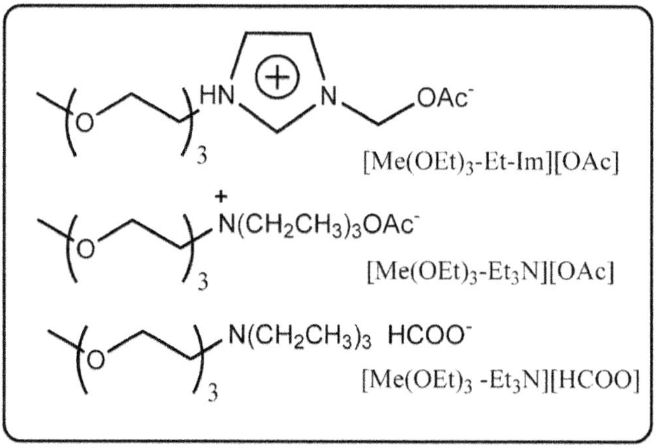

Figure 5. Novel triethyleneglycol-functionalized ionic liquids for biodiesel synthesis [64].

Zhang *et al.* have demonstrated the activity of *Penicillium expansum* lipase (PEL) in ionic liquids and reported the application of this system for the synthesis of corn oil-based biodiesel [66]. The ionic liquid [Bmim][PF$_6$] was found to be an optimal solvent for the transesterification reaction, giving a higher conversion (86%) compared to *t*-butanol (52%), and PEL was found to be more tolerant to the reaction conditions compared to other lipase enzymes routinely used. Furthermore, the biphasic ionic liquid-FAME reaction system facilitated separation of the biodiesel and recycling of the catalyst. Lai *et al.*, from the same group, have reported the use of PEL for the transesterification of triglycerides from the microalgae *Chlorella vulgaris*, the first example of ionic liquid mediated biocatalytic conversion of microalgal oil to biodiesel [67].

Ionic Liquid-Based Supported Liquid Membranes in Separation Processes

Selectively permeable membranes that separate bulk solvent phases have potential application in separation processes. For example, kinetic resolution employing enzyme catalysis followed by product separation using selectively permeable membranes allows the conversion and separation of the desired product in a single continuous process which is attractive from an environmental and industrial perspective. Supported liquid membranes (SLMs) have been identified as suitable materials for such continuous separation processes.

These materials are usually comprised of a solvent immobilized in the porous structure of polymeric or ceramic membrane [68]. The SLM serves to separate a feedstock solvent and a receiving solvent with the immobilized solvent mediating solute transport across the membrane. A critical feature of SLMs in continuous extraction processes is negligible loss of the immobilized solvent to either of the solvent phases and to the atmosphere.

The potential of using SLMs incorporating ionic liquids as the immobilized solvent for transport of organic molecules was first demonstrated by Branco *et al.*, who employed [bmim][PF$_6$] in a polyvinyl-idene fluoride membrane owing to its poor solubility in the feedstock and receiving solvents, and negligible vapor pressure [69]. This system was able to selectively transport diisopropylamine from a mixture of amines in the feed solvent to the receiving solvent.

Miyako *et al.* have subsequently demonstrated a lipase-facilitated transport of organic acids across SLMs employing ionic liquids as transport solvents (Scheme 9) [70]. Based on this technology, organic acids were esterified by *Candida rugosa* lipase (CRL) in the feedstock phase with the resulting ester partitioning into the ionic liquid phase in the SLM and then diffusing into a receiving phase that contained a porcine pancreatic lipase (PPL) to hydrolyze the ester back to the initial organic acid. Accumulation of organic acids in the receiving phase was not observed in the absence of CRL due to their poor solubility in the ionic liquid used to construct the SLM. The use of ionic liquids as transport solvents also enabled long-term stability of the SLM compared to conventional organic solvent.

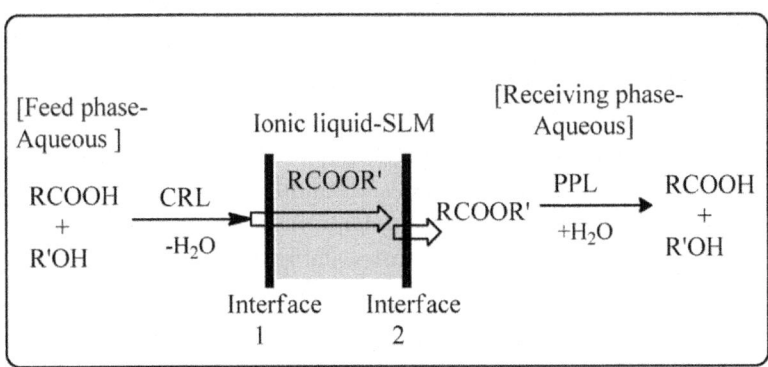

Scheme 9. Lipase-facilitated transport of carboxylic acids across an ionic liquid-based supported liquid membrane (SLM) [70].

This methodology was further extended to the resolution of (S)-ibuprofen through selective CRL-catalyzed esterification of the (S)-enantiomer in the feedstock, diffusion of the ester through the SLM ionic liquid and regeneration of (S)-ibuprofen in the receiving phase by PPL (Scheme 10) [71]. This process provided (S)-ibuprofen in up to 75% enantioselectivity.

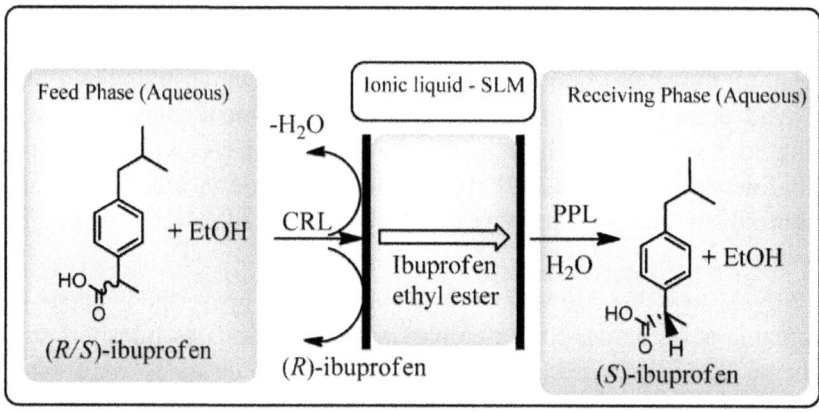

Scheme 10. Lipase-facilitated enantioselective transport of (S)-ibuprofen across an ionic liquid based supported liquid membrane (SLM) [71].

An ionic liquid-based SLM system for the kinetic resolution of racemic 2-pentanol and separation of the unreacted enantiomer has been developed by Hernández-Fernández *et al.* [72]. (S)-2-Pentanol is an interesting chiral intermediate in the synthesis of drug candidates for Alzheimer's disease [73]. Resolution and separation of (S)-2-pentanol from a racemic mixture was achieved by Novozym® 435 lipase-catalyzed enantioselective transesterification of vinyl esters resulting in (R)-ester and (S)-alcohol (Scheme 11). In the SLM system, the alcohols readily diffused across the membrane whilst the diffusion rate of the esters was relatively low. Thus, enantioselective transesterification resulted in (S)-2-pentanol selectively accumulating in the receiving phase, enabling its separation from the (R)-enantiomer. Of six ionic liquids tested, [bmim][BF$_4$] was found to be the most suitable as a transport solvent, giving 60–70-fold enrichment of the (S)-2-pentanol enantiomer in the receiving phase.

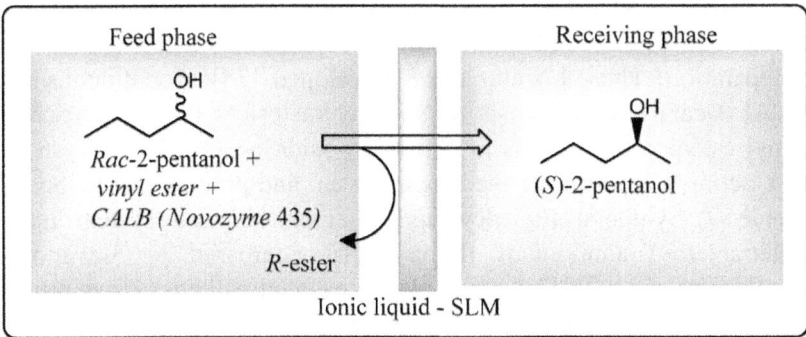

Scheme 11. Lipase-facilitated resolution of *rac*-2-pentanol using an ionic liquid-based supported liquid membrane (SLM) [72].

This methodology was further extended to the separation of (*S*)-phenylethanol, a commercially important chiral building block in the fine chemical and pharmaceutical industries [74]. The kinetic resolution of *rac*-1-phenylethanol was achieved by transesterification of a vinyl ester with the (*R*)-enantiomer, catalysed by Novozym® 435 lipase. (*S*)-1-Phenylethanol and the (*R*)-vinyl ester derivative were separated in the reactor using a [bmim] [BF$_4$]-based SLM to give the (*S*)-enantiomer in 99% enantioselectivity.

Biocatalysis in Biphasic Ionic Liquid-Supercritical Carbon Dioxide Systems

Some industrially important enzymes have been found to retain activity in a variety of non-aqueous solvents, including supercritical carbon dioxide (ScCO$_2$) with several types of enzyme catalyzed reactions documented in this alternative solvent [75,76]. ScCO$_2$ is an attractive industrial solvent owing to its non-flammable properties and low toxicity, and is relatively easily produced in terms of pressure and temperature compared to other supercritical fluids. A number of reports have demonstrated the value of combining ionic liquids and ScCO$_2$ for green biocatalytic processes [42,77]. Because ScCO$_2$ is immiscible with several different ionic liquids, it can be used to deliver substrates to, and extract products from, the ionic liquid phase. This particular use of ScCO$_2$ makes it an attractive candidate for continuous flow and batch industrial processes for reactions involving the use of enzymes in ionic liquids. A critical feature of such systems is efficient mass transfer of substrates and products between ScCO$_2$ and ionic liquids, which is dependent on ionic liquid hydrophobicity and viscosity.

The group of Iborra have carried out detailed work on the use of biphasic systems consisting of ionic liquid and ScCO$_2$ phases for combined biocatalysis and separation. Thus, Lozano *et al.* developed [78] a continuous process for CALB-catalyzed synthesis of short chain esters by transesteric-ation of vinyl esters with alcohols in an ionic liquid/Sc-CO$_2$ biphasic system with ScCO$_2$ acting as a substrate feedstock solvent and product extraction solvent (Scheme 12). Synthetic efficiency was greater in the biphasic system employing butyltetramethyl ammonium [btma][NTf$_2$] compared to 3-cyanopropyl-trimethylammonium [NTf$_2$] which was attributed to better mass transfer of substrates between ScCO$_2$ and the more hydrophobic [btma][NTf$_2$] ionic liquid.

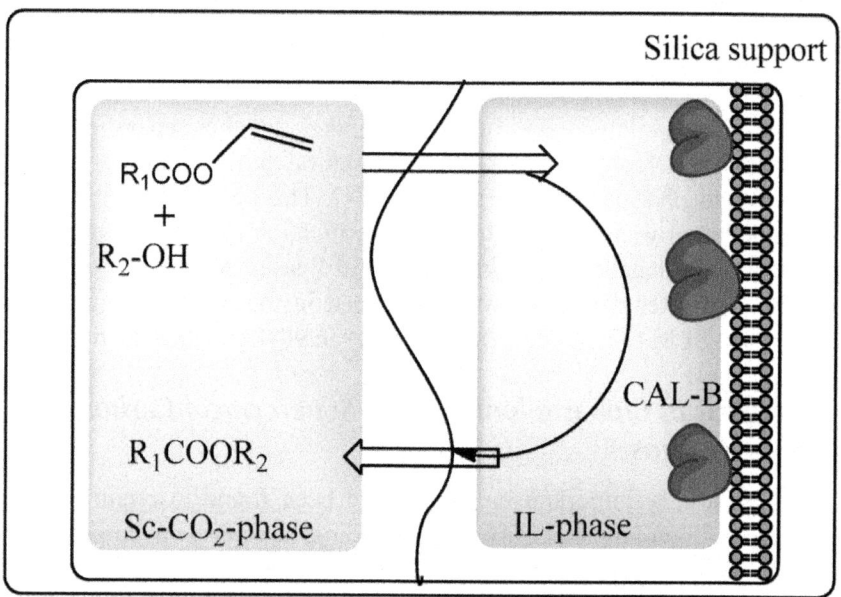

Scheme 12. Transesterification of vinyl esters in a biphasic ionic liquid-ScCO$_2$ system [78].

Immobilization of ionic liquids onto an insoluble solid support is an attractive strategy for continuous biocatalytic processes employing a mobile ScCO$_2$ phase since this approach minimizes the amount of ionic liquid used and enables easy product separation and catalyst recycling [79,80,81]. Lozano *et al* have synthesized immobilized ionic liquid phases by grafting or copolymerizing imidazolium-based ionic liquids to generate macroporous polymer monoliths with imidazolium loading ranging from 40% to 55% *w/w* [82]. The enzyme CALB was non-covalently adsorbed onto polymer monoliths from an aqueous solution. The systems were then used as heterogeneous catalysts for the continuous flow transesterification synthesis of citronellyl propionate from citronellol and vinyl propionate in ScCO$_2$ (Figure 6). Productivity was highest when a styrene/divinylbenzene/imidazolium-based monolith was used compared to a 2-hydroxyethyl/ethylene glycol dimethyl acrylate/imidazolium-based monolith which was attributed to better mass transfer properties due to its greater hydrophobicity and porosity. Yields of up to 93% were achieved with no enzyme leaching observed.

The same group later reported studies on the immobilization of CALB by adsorption onto silica gel supports containing covalently grafted [btma] or trioctyl-methyl ammonium [toma][NTf$_2$] ionic liquids [83]. The immobilized CALB was then assessed for its ability to kinetically resolve *rac*-1-phenylethanol by enantioselective transesterification of vinyl propionate in a continuous flow reactor using ScCO$_2$ as the mobile phase. A yield of up to 96% and an enantioselectivity of >99% was achieved for the *trans*-esterification with the (*R*)-enantiomer. The productivity of this reaction system was found to be up to six times greater when using ScCO$_2$ compared to hexane, attributed to better ability of ScCO$_2$ to transport solutes through the grafted ionic liquid phase.

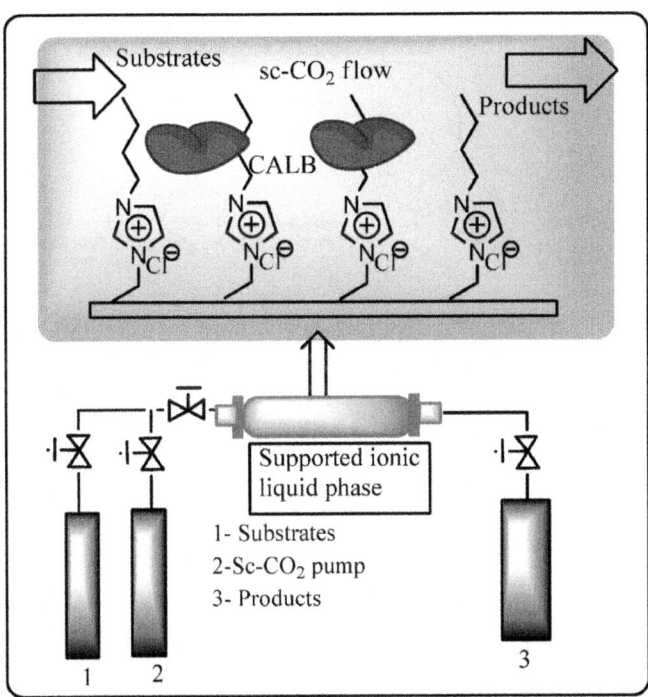

Figure 6. Continuous flow reactor setup employing CALB supported on an immobilized ionic liquid phase and sc-CO$_2$ as a mobile phase [82].

In a seminal study, Lozano *et al.* [84,85] developed a continuous dynamic kinetic resolution of *rac*-1-phenylethanol employing a mixture of Novozym® 435 lipase and a heterogeneous acidic zeolite in an ionic liquid-ScCO$_2$ biphasic system. The role of the lipase was to catalyze transesterification of vinyl propionate with (*R*)-1-phenylethanol while the acidic zeolite was included to catalyze the isomerization of (*S*)-1-phenylethanol to the (*R*)-enantiomer. Both catalysts were first coated with ionic liquids by adsorption to stabilize the enzyme from the strongly acidic sites of the zeolite and the mobile ScCO$_2$ phase. Overall, this catalytic system provided (*R*)-phenylethylpropionate in up to 98% yield and 97% enantioselectivity (Scheme 13) without any loss of catalyst activity after 14 days of operation. These findings clearly demonstrate the industrial potential of a combination of heterogeneous biological and chemical catalysis employing ionic liquids and ScCO$_2$ in continuous flow process for green production of commercially important enantiomers.

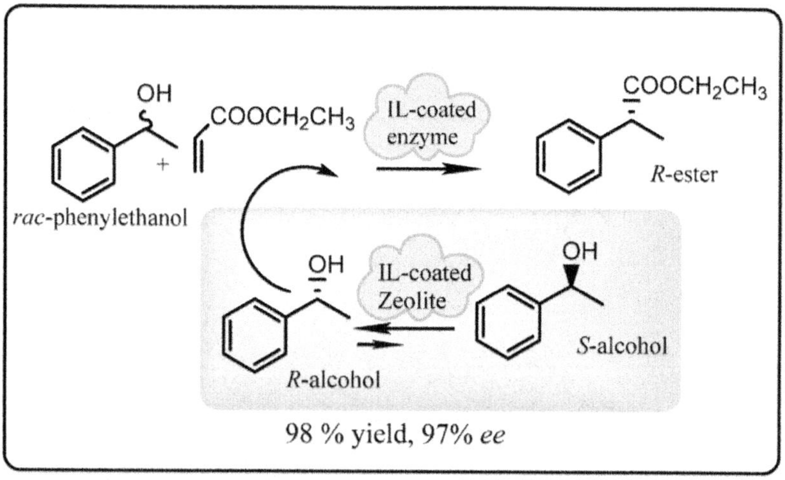

Scheme 13. Dynamic kinetic resolution of *R*-phenylethylpropionate with ionic liquid-coated enzyme and zeolite [85].

Use of Biodegradable Ionic-Based Deep Eutectic Solvents for Biocatalysis

Although ionic liquids are often claimed to be new generation green solvents due to their low flammability and vapor pressure, their component ions may be toxic, slow to biodegrade and may accumulate in marine organisms. Hence, the presence of ionic liquids in industrial waste streams may result in an environmental hazard. In light of this potential problem, the design and use of biodegradable ionic liquids that will not persist in the environment is a matter of importance. The building blocks used to synthesize such a class of ionic liquids are generally those regarded as being biocompatible, e.g., amines, amino acids, amino alcohols, essential oils, carbohydrates and carboxylic acids [86]. An added advantage is that these building blocks are obtained from renewable resources, which further enhances the sustainability profile of the ionic liquid.

An interesting class of designed biodegradable ionic liquids are physical mixtures of biological salts, such as choline chloride and uncharged biological hydrogen bond donors, e.g., urea or glycerol, which exist as deep eutectic solvents (DESs). Such eutectic solvents exist as a physical mixture with a specific component composition that imparts a melting temperature much lower than that of any of the individual components. Compared to earlier generation ionic liquids, ionic-based DESs are relatively easy to prepare in accord with green chemistry principles since they simply comprised of mixtures of compounds in defined ratios and hence do not require any purification [43].

Currently, there are only a few examples of enzyme catalysis in a DES compared to earlier generations of ionic liquids. The first report of enzymes used economically with a DES was a lipase-catalyzed transesterification of ethyl valerate with 1-butanol [87]. The specific activity of immobilized CALB (iCALB) was found to be 2–7 times higher in a choline chloride:glycerol (ChCl:Gly) DES compared to earlier generation ionic liquids [bmim][Tf$_2$N] or [bmim][BF$_4$], with conversion of 96% achieved. Remarkably, transesterification with the glycerol component of the DES was extremely low (<0.5%). iCALB-catalysed aminolysis of ethyl valerate with 1-butylamine also proceeded efficiently in ChCl:Gly with >90% conversion after 1 h. ChCl:Gly was also found to be a suitable co-solvent with aqueous buffers; a significant 20-fold enhancement in the rate of epoxide hydrolase-catalyzed hydrolysis of styrene oxide was observed in 25% v/vChCl:Gly/buffer compared to buffer alone.

The biocatalytic oxidation of cellulose-derived carbohydrates has potential application in bioelectrochemical synthesis and energy production in biofuel cells. Recently, the potential of a biocompatible hydrated choline dihydrogen phosphate (choline dhp) ionic liquid containing 35 wt % water as a solvent for the cellobiose dehydrogenase (CDH)-catalyzed oxidation of cellobiose to cellobionolactone (Scheme 14) was demonstrated [88]. Inter-electron transfer to a synthetic dye acceptor, DCIP, was observed, indicating the system may also have use in bioelectrochemical synthesis. Since CDH is known to be able to transfer electrons to electrodes, the CDH-DES system may have application in ionic liquid/carbohydrate-based bioenergy cells.

Recently, Zhao et al have employed DES-made from choline acetate and glycerol in the enzymatic preparation of biodiesel from Miglyol® oil 812 [89].

A eutectic mixture of choline acetate/glycerol (1:1.5) had a lower viscosity than choline chloride/urea eutectic mixtures, and was found to support CALB enzyme activity. Transesterification reactions performed in DES and 20% (*v/v*) MeOH and 1% water using Novozym® 435 gave a faster reaction rate compared to that in organic solvents or hydrophobic ionic liquids, and afforded 97% conversion. The biphasic separation of the resultant biodiesel allowed for its easy separation from the DES.

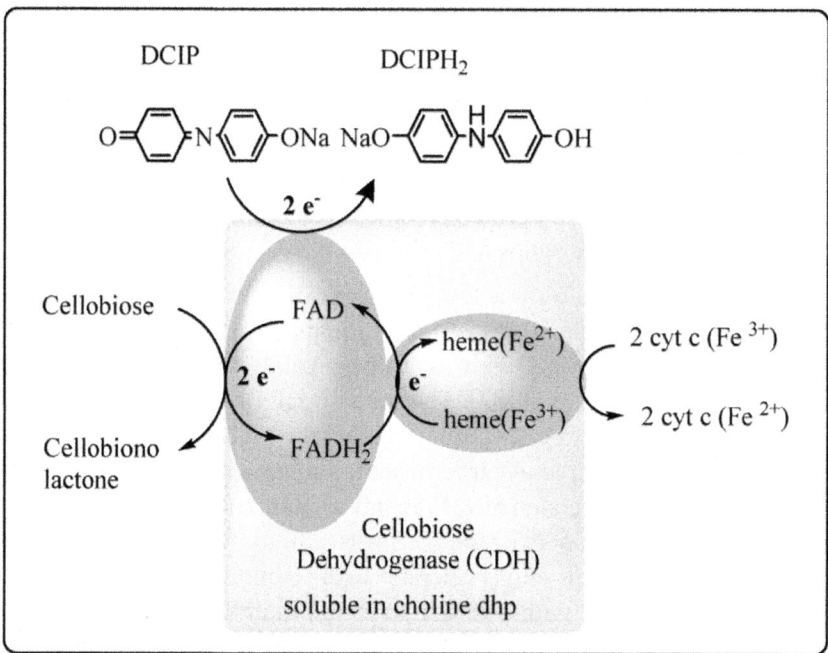

Scheme 14. CDH-catalyzed oxidation of cellobiose by inter and intra-electron transfer pathways [88].

Application of Ionic Liquids in Whole-Cell Biocatalysis: A Greener Perspective

Biocatalysis in organic synthesis employing recombinant enzymes requires an expensive set of enzyme purification unit operations, which themselves generate undesirable waste streams. Using whole cells to catalyze chemical transformations avoids this problem. However, technological challenges with whole-cell biocatalysis often include poor substrate and product solubility in the aqueous buffer/medium critical for cell survival. Furthermore, product accumulation in the cell culture medium can lead to cell toxicity. To overcome these problems, multiphase solvent systems have been developed using immiscible organic solvents, which act as substrate reservoirs and product extraction agents. However, the limitations of these systems are the flammability/toxicity of these solvents, and their deleterious effects on cell viability due to their destructive effects on cell membrane structure [77]. Hydrophobic ionic liquids are an attractive alternative to overcome these problems. Compared to conventional organic solvents, the ionic liquids [bmim][PF$_6$], [bmim][NTF$_2$] and methytrioctyl-ammonium [NTf$_2$] have been found not to affect cell membrane integrity of *Lactobacillus kefir*, *Escherichia coli* and *Saccharomyces cerevisiae* in biphasic systems, thus offering the advantage of preserving cell function and allowing intracellular co-factor regeneration in organic synthesis.

The first example of whole-cell biocatalysis employing a water-ionic liquid biphasic system was reported by Cull *et al.* using *Rhodococcus* R312 [90]. These cells contain nitrile hydratase which catalyzes the hydration of the nitrile group to an amide, a process of synthetic interest for producing fine chemicals and pharmaceutical intermediates. Toluene could be replaced by [bmim][PF$_6$] as a reservoir for the substrate 1,3-dicyanobenzene and as an extraction solvent for the product 3-cyano-benzamide in this biphasic system (Scheme 15). It was observed that the initial rate of 3-cyanobenz-amide production was lower in the [bmim][PF$_6$]-water biphasic system compared to the toluene-water system but that the overall yield was slightly higher. Since the specific activity of nitrile hydratase activity was found to be higher in the ionic liquid-water system and that the ionic liquid did not reduce cell viability, it was concluded the lower rate of product formation was due to reduced mass transfer of substrate from the more viscous ionic liquid phase. An advantage of [bmim][PF$_6$] in downstream processing was reduced cell aggregation at the phase interface, which made product recovery and material recycling easier.

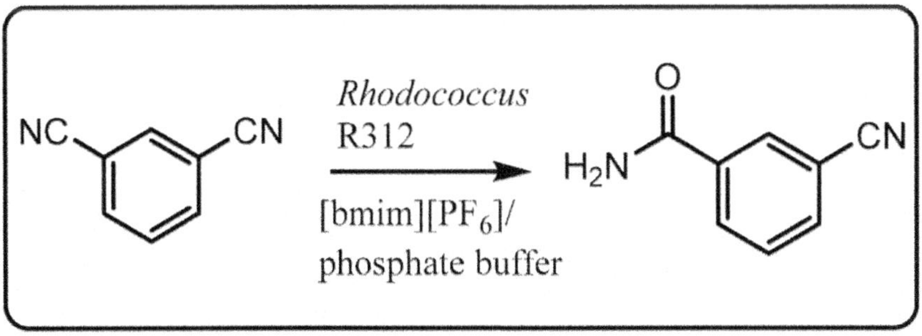

Scheme 15. Hydration of 1,3-dicyanobenzene to 3-cyanobenzamide using *Rhodococcus* R312 in biphasic system [90].

Asymmetric reduction of prochiral ketones to chiral alcohols is an important reaction in the development of chiral pharmaceuticals. Howarth *et al.* have reported whole-cell biocatalysis for asymmetric reduction of prochiral ketones in a biphasic ionic liquid-water system [91]. In this work, yeast was immobilized by encapsulation in low-cost calcium alginate beads and the reduction of several different prochiral ketones performed in a biphasic system of water and [bmim][PF$_6$] containing the substrate (Scheme 16). Yields ranged from poor to good depending on the chosen substrate, with some excellent enantioselectivities (>95%) obtained. The asymmetric reduction of prochiral ketones in ionic liquid-water biphasic systems has since been further developed [92,93].

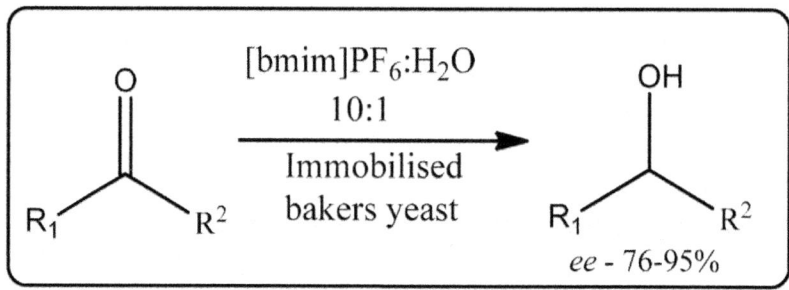

Scheme 16. Immobilized bakers yeast-catalyzed reduction of ketones in a biphasic ionic liquid-water system [91].

Asymmetric reduction of 4-chloro-acetophenone to (*R*)-1-(4-chlorophenyl) ethanol with *Lactobacillus kefir* was carried out as a representative example of a whole-cell-catalyzed multiphase process for the synthesis of fine chemicals (Scheme 17) [92]. In this biphasic system, the ionic liquids employed did not disrupt the necessary cellular cofactor generation systems required for performing the reaction, thus eliminating the need to add these expensive cofactors externally, and provided the product in 92.8% yield with 99.7% enantioselectivity. The potential industrial application of this system was demonstrated by scaling up the reaction volume nearly a hundred fold with the same yield and product purity. Furthermore, no emulsion formed when mixing the two phases, enabling their easy separation for subsequent product recovery and solvent recycling. Similar biphasic ionic liquid-water systems have been used for the *Saccharomyces cerevisiae*asymmetric reduction of 4-chloro acetoacetate to (*S*)-4-chloro-3-hydroxy-butanoate, an intermediate in the synthesis of cholesterol-reducing statin inhibitors, in good yield (80%) and enantioselectivity (84%). In another example of asymmetric reduction, Xu *et al.* [93] used immobilized *Acetobacter* sp. CCTCC M209061 cells in a deep eutectic comprised of chloline chloride and urea as a biocatalytic system to reduce 3-chloro-propiophenone to (S)-3chloro-1-phenylpropanol in 82% yield and >99% enantioselectivity. The reaction could be scaled up to 500 mL with comparable yield and enantioselectivity.

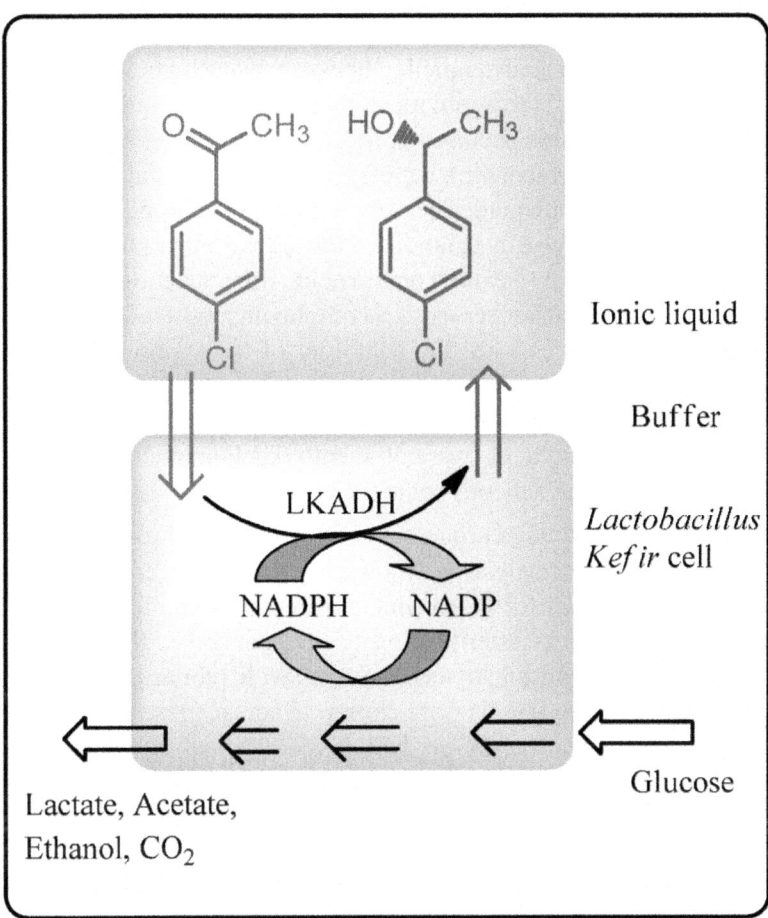

Scheme 17. Asymmetric reduction of 4-chloroacetophenone by *Lactobacillus kefir* in a biphasic ionic liquid-water system [92].

In addition to being employed in whole-cell biphasic systems for the production of fine chemicals or pharmaceutical intermediates, ionic liquids have also been assessed as post-fermentation extraction solvents of amino acids and antibiotics [90,93,94,95,96]. The partition coefficients of erythromycin in biphasic water-butyl acetate and water-[bmim][PF$_6$] were, for example, found by Pfruender *et al.* to be comparable at physiological pH, with about a 10-fold accumulation into the non-aqueous phase [93]. At pH 10 or above, partitioning of erythromycin into [bmim][PF$_6$] decreased to 2-fold while accumulation into butyl acetate increased. These results suggest a potential use of [bmim] [PF$_6$] as a non-flammable extraction solvent for erythromycin at physiological pH followed by back extraction with an alkaline aqueous solution to enable recycling of the ionic liquid. Similarly, a pH-dependent partitioning of the antibiotics amoxicillin and ampicllin between water and ionic liquids was observed by Soto *et al.*; the antibiotics accumulated 3–5-fold into ionic liquid at pH 8 while the reverse occurred at pH 4 [96]. Many different amino acids are prepared by cell fermentation but their isolation by liquid-liquid extraction technologies can be problematic due to their hydrophilic properties. Smirnova *et al.* assessed the ability of a solution of the crown ether dicyclohexano-18-crown-6 in [bmim][PF$_6$] to extract amino acids from aqueous solution [95]. In this system, the crown ether served as a complexing agent with the ammonium group at acidic pH to enhance the solubility of the amino acid in the ionic liquid. Excellent recoveries (>90%) of hydrophilic and hydrophobic amino acids were obtained from acidic aqueous solutions and fermentation broths. Furthermore, the amino acids were efficiently back-extracted from the ionic-liquid using an alkaline aqueous solution.

Although enzyme-based production of biodiesel is an attractive alternative to homogenous alkali catalysis, a whole-cell process for producing biodiesel would remove the need for an enzyme purification step. To this end, Arai *et al.*demonstrated the transesterification of soybean oil with methanol using fungi in an ionic liquid-oil biphasic system with [emim][BF$_4$] and [bmim] [BF$_4$] [97]. These ionic liquids were chosen to act as a reservoir for methanol and the transesterification by-product glycerol, both of which can have detrimental effects on whole cell catalysts, while allowing phase separation of the biodiesel for easy product separation. Four types of whole-cell biocatalysts supported on polyurethane biomass support particles were assessed in this study; wild type *Rhizopus oryzae* producing triacylglycerol lipase (w-ROL) and genetically engineered *Aspergillus oryzae* expressing either *Fusarium heterosporum* lipase (r-FHL), *Candida antarctica* lipase B (r-CALB) or *A. oryzae* mono- and diacylglycerol lipase (r-mdlB). Supported w-ROL was found to have the highest activity achieving 60% FAME content. Its use alone did not result in complete FAME production because of the inherent positional specificity of its triacylglycerol lipase. However, a higher FAME content of 90% was achieved using a mixture of supported w-ROL and r-mdlB. A difficulty associated with the long-term stability of w-ROL in the presence of ionic liquid was overcome by treating it with glutaraldehyde to cross-link its endogenous lipases. This greatly stabilized w-ROL and enabled its recycling. Overall, this research demonstrated the potential application of ionic liquids and whole-cell biocatalysis in developing a more sustainable and environmentally friendly route toward biodiesel production.

CONCLUSIONS

Ionic liquids as bulk solvent phases offer new opportunities as recyclable alternatives to conventional organic solvents for biocatalytic production of commercially important chemicals including asymmetric synthesis, with enhancement in certain cases of the rate of product formation and enantioselectivity. The application of benign ionic liquids that can both dissolve cellulose and maintain enzyme activity bodes well for future developments in utilizing cellulose as raw chemical resource. Similar strategies can be forecasted as apparent from our recent work with biphasic biocatalytic conversion of xylan, the second most abundant biomaterial found in nature after cellulose, using recombinantly expressed, thermostable xylanases, genetically engineered to have high activity in a variety of non-polar ionic liquids, leading to the production of a range of chemical intermediates that can be directly recovered from the aqueous phase in continuous processes [98]. Similar approaches and processes are equally germane in applications with biphasic ionic liquid-vegetable oil mixtures and show great potential as efficient reaction systems for producing biodiesel using enzyme or whole-cell-catalyzed transesterification. In addition to their use as bulk solvents, ionic liquids have been applied as enzyme coatings for the development of recyclable heterogeneous biocatalysts, as immobilized supports for enzymes in biphasic systems employing $ScCO_2$ for continuous reaction systems and as components of supported liquid membranes for continuous separation of reactants and products in enzyme catalyzed reactions. These are exciting developments as the amounts of ionic liquid required in these systems are much less than those for bulk solvent applications, which will reduce any unfavorable environmental impacts of synthesizing or using ionic liquids. The inherent biphasic nature of these systems also enables easier recycling of the ionic liquid component. The CDH-catalyzed oxidation of cellobiose in a biodegradable ionic-based DES demonstrates the potential of employing ionic liquids as components of bioelectrochemical energy cells. Overall, these developments highlight the potential of ionic liquid-based biocatalytic processes in designing more benign sustainable chemical processes for enhancing the quality of life of future generations.

ACKNOWLEDGMENTS

The support of the Australian Research Council is gratefully acknowledged.

REFERENCES AND NOTES

1. Anastas, P.T.; Kirchhoff, M.M. Origins, current status, and future challenges of green chemistry. *Acc. Chem. Res.* **2002**,*35*, 686–694.

2. Welton, T. Room-temperature ionic liquids. Solvents for synthesis and catalysis. *Chem. Rev.* **1999**, *99*, 2071–2083.

3. MacFarlane, D.R.; Forsyth, M.; Howlett, P.C.; Pringle, J.M.; Sun, J.; Annat, G.; Neil, W.; Izgorodina, E.I. Ionic liquids in electrochemical devices and processes: Managing interfacial electrochemistry. *Acc. Chem. Res.* **2007**, *40*, 1165–1173.

4. Dupont, J.; de Souza, R.F.; Suarez, P.A.Z. Ionic liquid (molten salt) phase organometallic catalysis. *Chem. Rev.* **2002**,*102*, 3667–3691.

5. Roosen, C.; Muller, P.; Greiner, L. Ionic liquids in biotechnology: Applications and perspectives for biotransformations. *Appl. Microbiol. Biotechnol.* **2008**, *81*, 607–614.

6. Wang, Z.S.; Koumura, N.; Cui, Y.; Miyashita, M.; Mori, S.; Hara, K. Exploitation of ionic liquid electrolyte for dye-sensitized solar cells by molecular modification of organic-dye sensitizers. *Chem. Mater.* **2009**, *21*, 2810–2816.

7. Riisager, A.; Fehrmann, R.; Haumann, M.; Wasserscheid, P. Supported ionic liquid phase (silp) catalysis: An innovative concept for homogeneous catalysis in continuous fixed-bed reactors. *Eur. J. Inorg. Chem.* **2006**, *2006*, 695–706.

8. Gruttadauria, M.; Riela, S.; Aprile, C.; Lo Meo, P.; D'Anna, F.; Noto, R. Supported ionic liquids. New recyclable materials for the l-proline-catalyzed aldol reaction. *Adv. Synth. Catal.* **2006**, *348*, 82–92.

9. Wasserscheid, P.; Keim, W. Ionic liquids—New "solutions" for transition metal catalysis. *Angew. Chem. Int. Ed.* **2000**,*39*, 3773–3789.

10. Fei, Z.F.; Geldbach, T.J.; Zhao, D.B.; Dyson, P.J. From dysfunction to bis-function: On the design and applications of functionalised ionic liquids. *Chem. -Eur. J.* **2006**, *12*, 2123–2130.

11. Lee, S.G. Functionalized imidazolium salts for task-specific ionic liquids and their applications. *Chem. Commun.* **2006**, 1049–1063.

12. De Maria, P.D. "Nonsolvent" applications of ionic liquids in biotransformations and organocatalysis. *Angew. Chem. Int. Ed.* **2008**, *47*, 6960–6968.

13. Van Rantwijk, F.; Sheldon, R.A. Biocatalysis in ionic liquids. *Chem. Rev.* **2007**, *107*, 2757–2785.

14. Miao, W.S.; Chan, T.H. Ionic-liquid-supported peptide synthesis demonstrated by the synthesis of leu[5]-enkephalin. *J. Org. Chem.* **2005**, *70*, 3251–3255.

15. Fraga-Dubreuil, J.; Bazureau, J.P. Grafted ionic liquid-phase-supported synthesis of small organic molecules.*Tetrahedron Lett.* **2001**, *42*, 6097–6100.

16. Deetlefs, M.; Seddon, K.R. Assessing the greenness of some typical laboratory ionic liquid preparations. *Green Chem.***2010**, *12*, 17–30.

17. Harjani, J.R.; Singer, R.D.; Garciac, M.T.; Scammells, P.J. Biodegradable pyridinium ionic liquids: Design, synthesis and evaluation. *Green Chem.* **2009**, *11*, 83–90.

18. Harjani, J.R.; Farrell, J.; Garcia, M.T.; Singer, R.D.; Scammells, P.J. Further investigation of the biodegradability of imidazolium ionic liquids. *Green Chem.* **2009**, *11*, 821–829.

19. Tao, J.H.; Zhao, L.S.; Ran, N.Q. Recent advances in developing chemoenzymatic processes for active pharmaceutical ingredients. *Org. Process Res. Dev.* **2007**, *11*, 259–267.

20. Straathof, A.J.J.; Panke, S.; Schmid, A. The production of fine chemicals by biotransformations. *Curr. Opin. Biotechnol.***2002**, *13*, 548–556.

21. Klibanov, A.M. Improving enzymes by using them in organic solvents. *Nature* **2001**, *409*, 241–246.

22. Klibanov, A.M. Enzymatic catalysis in anhydrous organic-solvents. *Trends Biochem. Sci.* **1989**, *14*, 141–144.

23. Magnuson, D.K.; Bodley, J.W.; Evans, D.F. The activity and stability of alkaline phosphatase in solutions of water and the fused salt ethylammonium nitrate. *J. Sol. Chem.* **1984**, *13*, 583–587.

24. Moniruzzaman, M.; Nakashima, K.; Kamiya, N.; Goto, M. Recent advances of enzymatic reactions in ionic liquids.*Biochem. Eng. J.* **2010**, *48*, 295–314.

25. Zhao, H. Methods for stabilizing and activating enzymes in ionic liquids. *J. Chem. Technol. Biotechnol.* **2010**, *85*, 891–907.

26. Moniruzzaman, M.; Kamiya, N.; Goto, M. Activation and stabilization of enzymes in ionic liquids. *Org. Biomol. Chem.***2010**, *8*, 2887–2899.

27. Yang, Z. Hofmeister effects: An explanation for the impact of ionic liquids on biocatalysis. *J. Biotechnol.* **2009**, *144*, 12–22.

28. Zhao, H. Effect of ions and other compatible solutes on enzyme activity, and its implication for biocatalysis using ionic liquids. *J. Mol. Catal. B Enzym.* **2005**, *37*, 16–25.

29. De Gonzalo, G.; Lavandera, I.; Durchschein, K.; Wurm, D.; Faber, K.; Kroutil, W. Asymmetric biocatalytic reduction of ketones using hydroxy-functionalised water-miscible ionic liquids as solvents. *Tetrahedron Asymmetry* **2007**, *18*, 2541–2546.

30. Lau, R.M.; Sorgedrager, M.J.; Carrea, G.; van Rantwijk, F.; Secundo, F.; Sheldon, R.A. Dissolution of candida antarctica lipase b in ionic liquids: Effects on structure and activity. *Green Chem.* **2004**, *6*, 483–487.

31. Erbeldinger, M.; Mesiano, A.J.; Russell, A.J. Enzymatic catalysis of formation of z-aspartame in ionic liquid—An alternative to enzymatic catalysis in organic solvents. *Biotechnol. Prog.* **2000**, *16*, 1129–1131.

32. Lau, R.M.; van Rantwijk, F.; Seddon, K.R.; Sheldon, R.A. Lipase-catalyzed reactions in ionic liquids. *Org. Lett.* **2000**, *2*, 4189–4191.

33. Harjani, J.R.; Naik, P.U.; Nara, S.J.; Salunkhe, M.M. Enzyme mediated reactions in ionic liquids. *Curr. Org. Synth.* **2007**, *4*, 354–369.

34. Yang, Z.; Pan, W.B. Ionic liquids: Green solvents for nonaqueous biocatalysis. *Enzym. Microb. Technol.* **2005**, *37*, 19–28.

35. Sureshkumar, M.; Lee, C.K. Biocatalytic reactions in hydrophobic ionic liquids. *J. Mol. Catal. B Enzym.* **2009**, *60*, 1–12.

36. Sheldon, R.A. Biocatalysis in ionic liquids. *RSC Catal. Ser.* **2014**, *15*, 20–43.

37. Lozano, P.; Bernal, J.M.; Garcia-Verdugo, E.; Vaultier, M.; Luis, S.V. *Biocatalysis in Ionic Liquids*; CRC Press: Boca Raton, FL, USA, 2015; pp. 31–66.

38. Stein, F.; Kragl, U. *Biocatalytic Reactions in Ionic Liquids*; John Wiley & Sons, Inc.: Hoboken, NJ, USA, 2014; pp. 193–216.

39. Klembt, S.; Dreyer, S.; Eckstein, M.; Kragl, U. Biocatalytic reactions in ionic liquids. In *Ionic Liquids in Synthesis*, 2nd ed.; Wasserscheid, P., Welton, T., Eds.; Wiley-VCH Verlags GmbH & Co KGaA: Weinheim, Germany, 2008; Volume 2, pp. 641–661.

40. Schöfer, S.H.; Kaftzik, N.; Wasserscheid, P.; Kragl, U. Enzyme catalysis in ionic liquids: Lipase catalysed kinetic resolution of 1-phenylethanol with improved enantioselectivity. *Chem. Commun.* **2001**, 425–426.

41. Itoh, T.; Akasaki, E.; Kudo, K.; Shirakami, S. Lipase-catalyzed enantioselective acylation in the ionic liquid solvent system: Reaction of enzyme anchored to the solvent. *Chem. Lett.* **2001**, *30*, 262–263.

42. Cantone, S.; Hanefeld, U.; Basso, A. Biocatalysis in non-conventional media-ionic liquids, supercritical fluids and the gas. *Green Chem.* **2007**, *9*, 954–971.

43. Gorke, J.; Srienc, F.; Kazlauskas, R. Toward advanced ionic liquids. Polar, enzyme-friendly solvents for biocatalysis.*Biotechnol. Bioprocess Eng.* **2010**, *15*, 40–53.

44. Mohile, S.S.; Potdar, M.K.; Harjani, J.R.; Nara, S.J.; Salunkhe, M.M. Ionic liquids: Efficient additives for candida rugosa lipase-catalyzed enantioselective hydrolysis of butyl 2-(4-chlorophenoxy)propionate. *J. Mol. Catal. B Enzym.* **2004**, *30*, 185–188.

45. Rasalkar, M.S.; Potdar, M.K.; Salunkhe, M.M. Pseudomonas cepacia lipase-catalysed resolution of racemic alcohols in ionic liquid using succinic anhydride: Role of triethylamine in enhancement of catalytic activity. *J. Mol. Catal. B Enzym.* **2004**, *27*, 267–270.

46. Lourenco, N.M.T.; Barreiros, S.; Afonso, C.A.M. Enzymatic resolution of indinavir precursor in ionic liquids with reuse of biocatalyst and media by product sublimation. *Green Chem.* **2007**, *9*, 734–736.

47. Mai, N.L.; Ahn, K.; Bae, S.W.; Shin, D.W.; Morya, V.K.; Koo, Y.-M. Ionic liquids as novel solvents for the synthesis of sugar fatty acid ester. *Biotechnol. J.* **2014**, *9*, 1565–1572.

48. Lee, J.K.; Kim, M.-J. Ionic liquid-coated enzyme for biocatalysis in organic solvent. *J. Org. Chem.* **2002**, *67*, 6845–6847.

49. Itoh, T.; Matsushita, Y.; Abe, Y.; Han, S.-H.; Wada, S.; Hayase, S.; Kawatsura, M.; Takai, S.; Morimoto, M.; Hirose, Y. Increased enantioselectivity and remarkable acceleration of lipase-catalyzed transesterification by using an imidazolium peg-alkyl sulfate ionic liquid. *Chem. Eur. J.* **2006**, *12*, 9228–9237.

50. Nakashima, K.; Kamiya, N.; Koda, D.; Maruyama, T.; Goto, M. Enzyme encapsulation in microparticles composed of polymerized ionic liquids for highly active and reusable biocatalysts. *Org. Biomol. Chem.* **2009**, *7*, 2353–2358.

51. Yu, C.-Y.; Wei, P.; Li, X.-F.; Zong, M.-H.; Lou, W.-Y. Using ionic liquid in a biphasic system to improve asymmetric hydrolysis of styrene oxide catalyzed by cross-linked enzyme aggregates (cleas) of mung bean epoxide hydrolases.*Ind. Eng. Chem. Res.* **2014**, *53*, 7923–7930.

52. Rehmann, L.; Ivanova, E.; Gunaratne, H.Q.N.; Seddon, K.R.; Stephens, G. Enhanced laccase stability through mediator partitioning into hydrophobic ionic liquids. *Green Chem.* **2014**, *16*, 1462–1469.

53. Turner, M.B.; Spear, S.K.; Huddleston, J.G.; Holbrey, J.D.; Rogers, R.D. Ionic liquid salt-induced inactivation and unfolding of cellulase from trichoderma reesei. *Green Chem.* **2003**, *5*, 443–447.

54. Toral, A.R.; de los Rios, A.P.; Hernandez, F.J.; Janssen, M.H.A.; Schoevaart, R.; van Rantwijk, F.; Sheldon, R.A. Cross-linked candida antarctica lipase b is active in denaturing ionic liquids. *Enzym. Microb. Technol.* **2007**, *40*, 1095–1099.

55. Zhao, H.; Baker, G.A.; Song, Z.; Olubajo, O.; Crittle, T.; Peters, D. Designing enzyme-compatible ionic liquids that can dissolve carbohydrates. *Green Chem.* **2008**, *10*, 696–705.

56. Vancov, T.; Alston, A.-S.; Brown, T.; McIntosh, S. Use of ionic liquids in converting lignocellulosic material to biofuels. *Renew. Energy* **2012**, *45*, 1–6.

57. Long, J.; Li, X.; Guo, B.; Wang, F.; Yu, Y.; Wang, L. Simultaneous delignification and selective catalytic transformation of agricultural lignocellulose in cooperative ionic liquid pairs. *Green Chem.* **2012**, *14*, 1935–1941.

58. Shi, J.; Gladden, J.M.; Sathitsuksanoh, N.; Kambam, P.; Sandoval, L.; Mitra, D.; Zhang, S.; George, A.; Singer, S.W.; Simmons, B.A.; *et al.* One-pot ionic liquid pretreatment and saccharification of switchgrass. *Green Chem.* **2013**, *15*, 2579–2589.

59. Sun, N.; Parthasarathi, R.; Socha, A.M.; Shi, J.; Zhang, S.; Stavila, V.; Sale, K.L.; Simmons, B.A.; Singh, S. Understanding pretreatment efficacy of four cholinium and imidazolium ionic liquids by chemistry and computation.*Green Chem.* **2014**, *16*, 2546–2557.

60. Moniruzzaman, M.; Ono, T. Ionic liquid assisted enzymatic delignification of wood biomass: A new green' and efficient approach for isolating of cellulose fibers. *Biochem. Eng. J.* **2012**, *60*, 156–160.

61. Melero, J.A.; Iglesias, J.; Morales, G. Heterogeneous acid catalysts for biodiesel production: Current status and future challenges. *Green Chem.* **2009**, *11*, 1285–1308.

62. Ha, S.H.; Lan, M.N.; Lee, S.H.; Hwang, S.M.; Koo, Y.-M. Lipase-catalyzed biodiesel production from soybean oil in ionic liquids. *Enzym. Microb. Technol.* **2007**, *41*, 480–483.

63. Gamba, M.; Lapis, A.A.M.; Dupont, J. Supported ionic liquid enzymatic catalysis for the production of biodiesel. *Adv. Synth. Catal.* **2008**, *350*, 160–164.

64. Zhao, H.; Song, Z.; Olubajo, O.; Cowins Janet, V. New ether-functionalized ionic liquids for lipase-catalyzed synthesis of biodiesel. *Appl. Biochem. Biotechnol.* **2010**, *162*, 13–23.

65. Yu, D.; Wang, C.; Yin, Y.; Zhang, A.; Gao, G.; Fang, X. A synergistic effect of microwave irradiation and ionic liquids on enzyme-catalyzed biodiesel production. *Green Chem.* **2011**, *13*, 1869–1875.

66. Zhang, K.P.; Lai, J.Q.; Huang, Z.L.; Yang, Z. Penicillium expansum lipase-catalyzed production of biodiesel in ionic liquids. *Bioresour. Technol.* **2011**, *102*, 2767–2772.

67. Lai, J.-Q.; Hu, Z.-L.; Wang, P.-W.; Yang, Z. Enzymatic production of microalgal biodiesel in ionic liquid [bmim][pf6].*Fuel* **2012**, *95*, 329–333.

68. Pellegrino, J.J.; Noble, R.D. Enhanced transport and liquid membranes in bioseparations. *Trends Biotechnol.* **1990**, *8*, 216–224.

69. Branco, L.C.; Crespo, J.G.; Afonso, C.A.M. Highly selective transport of organic compounds by using supported liquid membranes based on ionic liquids. *Angew. Chem. Int. Ed.* **2002**, *114*, 2895–2897.

70. Miyako, E.; Maruyama, T.; Kamiya, N.; Goto, M. Use of ionic liquids in a lipase-facilitated supported liquid membrane. *Biotechnol. Lett.* **2003**, *25*, 805–808.

71. Miyako, E.; Maruyama, T.; Kamiya, N.; Goto, M. Enzyme-facilitated enantioselective transport of (*S*)-ibuprofen through a supported liquid membrane based on ionic liquids. *Chem. Commun.* **2003**, 2926–2927.

72. Hernandez-Fernandez, F.J.; de los Rios, A.P.; Tomas-Alonso, F.; Gomez, D.; Villora, G. On the development of an integrated membrane process with ionic liquids for the kinetic resolution of rac-2-pentanol. *J. Membr. Sci.* **2008**, *314*, 238–246.

73. Audia, J.E.; Britton, T.C.; Droste, J.J.; Folmer, B.K.; Huffman, G.W.; John, V.; Latimer, L.H.; Mabry, T.E.; Nissen, J.S. Preparation of *N*-(phenylacetyl)di- and Tripeptide Derivatives for Inhibiting β-Amyloid Peptide Release. WO9822494, 28 May 1998.

74. Hernandez-Fernandez, F.J.; de los Rios, A.P.; Tomas-Alonso, F.; Gomez, D.; Villora, G. Kinetic resolution of 1-phenylethanol integrated with separation of substrates and products by a supported ionic liquid membrane. *J. Chem. Technol. Biotechnol.* **2009**, *84*, 337–342.

75. Kamat, S.V.; Beckman, E.J.; Russell, A.J. Enzyme-activity in supercritical fluids. *Crit. Rev. Biotechnol.* **1995**, *15*, 41–71.

76. Mesiano, A.J.; Beckman, E.J.; Russell, A.J. Supercritical biocatalysis. *Chem. Rev.* **1999**, *99*, 623–633.

77. Lozano, P. Enzymes in neoteric solvents: From one-phase to multiphase systems. *Green Chem.* **2010**, *12*, 555–569.

78. Lozano, P.; de Diego, T.; Gmouh, S.; Vaultier, M.; Iborra, J.L. Criteria to design green enzymatic processes in ionic liquid/supercritical carbon dioxide systems. *Biotechnol. Prog.* **2004**, *20*, 661–669.

79. Fonseca, G.S.; Scholten, J.D.; Dupont, J. Iridium nanoparticles prepared in ionic liquids: An efficient catalytic system for the hydrogenation of ketones. *Synlett* **2004**, 1525–1528.

80. Valkenberg, M.H.; de Castro, C.; Holderich, W.F. Immobilisation of ionic liquids on solid supports. *Green Chem.* **2002**,*4*, 88–93.

81. Mehnert, C.P. Supported ionic liquid phases. *Chem. -Eur. J.* **2004**, *11*, 50–56.

82. Lozano, P.; Garcia-Verdugo, E.; Piamtongkam, R.; Karbass, N.; De Diego, T.; Burguete, M.I.; Luis, S.V.; Iborra, J.L. Bioreactors based on monolith-supported ionic liquid phase for enzyme catalysis in supercritical carbon dioxide. *Adv. Synth. Catal.* **2007**, *349*, 1077–1084.

83. Lozano, P.; De Diego, T.; Sauer, T.; Vaultier, M.; Gmouh, S.; Iborra, J.L. On the importance of the supporting material for activity of immobilized candida antarctica lipase b in ionic liquid/hexane and ionic liquid/supercritical carbon dioxide biphasic media. *J. Supercrit. Fluids* **2007**, *40*, 93–100.

84. Lozano, P.; de Diego, T.; Mira, C.; Montague, K.; Vaultier, M.; Iborra, J.L. Long term continuous chemoenzymatic dynamic kinetic resolution of rac-1-phenylethanol using ionic liquids and supercritical carbon dioxide. *Green Chem.***2009**, *11*, 538–542.

85. Lozano, P.; de Diego, T.; Vaultier, M.; Iborra, J.L. Dynamic kinetic resolution of sec-alcohols in ionic liquids/supercritical carbon dioxide biphasic systems. *Int. J. Chem. React. Eng.* **2009**, *7*.

86. Imperato, G.; Konig, B.; Chiappe, C. Ionic green solvents from renewable resources. *Eur. J. Org. Chem.* **2007**, *2007*, 1049–1058.

87. Gorke, J.T.; Srienc, F.; Kazlauskas, R.J. Hydrolase-catalyzed biotransformations in deep eutectic solvents. *Chem. Commun.* **2008**, 1235–1237.

88. Fujita, K.; Nakamura, N.; Igarashi, K.; Samejima, M.; Ohno, H. Biocatalytic oxidation of cellobiose in an hydrated ionic liquid. *Green Chem.* **2009**, *11*, 351–354.

89. Zhao, H.; Baker, G.A.; Holmes, S. New eutectic ionic liquids for lipase activation and enzymatic preparation of biodiesel. *Org. Biomol. Chem.* **2011**, *9*, 1908–1916.

90. Cull, S.G.; Holbrey, J.D.; Vargas-Mora, V.; Seddon, K.R.; Lye, G.J. Room-temperature ionic liquids as replacements for organic solvents in multiphase bioprocess operations. *Biotechnol. Bioeng.* **2000**, *69*, 227–233.

91. Howarth, J.; James, P.; Dai, J. Immobilized baker's yeast reduction of ketones in an ionic liquid, [bmim]PF6 and water mix. *Tetrahedron Lett.* **2001**, *42*, 7517–7519.

92. Pfruender, H.; Midjojo, M.; Kragl, U.; Weuster-Botz, D. Efficient whole-cell biotransformation in a biphasic ionic liquid/water system. *Angew. Chem. Int. Ed.* **2004**, *43*, 4529–4531.

93. Xu, P.; Xu, Y.; Li, X.-F.; Yi Zhao, B.-Y.; Zong, M.-H.; Lou, W.-Y. Enhancing Asymmetric Reduction of 3-Chloropropiophenone with Immobilized Acetobacter sp. CCTCC M209061 Cells by Using Deep Eutectic Solvents as Cosolvents. *ACS Sustain. Chem. Eng.* **2015**, *3*, 718–724.

94. Pfruender, H.; Jones, R.; Weuster-Botz, D. Water immiscible ionic liquids as solvents for whole cell biocatalysis. *J. Biotechnol.* **2006**, *124*, 182–190.

95. Smirnova, S.V.; Torocheshnikova, I.I.; Formanovsky, A.A.; Pletnev, I.V. Solvent extraction of amino acids into a room temperature ionic liquid with dicyclohexano-18-crown-6. *Anal. Bioanal. Chem.* **2004**, *378*, 1369–1375.

96. Soto, A.; Arce, A.; Khoshkbarchi, M.K. Partitioning of antibiotics in a two-liquid phase system formed by water and a room temperature ionic liquid. *Sep. Purif. Technol.* **2005**, *44*, 242–246.

97. Arai, S.; Nakashima, K.; Tanino, T.; Ogino, C.; Kondo, A.; Fukuda, H. Production of biodiesel fuel from soybean oil catalyzed by fungus whole-cell biocatalysts in ionic liquids. *Enzym. Microb. Technol.* **2010**, *46*, 51–55.

98. Lim, A.; Zhang, C.; Oktavianawati, I.; Hearn, M.T.W. Continuous enzymatic conversion of xylan with product recovery by liquid-liquid two-phase extraction. Clayton, Victoria, Australia, Unpublished work . 2015.

Chapter 5

RECENT ADVANCES IN ENZYME PROMISCUITY

Rinkoo Devi Gupta

Faculty of Life Sciences and Biotechnology, South Asian University, New Delhi 110021, India

ABSTRACT

Enzyme promiscuity is defined as the capability of an enzyme to catalyze a reaction other than the reaction for which it has been specialized. Although, enzyme is known for its specificity, many enzymes are reported to be promiscuous in nature. However, the promiscuous function may not be relevant in physiological conditions. The reasons could be either very low level of catalytic activity or unavailability of the substrates in the cell. Hitherto, the enzyme promiscuity is of great importance because they are the starting point for the evolution of new functions in the nature. In addition, the promiscuous activities are utilized for the development of new catalytic functions by applying directed laboratory evolution and protein engineering techniques. The aim of this review is to provide recent developments on the understanding of the mechanism of catalytic promiscuity, evolvability of promiscuous functions and the applications of enzyme promiscuity in the designing of enhanced or new functional biocatalysts.

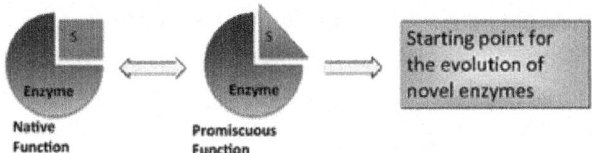

BACKGROUND

Enzymes are well known for catalyzing specific reactions and hence the specificity is a key feature of enzymes. However, there are many exceptions of this universally accepted enzyme specificity, which have been intensive area of research since last two decades. These enzymes are capable of binding with more than one substrates and catalyzing reactions other than its physiological substrate. The secondary function of enzyme is referred as promiscuous function, moonlighting function, substrate ambiguity or some time as cross reactivity of enzymes [1, 2]. Generally, enzymes catalyze a range of similar substrates of the same class by identical reaction mechanism, which is referred as broad substrate specificity. Hence, commonly occurring broad substrate specificity, for example lipase catalyzes a broad range of carboxylic acids and alcohols, are not considered as enzyme promiscuity. However, it also catalyzes C–C bond formation reactions, which is a promiscuous function of lipase. Other examples are Glutathione S transferase, cytochrome P450 and serum paraoxonase, which are known to neutralize a broad range of related substrates, are not promiscuous functions. However, they also catalyze dissimilar substrates that are promiscuous functions of these enzymes. For better distinction between promiscuous function and broad substrate specificity, the function of enzyme, which has no physiological relevance, can be categorized as promiscuous function.

If not all, several enzymes are promiscuous in nature. The physiologically irrelevant promiscuous function of an enzyme is hypothesized to be the starting point for the evolution of new enzymes [1]. This hypothesis was proven by the excellently designed experiments in the laboratory by applying directed laboratory evolution technique on the targets like serum paraoxonase, bacterial phosphotriesterase and several other enzymes [3–6]. Subsequently, many other promiscuous enzymes were studied for the evolution of new enzymes [7, 8], and to appreciate the mechanistic and evolutionary prospective [1, 2, 9–11]. Previous reviews highlight the importance of promiscuity, and also examined the possible mechanisms [1, 6, 12]. Other more recent reviews have focused on the practical implications of enzyme promiscuity in the development of biocatalysts involved in organic synthesis [13–15], or divergence in certain enzyme families [16–19]. The aim of writing this review is to encapsulate the advances in the field of enzyme promiscuity. Focus of the review is to highlight the recent findings on mechanism of catalytic promiscuity, evolutionary potential of promiscuous functions and applications of enzyme promiscuity in the designing of novel biocatalysts.

Mechanism of Catalytic Promiscuity

Promiscuous nature of enzymes raises many questions in our mind for e.g., what makes an enzyme promiscuous? How does an active site of enzyme accommodate a substrate other than its physiological one? Previous studies showed that the mechanism of catalysis of promiscuous substrate is different from the physiological reaction mechanism. These mechanisms underlying enzyme promiscuity are based on the conformational diversity of enzyme's active site, substrate ambiguity, different protonation states, different subsites within the same active sites and cofactor ambiguity [1, 2]. In my view, it can be categorized in three major groups based on the different factors responsible for enzyme promiscuity: (1) active site plasticity of enzymes, (2) substrate ambiguity and (3) cofactor ambiguity (Fig. 1) which have been deliberated below.

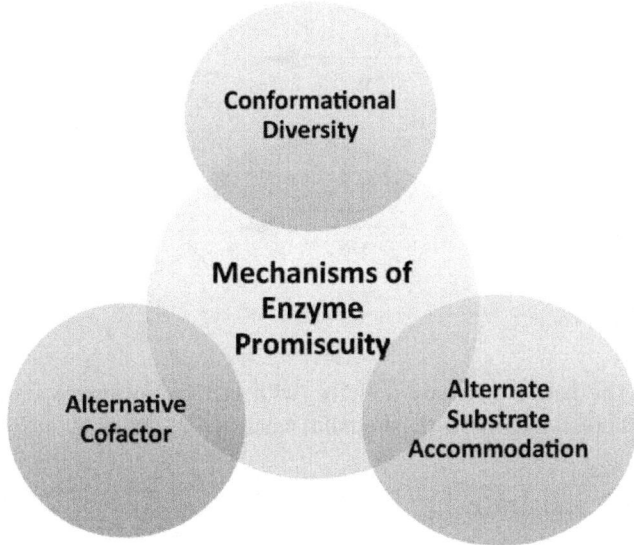

Figure. 1: Enzyme promiscuity can be categorized in three major groups based on the different factors responsible for enzyme promiscuity. These are: active site plasticity of enzymes, substrate ambiguity and cofactor ambiguity.

Active Site Plasticity

Active site plasticity of enzymes facilitates the binding with promiscuous substrates and thereby the catalysis. The plasticity allows enzyme active site to accommodate promiscuous substrate and thus the enzyme-substrate complex acquires diverse conformations. Hence, the native and the promiscuous functions are facilitated by different active-site conformations. For examples, β-lactamase and sulfo-transferase display increased plasticity and altered substrate hydrolysis [20, 21]. Other than active-site plasticity, many enzymes possess a loop like structure passing through or over the active site of enzyme, which determines the specificity [22, 23]. For example, isopropylmalate isomerase exhibits dual-substrate specificity, where the conformation of a loop differs on the specific substrate present [24]. In the above-mentioned examples as well as many other promiscuous enzymes, both the native and promiscuous activities reside within the same active site. However, there are evidences of different sub-sites within the same active site of enzymes. For example serum paraoxonase has native lactonase with promiscuous phosphotriesterase activity where hydrolysis of lactone is mediated by histidine dyad while phosphotriesterse activity is mediated by another set of residues in the same active site groove [1, 3] (Fig. 2).

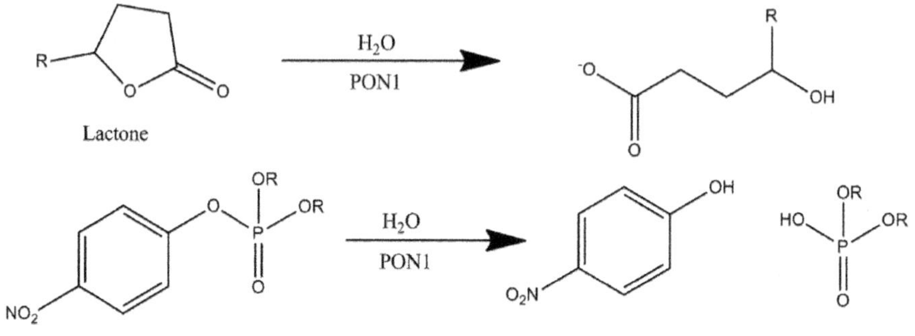

Figure. 2: The native lactonase activity (with lactone substrate) and promiscuous phosphotriesterase activity (with paraoxon) catalyzed by serum paraoxonase (PON1).

Ambiguous Substrate

Several enzymes can accommodate ambiguous substrates by different modes of interactions in the very same active site. Some of these might be employed by promiscuous substrates. For example, cytochromes P450 (CYP) represents a diverse group of heme-thiolate proteins, which share common protein fold however they differ in substrate selectivity. The mechanism of catalysis is through activation of molecular oxygen on varied monooxygenation reactions. Cytochrome CYP3A4 shows remarkably extreme promiscuity in substrate specificity and cooperative substrate binding. Sevrioukova has reviewed extensively the promiscuous nature of cytochromes P450 [25]. Similarly, salicylic acid binding protein has promiscuous esterase activity towards a series of substrates. The highest activity was reported for its native substrate, methyl salicylate that is mediated by substrate-assisted catalysis involving the hydroxyl group from methyl salicylic acid [26].

Cofactor Ambiguity

In many enzymes, changes in cofactor alter enzyme specificity. Active sites of metalloenzymes harbor such amino acid side chain ligands, which create metal binding sites, are often capable of binding a range of metals and other chemical ligands. The binding of a variety of metals may create metal-binding promiscuity and thus catalytic promiscuity [27, 28]. Metal-binding promiscuity can be employed in artificial metalloenzyme design by using the native scaffold [29]. Similarly, farnesyl diphosphate synthase, a key enzyme of isoprenoid biosynthetic pathway, displays its activity in the presence of metal cofactors. Mg^{2+} ions lead to the production of farnesyl diphosphate while the presence of Co^{2+} ions lead to geranyl diphosphate production. This metal ion-dependent enzyme promiscuity dictates a regulatory mechanism which allows a single enzyme to specifically control the metabolites it produces, thus potentially changing the flow of carbon into different metabolic pathways [30]. A recently identified enzyme named NDM-1, from *Klebsiella pneumoniae* from a patient sample, exemplifies extreme enzyme promiscuity. NDM-1 hydrolyzes and degrades nearly all known β-lactam-based antibiotics. NDM-1 can employ different metal cofactors and different reaction mechanism for its promiscuous activities [31].

Evolutionary Potential of Promiscuous Function

Many scientists have explored the evolutionary aspect of promiscuity [1, 2, 32, 33]. It is proposed that a new function can evolve without negative trade-offs in the native activity, leading to a generalist enzyme. Later on, this generalist enzyme can become a specialist for a new catalytic activity. It is well stated that the process of natural evolution is gradual and slow. Owing to the weaker promiscuous activity as compare to native activity, it could be a better starting point for the evolution of new function. It also provides an immediate advantage to the gene under selection for a new function. Initially, weaker promiscuous activity can be compensated by higher expression level of the protein. Further, gradual improvement in the weaker promiscuous activities is achievable by accumulation of a few beneficial mutations or neutral mutations [34, 35]. Small variations in the active site can lead to the emergence of new functions in existing protein folds. For evidence, bacterial and mammalian 6-pyruvoyltetrahydropterin synthase homologs catalyze distinct reactions using the same 7, 8-dihydroneopterin triphosphate as substrate. The bacterial enzyme catalyzes the formation of 6-carboxy-5, 6, 7, 8-tetrahydropterin, whereas the mammalian enzyme converts 7, 8-dihydroneopterin triphosphate to 6-pyruvoyltetrahydropterin [36], which indicates that small variation in active site can lead to the emergence of new enzyme. This hypothesis has been utilized for the construction of an efficient and thermostable phosphotriesterase from lactonase by creating simple double mutations His250Ile and Ile263Trp [37].

When complicated and challenging enzyme activities are sought by applying directed evolution, smart gene libraries are required. A neutral drift of the gene can help in achieving such libraries. Neutral drift is a gradual accumulation of mutations under selection to maintain a protein's original function and structure, where each and every variant is folded and functional and maintains a certain degree of the enzyme's existing function. For making neutral libraries, several iterative rounds of mutagenesis and selection are applied to maintain the protein's original function and structure [34, 35]. This results a library of highly diversified mutants with different polymorphic characteristics. The most significant point regarding neutral libraries is that all the mutants are stable and functional which may not be available in the wild-type starting point. We have described the generation of neutral libraries for directed enzyme evolution using serum paraoxonase (PON1) as a model. This allowed the selection of mutants presenting improved activity, as well as the identification of mutants displaying higher specific activity. The resulting neutral libraries confined to only a few hundreds of variants. These neutral mutants were used as starting point for the directed evolution, which was resulted in array of new variants, including PON1 variants capable of degrading V-type organophosphates and sterically hindered organophosphates. This method is generally applicable methods for the preparation of such libraries starting from the wild-type gene [34, 35].

Neutral drift happening over billions of year results in considerable sequence divergence among proteins that binds with same molecule or catalyze the same reaction. Natural selection maintains the primary activity of these proteins; thereby maintain physiologically irrelevant promiscuous activities. Thus, the levels and the evolvability of promiscuous activities may vary among these orthologous proteins. For example, there are nine gamma-glutamyl phosphate reductase (ProA) orthologs, which display different level of promiscuity varying by about 50-fold [38]. A single amino acid change from glutamine to alanine near the active site seemed to be critical for enhanced promiscuous activity in these orthologous genes. The improvement in the promiscuous activity has been shown wide-ranging from 50- to 770-fold. This fold improvement was not correlated with the original level of the promiscuous activity. Similarly, the decrease in the native activity varied from 190- to 2100-fold [38]. These results suggest that evolution of a novel enzyme may be possible with some orthologous enzymes, but not with all [39, 40]. Hence, it is suggested to use several orthologous enzymes, instead of only one, as a starting point for directed laboratory evolution.

Applications of Promiscuous Functions

As we have discussed above, the promiscuity of enzymes can be a starting point for the divergent evolution. The low promiscuous activity towards a physiologically irrelevant substrate might turn the enzyme into a much more proficient catalyst which is possible by accumulation of just one or more beneficial mutations affording a survival benefit to the organism. Furthermore, a promiscuous function can be the starting point for the creation of a new enzyme activity by applying rational, semi-rational protein engineering methodologies or by applying directed laboratory evolution [41, 42]. Certain recent reports are discussed below where significant improvements in promiscuous activities and/or change in substrate specificity have been established (Table 1).

Table 1: Examples of fold improvements in promiscuous functions

Enzymes	Native function/substrate	Promiscuous functions/substrate	Improvement by directed laboratory evolution and/or rational designing	References
Mammalian serum paraoxonase	Lipo-lactonase	Phosphotriesterase CMP-Coumarine (racemic mixture)	77-fold higher	[35]
		CMP-Coumarine (sp isomer)	86,000-fold	[39]
Phosphotriesterase (PTE) from bacteria	Paraoxon hydrolysis	Racemic nerve agent VX	230-fold	[42]
		DEVX	78-fold	
		Malathion	5000-fold	[43]
PLL scaffold (Dr0930) from *Deinococcus radiodurans*	Lactonase	OP hydrolase	69,000-fold	[52]
Sortase A from *Staphylococcus aureus*	Transpeptidation of LPXTG	LAXTG and LPXSG	51,000-fold	[44]
Human glutathione transferase (GST) M2-2	2-cyano-1,3-dimethyl-1-nitrosoguanidine	Azathioprine	30-fold	[53]

Organophosphates (OPs) poisoning poses great danger to both military and civilian populations. A timely and effective control with pharmacological agents can minimize the damage. Rapid in vivo organophosphate clearance requires bio-scavenging enzymes with catalytic efficiencies of more than 10^7 M^{-1} min^{-1}. Hydrolysis of OPs by mammalian enzymes at high efficiency has been a challenge, specially the more toxic stereoisomer *Sp*, with a required catalytic efficiency. Though mammalian PON1 is a foremost candidate for such a treatment, it hydrolyzes the toxic *Sp* isomers of G-agents with very slow rates. By performing three rounds of directed evolution of PON1, a new mutant for *Sp* isomer with a catalytic efficiency of the order of 10^5 $M^{-1}min^{-1}$, which was ~86,000 fold improved over wild-type PON1 G3C9 was isolated (Fig. 3) [3]. Based on the mutations appeared in the variants obtained by directed evolution experiment and PON1's active sites, we designed a library by incorporating synthetic oligonucleotides via gene reassembly. Following shuffling of the improved variants, a screening of the random mutagenesis library resulted in a range of variants with the catalytic efficiency of up to 10^7 $M^{-1}min^{-1}$. We characterized several such mutants with broader stereo-specificity and also with different leaving groups, which could be used for the in vivo detoxification of real nerve agents. These in vivo prophylactic activity of evolved variants and the newly developed screens provide the foundation for designing PON1 and other OP degrading enzymes for prophylaxis against other G-type agents [3, 43].

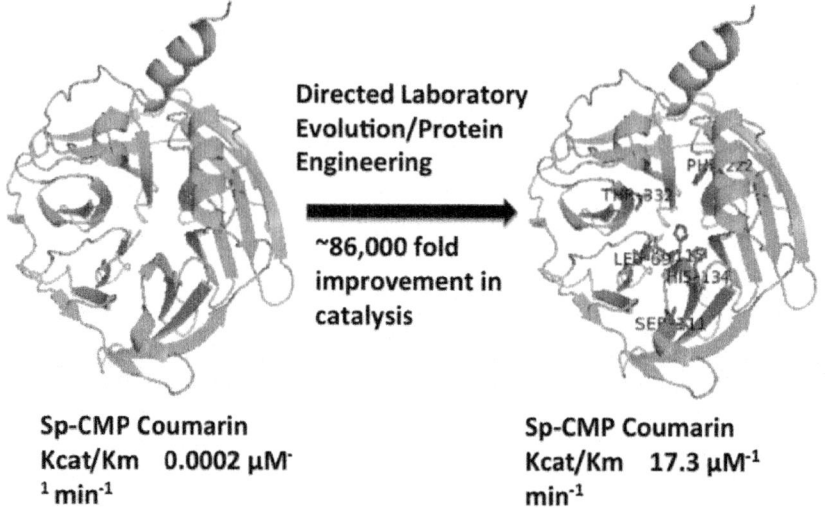

Figure. 3 Improvement in promiscuous function of mammalian serum PON1. Several rounds of mutagenesis and selection by applying directed laboratory evolution and ra-

tional protein designing approaches resulted in the PON1 mutant exhibiting ~86,000-fold improvement in $Kcat/Km$ values for the hydrolysis of stereospecific G-type nerve agent analogue Sp-CMP Coumarin. The selected clone had following mutations as compare to the starting gene -Ll69G, S111T, H115W, H134R, F222S and T332S, which are displayed in the figure.

Another category of nerve agent is V-type (VX and VR), which is known to be one of the most toxic substances. Promiscuous bacterial phosphotriesterases (PTEs) are capable of hydrolyzing VX but with very low level of activity [44]. PTE mutant library was created by mutating 12 active-site residues of PTE in order to enhance its catalytic efficiency. The library was screened for catalytic activity against a new VX analogue, DEVX, which contains the same thiolate-leaving group of VX attached to a diethoxyphosphate center rather than the ethoxymethylphosphonate center of VX. The evolved mutants showed up to 26-fold enhanced catalytic activity with DEVX relative to wild-type PTE. For further improvement, loop7 of Mutant PTE was mutated by error-prone PCR, which resulted in up to 78-fold increase in the rate of DEVX hydrolysis and 230-fold improvement in hydrolyzing racemic nerve agent VX as compare to wild-type wild-type PTE. However, stereo-selectivity for the hydrolysis of the two enantiomers of VX was relatively low [44]. The catalytic activity of PTE from *A. radiobacter*, for the common organophosphorous insecticide malathion was enhanced by changing serine to leucine at position 308 and tyrosine to alanine at position 309, which resulted in 5000-fold increase in $Kcat/Km$ value. X-ray crystal structures for the above variant demonstrate that the access to the binding pocket was enhanced by the replacement of the bulky tyrosine residue with the smaller alanine residue [45].

One example of creating highly active and specific enzyme is reported with sortase A from *Staphylococcus aureus,* which catalyzes the transpeptidation of an LPXTG peptide acceptor and a glycine-linked peptide donor. This has proven to be a powerful tool for site-specific protein modification due to its transpeptidation activity but its high specificity limits its broader utility. Recently, two orthogonal sortase A variants have been developed by applying directed laboratory evolution technique that recognize each of two different substrates, LAXTG and LPXSG, with high activity and specificity. The evolved sortases exhibit changes in specificity up to 51,000-fold, relative to the starting sortase without much loss of catalytic activity [46].

Lipases are well studied due to high industrial value and applications. *Candida antarctica* lipase B (CALB) is a well-known for its promiscuous activity which can be utilized for designing new catalysts for important organic reactions [47]. The active site of CALB is similar to that of soluble epoxide hydrolase, which are formed by a nucleophile-histidine-acid catalytic triad and an oxyanion hole similar to α/β hydrolases. Recently, a S105D variant of CALB was identified as a new catalyst for epoxide hydrolysis. The hydrolysis of the trans-diphenylpropene oxide is studied in particular and suggested that the mutant CALB is a good protein scaffold to be used for the biosynthesis of chiral compounds [48]. Similarly, lipase SrLip from *Streptomyces rimosus* (Q93MW7) revealed lipase, phospholipase, esterase and thioesterase activities [49]. Porcine pancreatic lipase (PPL), one of the best enzymes identified for biocatalytic aldol addition at lower temperature but much accelerated activity at elevated temperature. Recently, a novel peptidase is reported from thermophilic archaea *Sulfolobus tokodaii* for its catalytic promiscuity of aldol addition, which shows comparable activity as PPL. The catalytic efficiency of this enzyme at 55 °C adds up to 140 times higher than that of PPL at its optimum physiological temperature. This study is an example, which signifies the importance to identify a new enzyme with catalytic promiscuity and demonstrates the application of novel biocatalyst from thermophile microorganisms [50].

A recent report describes the designing of an enzyme capable of discriminately etherifying the parahydroxyl of coniferyl alcohol in the presence of excess sinapyl alcohol. The designed enzyme mutant has a considerably smaller substrate-binding pocket that forces a clear steric hindrance thus excluding larger lignin precursors. Lignin is derivative of three monolignols, which are polymerized by oxidative reactions. The composition of the monolignol monomers determines the degree of lignin condensation and thus the degradability of plant cell walls. Guaiacyl lignin is considered as the condensed structural unit [50]. The active site of a monolignol 4-O-methyltransferase (MOMT5) was precisely remodeled to create an enzyme that specifically methylates the condensed guaiacyl lignin precursor coniferyl alcohol. They started with the promiscuous engineered enzyme, MOMT5 with mutations T133L, E165I, F175I, F166W and H169F. Together with the mutant information, crystal structural information with combinatorial active site saturation mutagenesis were applied and remodeled its substrate binding pocket by the addition of four substitutions, i.e., M26H, S30R, V33S, and T319M [51]. Type I plant nucleases are known to play an important role in apoptotic processes and cell senescence. The first structure of tomato nuclease showed its oligomerization and activity profiles resulted in unexpected promiscuous activity towards phospholipids. Solving the crystal structure of this protein identified possible binding sites for double stranded DNA and other nucleic acids. Essentially, the phospholipase activity of tomato nuclease I significantly broaden the substrate promiscuity of the enzyme, and resulted in the release of diacylglycerol. The diacylglycerol is an important second messenger that can be related to the role of tomato nuclease I in apoptosis [52].

CONCLUSIONS

The above observations support the hypothesis that promiscuous functions are not exceptions but inherent features of proteins in general. However, the mechanisms of exhibiting promiscuity are different in different proteins, enzymes in particular. It can be categorized in three major groups based on the different factors responsible for enzyme promiscuity: (1) active site plasticity of enzymes, (2) substrate ambiguity and (3) cofactor ambiguity. Owing to the weaker promiscuous activity as compare to native activity, it could be a better starting point for the evolution of new function. It also provides an immediate advantage to the gene under selection for a new function. However, the importance of promiscuous functions especially with physiologically irrelevant substrates is yet to be understood. The promiscuous functions can also be utilized as a starting point for the evolution of new proteins in laboratory conditions for biotechnological applications.

ACKNOWLEDGEMENTS

Financial supports from Department of Science and Technology (SERB/F/1424) and South Asian University are gratefully acknowledged.

REFERENCES

1. Khersonsky O, Tawfik DS (2010) Enzyme promiscuity: mechanistic and evolutionary perspective. Annu Rev Biochem 79:471–505

2. Copley SD (2014) An evolutionary perspective on protein moonlighting. Biochem Soc Trans 42(6):1684–1691

3. Gupta RD, Goldsmith M, Ashani Y, Simo Y, Mullokandov G, Bar H, Ben-David M, Leader H, Margalit R, Silman I, Sussman JL, Tawfik DS (2011) Directed evolution of hydrolases for prevention of G-type nerve agent intoxication. Nat Chem Biol 7(2):120–125

4. Jackson CJ, Foo JL, Tokuriki N, Afriat L, Carr PD, Kim HK, Schenk G, Tawfik DS, Ollis DL (2009) Conformational sampling, catalysis, and evolution of the bacterial phosphotriesterase. Proc Natl Acad Sci USA 106(51):21631–21636

5. Bigley AN, Mabanglo MF, Harvey SP, Raushel FM (2015) Variants of phosphotriesterase for the enhanced detoxification of the chemical warfare agent VR. Biochemistry 54(35):5502–5512

6. Nobeli I, Favia AD, Thornton JM (2009) Protein promiscuity and its implications for biotechnology. Nat Biotechnol 27:157–167

7. Copley SD (2009) Evolution of efficient pathways for degradation of anthropogenic chemicals. Nat Chem Biol 5(8):559–566

8. López-Iglesias M, Gotor-Fernández V (2015) Recent advances in biocatalytic promiscuity: hydrolase-catalyzed reactions for nonconventional transformations. Chem Rec 15(4):743–759

9. Baier F, Tokuriki N (2014) Connectivity between catalytic landscapes of the metallo-β-lactamase superfamily. J Mol Biol 426(13):2442–2456

10. Parera M, Martinez MA (2014) Strong epistatic interactions within a single protein. Mol Biol Evol 31(6):1546–1553

11. Copley SD (2015) An evolutionary biochemist's perspective on promiscuity. Trends Biochem Sci 40(2):72–78

12. Atkins WM (2015) Biological messiness vs. biological genius: mechanistic aspects and roles of protein promiscuity. J Steroid Biochem Mol Biol 151:3–11

13. Arora B, Mukherjee J, Gupta MN (2014) Enzyme promiscuity: using the dark side of enzyme specificity in white Biotechnology. Sustainable Chemical Processes 2:25

14. Penning TM, Chen M, Jin Y (2015) Promiscuity and diversity in 3-ketosteroid reductases. J Steroid Biochem Mol Biol 151:93–101

15. Miao Y, Rahimi M, Geertsema EM, Poelarends GJ (2015) Recent developments in enzyme promiscuity for carbon-carbon bond-forming reactions. Curr Opin Chem Biol 25:115–123

16. Matange N, Podobnik M, Visweswariah SS (2015) Metallophosphoesterases: structural fidelity with functional promiscuity. Biochem J 467(2):201–216

17. Noda-García L, Juárez-Vázquez AL, Ávila-Arcos MC, Verduzco-Castro EA, Montero-Morán G, Gaytán P, Carrillo-Tripp M, Barona-Gómez F (2015) Insights into the evolution of enzyme substrate promiscuity after the discovery of (βα) isomerase evolutionary intermediates from a diverse metagenome. BMC Evol Biol 15:107

18. Huang H, Pandya C, Liu C, Al-Obaidi NF, Wang M, Zheng L, Toews Keating S, Aono M, Love JD, Evans B, Seidel RD, Hillerich BS, Garforth SJ, Almo SC, Dunaway-Mariano PS, Mariano D, Allen KN, Farelli JD (2015) Panoramic view of a superfamily of phosphatases through substrate profiling. Proc Natl Acad Sci USA 112(16):e1974

19. Mashiyama ST, Malabanan MM, Akiva E, Bhosle R, Branch MC, Hillerich B, Jagessar K, Kim J, Patskovsky Y, Seidel RD, Stead M, Toro R, Vetting MW, Almo SC, Armstrong RN, Babbitt PC (2014) Large-

scale determination of sequence, structure, and function relationships in cytosolic glutathione transferases across the biosphere. PLoS Biol 12(4):e1001843

20. Pratap S, Katiki M, Gill P, Kumar P, Golemi-Kotra D (2015) Active-site plasticity is essential to carbapenem hydrolysis by OXA-58 Class D β-lactamase of *Acinetobacter baumannii*. Antimicrob Agents Chemother 60:75–86

21. Alcolombri U, Elias M, Tawfik DS (2011) Directed evolution of sulfotransferases and paraoxonases by ancestral libraries. J Mol Biol 411(4):837–853

22. Kraus ML, Grimm C, Seibel J (2015) Redesign of the active site of sucrose phosphorylase by a clash induced cascade of loop shifts. Chem Bio Chem. doi:10.1002/cbic.201500514

23. Afriat-Jurnou L, Jackson CJ, Tawfik DS (2012) Reconstructing a missing link in the evolution of a recently diverged phosphotriesterase by active-site loop remodeling. Biochemistry 51(31):6047–6055

24. Yasutake Y, Yao M, Sakai N, Kirita T, Tanaka I (2004) Crystal structure of the *Pyrococcus horikoshii* isopropylmalate isomerase small subunit provides insight into the dual substrate specificity of the enzyme. J Mol Biol 344:325–333

25. Sevrioukova IF, Poulos TL (2013) Understanding the mechanism of cytochrome P450 3A4:recent advances and remaining problems. Dalton Trans 42(9):3116–3126

26. Yao J, Guo H, Chaiprasongsuk M, Zhao N, Chen F, Yang X, Guo H (2015) Substrate-assisted catalysis in the reaction catalyzed by salicylic acid binding protein 2 (SABP2), a potential mechanism of substrate discrimination for some promiscuous enzymes. Biochemistry 54(34):5366–5375

27. Baier F, Chen J, Solomonson M, Strynadka NC, Tokuriki N (2015) Distinct metal isoforms underlie promiscuous activity profiles of metalloenzymes. ACS Chem Biol 10(7):1684–1693

28. Marschner A, Klein CD (2015) Metal promiscuity and metal-dependent substrate preferences of *Trypanosoma brucei* methionine aminopeptidase 1. Biochimie 115:35–43

29. Pordea A (2015) Metal-binding promiscuity in artificial metalloenzyme design. Curr Opin Chem Biol 25:124–132

30. Rivera-Perez C, Nyati P, Noriega FG (2015) A corpora allata farnesyl diphosphate synthase in mosquitoes displaying a metal ion dependent substrate specificity. Insect Biochem Mol Biol 64:44–50

31. Kim Y, Cunningham MA, Mire J, Tesar C, Sacchettini J, Joachimiak A (2013) NDM-1, the ultimate promiscuous enzyme: substrate recognition and catalytic mechanism. FASEB J 27(5):1917–1927

32. Tokuriki N, Tawfik DS (2009) Protein dynamism and evolvability. Science 324:203–207

33. Kaltenbach M, Tokuriki N (2014) Dynamics and constraints of enzyme evolution. J Exp Zool B Mol Dev Evol 322(7):468–487

34. Amitai G, Gupta RD, Tawfik DS (2007) Laten evolutionary potentials under the neutral mutational drift of an enzyme. HFSP J 1(1):67–78

35. Gupta RD, Tawfik DS (2008) Directed enzyme evolution via small and effective neutral drift libraries. Nat Methods 5(11):939–942

36. Miles ZD, Roberts SA, McCarty RM, Bandarian V (2014) Biochemical and structural studies of 6-carboxy-5, 6, 7, 8-tetrahydropterin synthase reveal the molecular basis of catalytic promiscuity within the tunnel-fold superfamily. J Biol Chem 289(34):23641–23652

37. Luo XJ, Kong XD, Zhao J, Chen Q, Zhou J, Xu JH (2014) Switching a newly discovered lactonase into an efficient and thermostable phosphotriesterase by simple double mutations His250Ile/Ile263Trp. Biotechnol Bioeng 111(10):1920–1930

38. Khanal A, Yu McLoughlin S, Kershner JP, Copley SD (2015) Differential effects of a mutation on the normal and promiscuous activities of orthologs: implications for natural and directed evolution. Mol Biol Evol 32(1):100–108

39. de Visser JA, Krug J (2014) Empirical fitness landscapes and the predictability of evolution. Nat Rev Genet 15(7):480–490

40. Harms MJ, Thornton JW (2013) Evolutionary biochemistry: revealing the historical and physical causes of protein properties. Nat Rev Genet 14(8):559–571

41. Renata H, Wang ZJ, Arnold FH (2015) Expanding the enzyme universe: accessing non-natural reactions by mechanism-guided directed evolution. Angew Chem Int Ed Engl 54(11):3351–3367

42. Colin PY, Kintses B, Gielen F, Miton CM, Fischer G, Mohamed MF, Hyvönen M, Morgavi DP, Janssen DB, Hollfelder F (2015) Ultrahigh-throughput discovery of promiscuous enzymes by picodroplet functional metagenomics. Nat Commun 6:10008

43. Meier MM, Rajendran C, Malisi C, Fox NG, Xu C, Schlee S, Barondeau DP, Höcker B, Sterner R, Raushel FM (2013) Molecular engineering of

organophosphate hydrolysis activity from a weak promiscuous lactonase template. J Am Chem Soc 135(31):11670–11677

44. Bigley AN, Xu C, Henderson TJ, Harvey SP, Raushel FM (2013) Enzymatic neutralization of the chemical warfare agent VX: evolution of phosphotriesterase for phosphorothiolate hydrolysis. J Am Chem Soc 135(28):10426–10432

45. Naqvi T, Warden AC, French N, Sugrue E, Carr PD, Jackson CJ, Scott C (2014) A 5000-fold increase in the specificity of a bacterial phosphotriesterase for malathion through combinatorial active site mutagenesis. PLoS One. 9(4):e94177

46. Dorr BM, Ham HO, An C, Chaikof EL, Liu DR (2014) Reprogramming the specificity of sortase enzymes. Proc Natl Acad Sci USA 111(37):13343–13348

47. Sharma UK, Sharma N, Kumar R, Kumar R, Sinha AK (2009) Biocatalytic promiscuity of lipase in chemoselective oxidation of aryl alcohols/acetates: a unique synergism of CAL-B and [hmim] Br for the metal-free H_2O_2 activation. Org Lett 11(21):4846–4848

48. Bordes I, Recatalá J, Świderek K, Moliner V (2015) Is promiscuous CALB a good scaffold for designing new epoxidases? Molecules 20(10):17789–17806

49. Leščić Ašler I, Ivić N, Kovačić F, Schell S, Knorr J, Krauss U, Wilhelm S, Kojić-Prodić B, Jaeger KE (2010) Probing enzyme promiscuity of SGNH hydrolases. Chem Bio Chem 11(15):2158–2167

50. Li R, Perez B, Jian H, Jensen MM, Gao R, Dong M, Glasius M, Guo Z (2015) Characterization and mechanism insight of accelerated catalytic promiscuity of *Sulfolobus tokodaii* (ST0779) peptidase for aldol addition reaction. Appl Microbiol Biotechnol 99:9625–9634

51. Cai Y, Bhuiya MW, Shanklin J, Liu CJ (2015) Engineering a Monolignol 4-O-methyltransferase with High Selectivity for the Condensed Lignin Precursor Coniferyl Alchohol. J Biol Chem. 290:26715–26724

52. Koval' T, Lipovová P, Podzimek T, Matoušek J, Dušková J, Skálová T, Stěpánková A, Hašek J, Dohnálek J (2013) Plant multifunctional nuclease TBN1 with unexpected phospholipase activity: structural study and reaction-mechanism analysis. Acta Crystallogr D Biol Crystallogr 69(Pt 2):213–226

53. Norrgård MA, Mannervik B (2011) Engineering GST M2-2 for high activity with indene 1,2-oxide and indication of an H-site residue sustaining catalytic promiscuity. J Mol Biol 412(1):111–120

Chapter 6

RECENT ADVANCES ON BIOBUTANOL PRODUCTION

Luiz J Visioli, Heveline Enzweiler, Raquel C Kuhn, Marcio Schwaab and Marcio A Mazutt

Department of Chemical Engineering, Federal University of Santa Maria, 97105-900 Santa Maria, Brazil

ABSTRACT

Recent studies have shown that butanol is a potential gasoline replacement that can also be blended in significant quantities with conventional diesel fuel. However, biotechnological production of butanol has some challenges such as low butanol titer, high cost feedstocks and product inhibition. The present work reviewed the technical and economic feasibility of the main technologies available to produce biobutanol. The latest studies integrating continuous fermentation processes with efficient product recovery and the use of mathematical models as tools for process scale-up, optimization and control are presented.

INTRODUCTION

During the last decade the interest in the production of chemicals and fuels from renewable resources has increased. Reasons for this trend include growing concerns about global warming and climatic change, volatility of oil supply, increasing price of crude oil and legislation restricting the use of nonrenewable energy sources. Furthermore, the generation of biofuels may improve the local employment opportunities and contribute to the reduction of CO_2 emissions [1–3]. Among the alternative fuels, biobutanol has shown promise as its properties are similar to gasoline [4] and, in comparison with ethanol, it has a longer carbon chain length as well as higher volatility, polarity, combustion value, octane rating [5] and is less corrosive [6]. It can also be a substitute for gasoline without alteration in current vehicle or engine technologies [7]. In addition, it has less ignition problems since the heat of vaporization of butanol is less than half of that of ethanol, hence an engine running on butanol should be easier to start in cold weather than the one running on ethanol or methanol [8].

Commercial butanol fermentation processes have been developed by some companies [2]. There is an expectation that the number of companies devoted to biobutanol production will increase worldwide as well as the development of new technologies to increase the yield [9]. A difficulty in butanol fermentation is the inhibition caused by the product as butanol concentrations around 20 g/L inhibit microbial growth [5]. In addition, the clostridium species are strictly anaerobes [10] and the anaerobic conditions need to be established before the beginning of the fermentation and the reactor must be remain closed during the process [11].

The cost of the plant for butanol production depends on the price of the feedstock and is extremely sensitive to any price fluctuation [12]. Thereby, the commodity price is still very dependent of feedstock price and an expensive raw material generates an expensive product. Agricultural residues and wastes are demonstrated to be cheaper than other sources [13]. However, the hydrolysis of these materials can generate fermentation inhibitors, which is another problem to be solved [14].

Another important point in butanol production is the separation techniques and their application, mainly for in situ continuous recovery [15]. Distillation is the unit operation widely used in separation of aqueous solution from butanol fermentation. However, the problem in this process is the formation of an azeotrope that increases the energy cost [16]. Alternative methods are reported with the objective to promote a cheaper and efficient separation. More recently, mathematical models have been developed to design the process as well as to simulate its behavior on an industrial scale without the need to carry out experiments to optimize the operational conditions of the reactor [17–19].

Although there are excellent reviews available in the literature concerning butanol production [12, 20–29], the present work is focused on the presentation of the technical and economic feasibility of the main technologies that have been developed to produce biobutanol, complementing the existing literature about the topic. For this purpose, the latest studies reporting the microorganism used in butyric fermentation, as well as the integration of continuous fermentation processes with efficient product recovery and the use of mathematical models as tools for process, optimization and control are reviewed.

Microorganism

According to Liu et al. [30], the most common microbial strains employed for butanol fermentation are the mesophiles *Clostridium acetobutylicum* and *Clostridium beijerinckii,* where *Clostridium acetobutylicum* is the most reported in acetone–butanol–ethanol (ABE) fermentation, which are the major products obtained in the process [7]. Furthermore, *Clostridium acetobutylicum* was the first bacterium used for ABE fermentation [2]. However, other *clostridium* sp. have also been reported, for example; *C. pasteurianum* [31] , *C. sporogenes* [32] , *C. saccharoperbutylacetonicum* [33] and *C. saccharobutylicum* [26]. The main strains used for biobutanol production are reported in Table 1.

Table 1: Microorganism, substrate, yield/production and main aspects in the butanol production reported

Microorganism	Substrate	Yield/ Production	Technology	Reference
C. acetobutylicum (immobilized)	Cheese whey (lactose)	Yield: 15% to 0.54 h^{-1} of dilution and 28% to 0.97 h^{-1} of dilution	Reactor (PBR) with immobilized clostridium	[34]
C. beijeirinckii ATCC 55025	Hydrolysate of wheat bran	Yield: 32%/Production: 8.8 g.l^{-1} of biobutanol	Acid hydrolysis	[30]
C. beijerinckii	Cassava flour	Production: 23.98 g.l^{-1} of butanol	Enzymatic treatment with yield of 9.12% to Reducing sugar	[35]
C. beijerinckii P260	Wheat straw	Yield: 42%	Acid pretreatment and enzymatic hydrolysis	[14]
	Barley straw	Yield: 43%/Production: 26,64 g.l^{-1} of total solvents	Dilute sulfuric acid hydrolysis/overliming	[36]
	Corn stover	Yield: 43%/Production: 18.04 g.l^{-1} of total solvents	Acid and enzymatic steps of hydrolysis/overliming	[37]
	Switchgrass	Yield: 37%/Production: 8.91 g.l^{-1} of total solvents		
	Glucose	Production: 17.54 g.l^{-1} of butanol	Intermittent vacuum application	[10]
C. saccharobutylicum DSM 13864	Sago starch	Yield: 29%	Free microorganism fermentation	[26]

C. acetobutyli-cum	Cassava bagasse	Yield: 32%/Pro-duction: 76.4 g.l^{-1} of butanol	Hydrolyze by enzymes fibrous bed bioreactor/ Gas stripping	[38]
	Palm empty fruit bunches	Production: 1.262 g.l^{-1} of butanol	Acid pretreatment/enzy-matic hydrolysis	[39]
C. beijerinckii BA101	Liquefied corn starch	Butanol pro-duction: 81, 3 g.l^{-1} (with gas stripping)/18.6 g.l^{-1} (without gas stripping)	Bath reactor/gas stripping/enzymatic hydrolysis	[40]
C. acetobuty-licum ATCC 824 and Bacillus sub-tilis DSM 4451	Spoilage date palm fruits	Yield: 42%/pro-duction: 21.56 g.l^{-1} of Solvents	Bacterial consortium (anaerobic conditions)	[41]
C. beijerinckii NCIMB 8052	Tropical maize stalk juice	Production: 0.27 g-butanol/g-sugar	Optimization of pH, agitation, sugar concen-tration	[42]
C. acetobutyli-cum ATCC824	Sugar maple Hemicellu-losic material	Production: 7 g.l^{-1} of butanol	Alkali pretreatment/acid hydrolysis/overliming	[43]
C. saccharoper-butylacetonicum N1-4	Rice bran	Yield: 57% to sugar generated.	Acid hydrolysis	[44]
	De-oiled rice bran	Yield: 44% to sugar generated.	Acid pretreatment/enzy-matic Hydrolysis	
C. acetobutyli-cum XY16	Glucose	Production: 20.3 g.l^{-1} of butanol	pH steps in the fermen-tation	[45]
C. sporogenes BE01	rice straw	Production of 3.49 g/L and 5.32 g/L of butanol and total solvents respec-tively	Acid pretreatment/ enzymatic Hydrolysis/ Overliming	[32]
C. saccharoper-butylacetonicum N1-4	rice straw	Maximum butanol production of 6.6 g/L and buta-nol yield 0.2 g/g of total sugar.	Absence of pretreat-ment/enzymatic hydrolysis/Non-sterile conditions	[33]
C. pasteurianum	Glycerol	Maximum butanol production of 8.8 g/L and buta-nol yield 0.35 g/g of glycerol at initial substrate concentration of 25 g/L.	Immobilized cells/Bath fermentation	[31]

C. acetobutyli-cum NCIM 2337	Rice straw	Butanol production of 13.5 g/L and butanol yield 0.34 g/g of total sugar generated.	Acid treatment with shear stress	[46]
C. acetobutyli-cum MTCC 481	Rice straw	Butanol production of 1.72 g/L.	Steam explosion	[47]
		Butanol production of 1.6 g/L	Acid treatment	
		Butanol production of 2.1 g/L	Acid pre-treatment/enzymatic hydrolysis	
C beijerinckii NCIMB 8052	Corncob	Butanol production of 8.2 g/L	Alkali pre-treatment/ enzymatic hydrolysis/ overliming	[48]
C. acetobutyli-cum JB200	Glucose	Yield: 21%/Production: 172 g.l^{-1} of solvents	Gas stripping	[49]
C. beijerinckii ATCC 10132	Glucose	Production: 20 g.l^{-1} of butanol	Bath reactor	[50]
C. acetobutyli-cum CICC 8008	Corn straw	Production: 6.20 g.l^{-1} of butanol	Enzymatic hydrolysis/ bath reactor	[51]
C. acetobutyli-cum P262	Whey permeate medium	Yield: 44%/Production: 98.97 g.l^{-1} of solvents	Perstraction/bath reactor	[52]

ABE fermentation is one of the oldest known industrial fermentations with a history of more than 100 years [4]. However, this fermentation is not widely used as butanol is highly toxic to microorganisms and, for this reason, less than 13 g/L of butanol are produced during batch fermentation. In general, fermentation using Clostridia sp. results in the ABE production of around 15–25 g/L with a yield of 0.25–0.4 g ABE/g sugar [2]. Substrate inhibition has not been the major concern in ABE fermentation when glucose is used as a carbon source [53]. However, Chen et al. [54] reported inhibition of butanol production at high substrate concentrations. Ezeji et al. [55] also reported that *Clostridium* sp. showed a catabolic inhibition for sugar concentration higher than 162 g/L.

The pH of the fermentation broth, initially at 6.8–7.0, drops to 4.5–5.0 during the acidogenic phase. This phase is associated with the fast growth of cells and the secretion of the carboxylic acids, acetate, and butyrate [4]. According to Napoli et al. [34], the pH varies between 4.0 - 5.0 for butanol production. Li et al. [53] verified that pH 4.3, maintained constant during fermentation, was optimal for butanol production using *C. acetobutylicum*. Similar results were also reported previously by Bahl et al. [56]. On the other hand, Qureshi [14] reported that the pH is self controlled at approximately 5.2 ± 0.2 during the solventogenic stage of *C. beijerinckii*. These different ranges of pH are due to different clostridium species used in the process.

The metabolism of Clostridia strains has two distinct phases, acidogenesis and solventogenesis. The acidogenesis is characterized by substrate conversion into acids (acetic and butyric acids) and exponential cell growth with ATP formation. This is a fundamental step of fermentation, without which the number of viable cell would be greatly reduced, making the normal solvents production difficult. The solventogenesis phase is characterized by conversion of substrate and acids into solvents (ABE) [23, 34]. Solventogenic clostridia can utilize a wide range of carbon sources, such as starch, sucrose, glucose, fructose, galactose, cellobiose, xylose, arabinose, glycerol, and syngas as fermentation substrates for the ABE production.

According to Jang et al. [57], it is very important to have a better understanding of the genes that are the basis of microbial metabolism for the production of butanol, since it will be possible to obtain modified strains able to improve biomass conversion [4], increase oxygen tolerance, increase the cell density, prolong cell viability, direct the utilization of cellulosics and provide high solvent tolerance and high butanol selectivity [58]. Genetic modification of *Clostridium* is widely used, by inserting some heterogenetic genes or over expressing or knocking out/down some relative endogenous genes, to improve butanol production. Some researchers are working with genetic tools which are being used to manipulate their metabolism by introducing the genes that are responsible for butanol production from *Clostridium acetobutylicum* into *E. coli* and yeast (commonly *Saccharomyces cerevisiae*) [24]. This manipulation can enable the microbial strain to increase the production of butanol in the medium, without inhibition by products. The details of genetic engineering for butanol production can be observed in these papers e.g. [59, 60].

According to Lütke-Eversloh and Bahl [61], the modifications in the strains of the genus *Clostridium* can be achieved in the following ways: disruption of the pathway that synthesizes the unwanted products or changing the pathway for formation of acetate and butyrate (acidogenic stage). The disruption of the pathway for acetone production increased the butanol production from 71% to 80% [62]. Conversely Isar and Rangaswamy [50] reported an increase in the tolerance of solvents from 18 g/L to 25 g/L using *Clostridium beijerinckii* adapted to solvent. It indicated that the strain had adapted to butanol and become solvent tolerant in the absence of any mutation.

In the study of Abd-Alla and El-Enany [41], the authors described an alternative method to maintain the anaerobic medium during the clostridial fermentation. The culture of *Bacillus subtilis* DSM 4451 was used to maintain strict anaerobic conditions for *C. acetobutylicum* ATCC 824. Thus, fermentation does not need N_2 flushing to remove the oxygen so the costs decreased. The highest butanol production obtained was 21.7 g/L (this is similar to that reported by clostridium production) just using the consortium of microorganism to maintain the anaerobic medium. The main objective of using the microorganism in the butanol fermentation is to attempt to create an engineered microbe that can overcome the limitations of *Clostridium*. The increase of solvent tolerance, butanol titer, and tolerance towards traces of oxygen are the most desirable characteristics in the engineered microorganism for it to be viable for industrial process application.

Substrates

The prices of substrates for biobutanol production influence the economic competition with the petrochemical industry [34]. The cost of feedstock represents over 70% of the total production costs of biobutanol [63]. At the beginning of butanol fermentation, substrates based on sugars and starch were used, but these are expensive and the process becomes unfeasible. One of the strategies to decrease the production cost is to use cheap and renewable feedstocks, such as lignocellulosic materials (e.g. agricultural waste, paper waste, wood chips), which are abundant. The production of alcohol using lignocelluloses follows an integrated process involving basically three steps: pre-treatment, hydrolysis and fermentation [7]. The main substrates used for biobutanol production are reported in Table 1.

The molasses are used for biobutanol production. However, these kinds of substrate are more expensive than agricultural residues. On the other hand, molasses can be used directly in the fermentation. Thus, it is not possible to assert that the cellulosic residues will be cheaper than molasses. Van der Merwe et al. [64] reported analyses of the energy efficiency and economics of biobutanol production using sugarcane molasses. Another important point to be analyzed related to the choice of substrate and its availability throughout the year. The major sources of this kind of raw material are agricultural residues and wastes, such as rice straw, wheat straw, wood (hardwood), byproducts left over from the corn milling process (corn fiber), annual and perennial crops, waste paper [14] and sweet sorghum [65]. These raw materials consist of three types of polymers: cellulose, hemicellulose, and lignin. Cellulose has strong physical-chemical interaction with hemicelluloses and lignin. Cellulose, a linear glucose polymer (that is broken in the hydrolysis), is a highly ordered polymer formed of glucose representing about 50% of the wood mass. Hemicellulose is a short, highly branched heteropolymer formed mainly of xylose, plus glucose, mannose, galactose and arabinose and sometimes uronic acids. Lignin consists of phenylpropanoid units derived from the corresponding p-hydroxycinnapyl alcohols. Lignin is hydrophobic and highly resistant to chemical and biological degradation [66]. *Clostridium beijerinckiii* is being explored as a promising strain to produce biobutanol from cellulosic materials [26].

The problems in the use of cellulosic or lignocellulosic materials for butanol production are the processes for production of these hydrolysates, resulting in the generation of chemical byproducts that inhibit cell growth and fermentation. Such inhibitors include salts, furfural, hydroxymethyl furfural, acetic, ferulic, glucuronic, r-coumaric acids, and phenolic compounds. Lignocellulosic materials are difficult to hydrolyse biologically [39]. Furthermore, the hydrolytic process can generate significant amounts of waste and hence increase the cost of the butanol produced [36]. Moreover, with the fermentation of any of these substrates (mainly cellulosic and starchy after a hydrolysis treatment) there is the need for nutritional supplementation. Lee et al. [67] reported the use of KH_2PO_4, K_2HPO_4, ammonium acetate, para-aminobenzoic acid, thiamin, biotin, $MgSO_4 \cdot 7H_2O$, $MnSO_4 \cdot H_2O$, $FeSO_4 \cdot 7H_2O$, NaCl, and yeast extract as supplements for biobutanol production.

The pretreatment for starchy and cellulosic materials is a limiting step and needs to be optimized for a satisfactory production of butanol. Liu et al. [30] pretreated wheat bran using sulfuric acid at high temperature followed by neutralization with $Ca(OH)_2$ for biobutanol production by *C. beijerinckii 55025*. This procedure increased the cost of the butanol produced, but this cost can be considerably decreased if a large amount of a cheap source of raw material is used. Lépiz-Aguilar [35] used HCl 1 M combined with high temperature for 2 h or enzymatic hydrolysis (using α-amylase and β-glucoamylase) to hydrolyze the cassava flour. The best results in terms of butanol production were 23.98 and 13.78 g.L^{-1} using enzymatic and acid hydrolysis, respectively. Qureshi et al. [7] studied the pretreatment of wheat straw with a mix of enzymes (cellulose, β-glucosidase and xylanase) at pH 5.0, 45°C for 72 h and 80 rpm, obtaining a butanol production of 12.0 g/L. Alternative technologies to hydrolyze the raw material such as microwave-assisted pre-treatment processes, steam explosion, ozonolysis, oxidative delignification, pulsed-electric-field pretreatment were also reported [68–70].

Qureshi et al. [36] believed that barley straw can be used for butanol production. However, there is the presence of inhibitors in this substrate and hence pretreatment (with lime called overliming) is necessary for an effective fermentation. After the pretreatment, the production of butanol was higher than when using glucose as substrate. Similarly, Qureshi et al. [37] evaluated corn stover and switchgrass hydrolysates as substrates for butanol production. The production of butanol using corn stover hydrolysates was similar to that presented in previous work using barley [36].

Al-Shorgani et al. [44] reported the formation of inhibitors during the acid pretreatment of cellulosic raw material (rice bran and de-oiled rice bran). Similar to other studies, the authors used overliming treatment and extraction of inhibitors with a nonionic polymeric adsorbent resin. These procedures improved butanol production and yield. Qureshi et al. [71] concluded that the formation of fermentation inhibitors after hydrolysis of cellulosic raw material is substrate and pretreatment dependent. Thus, it is necessary for a specific study to be carried out for each substrate and treatment.

Lin et al. [51] reported the use of corn straw as a raw material for butanol production after hydrolysis using alkali pre-treatement. The sugar concentration obtained was around 44 g/L and this represents approximately 400 grams of sugar per kg of corn straw, producing 6.54 g/L (65 g/kg of corn straw) of butanol in the fermentation. Using another residue from corn production (corncob), Zhang et al. [48] reported production of 16 g/L of solvents using the enzymatic hydrolyzed corncob pre-treated and detoxified with Ca (OH)$_2$. Several authors have stated that biobutanol production will only be feasible industrially if a low cost substrate can be employed. However, it is important to consider the total cost involved in the substrate utilization. In these scenarios, the tendency is for the diversification of substrates and the use of regional crops (molasses, starch or cellulosic one) for butanol production.

Bioreactors for Biobutanol Production

According to Kumar and Gayen [26], the operation of bioreactors for biobutanol production can be accomplished in batch, fed-batch, and continuous modes. Continuous processes offer various advantages such as reduction in sterilization and inoculation time, high productivity, and reduction in butanol inhibition, but this reactor presents high product recovery costs due to low concentration of biofuel [26]. Fed-batch fermentation is started with a low substrate concentration. When the fermentation culture consumes the substrate, more substrate is added to maintain the fermentation process while not exceeding the detrimental substrate level [53]. The usage of a continuous packed bed reactor (PBR) is reported as an alternative for fermentation using immobilized microorganism [34]. Lu et al. [38] used a fibrous bed bioreactor (FBB). This reactor is interesting because the microorganism is immobilized in the bed enabling the process to recover products in situ without the loss of cells. However, the reactor most reported is the batch one (as can be verified in table 1). This preference can be explained because it is easy to handle, maintain the anaerobic medium, control the temperature and pH, and take samples.

Furthermore, this reactor has less difficulties when coupling to a separation unit.

Mariano et al. [15] reported the use of batch-bioreactor containing 7 litres of medium. The anaerobic medium was maintained by oxygen free nitrogen. Parekh et al. [72] reported the use of a pilot-scale of 200 litres using corn steep liquor as the raw material, obtaining 17.8 g/L of butanol. The same size of bioreactor was used by Lee et al. [59]. These bioreactors are larger than the others reported, thus, these studies are very important to predict the behavior of clostridial fermentation after the scale up.

Separation

The main separation process used for purification of biobutanol from the fermentation broth is the distillation. However, the butanol-water system at 101.3 kPa has an azeotrope at 55.5 wt% butanol. The greatest difficulty in this process is the low solubility of butanol in water (maximum of 7.7 wt%). As the azeotrope occurs above this solubility limit, two liquid phases are formed at the azeotrope. The upper phase contains 79.9 wt% butanol whereas the lower phase contains 7.7 wt% butanol [16], which boils at a lower temperature [10]. The recovery of low concentration butanol by traditional distillation is energy intensive and thus, economically unfeasible [73].

Mariano et al. [74], Secuianu et al. [75] and Ezeji [3] reported some of the most commonly used techniques for continuous recovery of butanol from the fermentation broth, namely adsorption, gas stripping, ionic liquids, liquid-liquid extraction, pervaporation, aqueous two-phase separation, supercritical extraction, and flash fermentation. Adsorption should allow separation of butanol from the bulk aqueous fermentation broth. Hydrophobic adsorbents potentially show high selectivity for butanol over water [5]. In adsorption, alcohol is preferentially transferred from the feed liquid to a solid adsorbent material [16]. Dhamole et al. [76] used a non-ionic surfactant to decrease butanol toxicity and its separation from the non-ionic surfactant micelle aqueous solution by cloud point extraction. Thus, the fermentation is not inhibited and butanol concentration is increased in the micelles.

Ezeji [3] used gas stripping for *in situ* separation because it is a simple technique that is free of emulsion formation and it does not require a membrane or expensive chemicals. The production of ABE using gas stripping in the fermentation of 500 g/L of sugar using the strain *C. beijerinckii* BA101 was 232.8 g/L compared to a control batch reactor, where 17.6 g/L ABE was produced. Moreover, gas stripping was more selective in removing butanol than acetone and ethanol [38]. Gas stripping is more efficient when butanol concentration in the fermentation broth is higher than 8 g/L [49]. According to

Ezeji et al. [40], the production of ABE increased from 18.6 g/L to 81.3 g/L, whereas sugar consumption increased about 487% (compared with the control of the same substrate without gas stripping) using gas stripping in the fermentation of liquefied corn starch with *C. beijerinckii* BA101.

Pervaporation technique using membranes with high product selectivity is one of the most promising alternatives to conventional distillation. Without heating energy, the pervaporation process enables the efficient separation and concentration of the product in a single step, and maintains the productivity of the microorganism as a result of preventing product inhibition [77]. Yen et al. [78] tested a membrane of poly (ether-block-amide) and 5% and 10% (w/v) of carbon nanotubes for pervaporation. The productivity and yield increased about 20% in comparison with the pervaporation using a poly (ether-block-amide) membrane.

In the perstractive separation, the fermentation broth and the extractant are separated by a membrane. The membrane contactor provides a surface area where the two immiscible phases can exchange butanol, thus, the toxicity of solvent for the cells does not occur [79]. According to Qureshi and Maddox [52], in the control experiment 28.6 g/L lactose was used while in the fermentation perstraction experiment 227 g/L lactose was utilized in ABE fermentation.

The removal of butanol or ABE from the fermentation broth by liquid–liquid extraction is considered an important technique. Usually, a water-insoluble organic extractant is mixed with the fermentation broth. The main problem concerned with the use of this technique is related to the toxicity of the solvent to the cells [79]. The major limitation is that the extractant with high partition coefficient often leads to microbial toxicity because of direct contact between the fermentation broth and the extractant [80].

In the membrane-assisted extractive fermentation, two phases of extractant and fermentation broth are separated by a porous membrane. The membrane can be either hydrophilic or hydrophobic and the interface is immobilized by the impregnation of its pores with one of the two phases depending on the membrane affinity. Thus, the microbial toxicity of the extractant can be reduced. Tanaka et al. [80] using this approach reported an increase in the glucose consumption from 66% to 100% due to the absence of inhibition for butanol in the medium.

Mariano et al. [15] reported the use of a cyclic vacuum applied in a bioreactor during fermentation. Using a vacuum is a good method to remove butanol from the fermentation medium, resulting in decreased product inhibition. Furthermore, in this study the continuous and intermittent vacuum has been tested. The use of intermittent vacuum showed a reduction of 39% in the energy expenditure without product inhibition because of low butanol concentration. Moreover, this process results in pre-concentration of the aqueous solution of butanol, which decreases the energy expenditure in the purification of the butanol. The same authors reported the use of a flash fermentation for *in situ* butanol recovery. Flash fermentation is a good way to decrease the butanol concentration in the fermentation broth. In this technology, a partial separation of the solvents and water occurs in the flash tank separator, where the liquid fraction returns to the fermentor and the vapor fraction (after condensation) plus the purge and permeate streams will compose the final stream that is sent to distillation. Thus, butanol concentration is always less than the critical concentration (for inhibition by product) [17]. Furthermore, they developed a mathematical model to predict the behavior of the process. This consisted of a batch fermentation reactor, and a vacuum flash vessel (with a filter to remove any solids before it gets in the flash vessel). The schematic design of the process is showed in the Figure 1. For the development of the model the differential equations for the batch reactor assume constant volume and factor in the removal of butanol during the process. The objective of this work was to demonstrate that the use of flash fermentation was able to decrease product inhibition [17]. The Figure 1 demonstrates the cyclic process that can be generated using this technology. The volatile compounds are removed in the flash tank and the liquid fraction returns to the fermentation. The feed is used to control the sugar concentration in the reactor, the purge is used to control the level and allow renewal of cell mass with removal of old cells and the filter is used to prevent solids entering the flash tank.

Mariano et al. [81] proposed a mathematical model for a continuous flash fermentation and used this model to optimize the process using response surface techniques. In other work, Mariano et al. [82] used the same model proposed by Mariano et al. [81], but the process was optimized using the method of particle swarm optimization to obtain the best operating conditions for butanol production. The same authors proposed the utilization of a servo control in flash fermentation [18]. This work was carried out because in previous studies it has been proved that this process can be used to improve butyric fermentation. So it becomes necessary to control the removal of butanol, since natural oscillations can occur during the dynamic process. The mathematical modeling is similar to the one used in [17] with some minor changes in the differential equations due to the alteration of the reaction volume. The objective of the control was to keep sugar and butanol concentrations constant in the fermentor. The controller was shown to efficiently regulate the operating conditions. Thus, the use of a controller in flash fermentation is able to enlarge the process, and suit this to an industrial application.

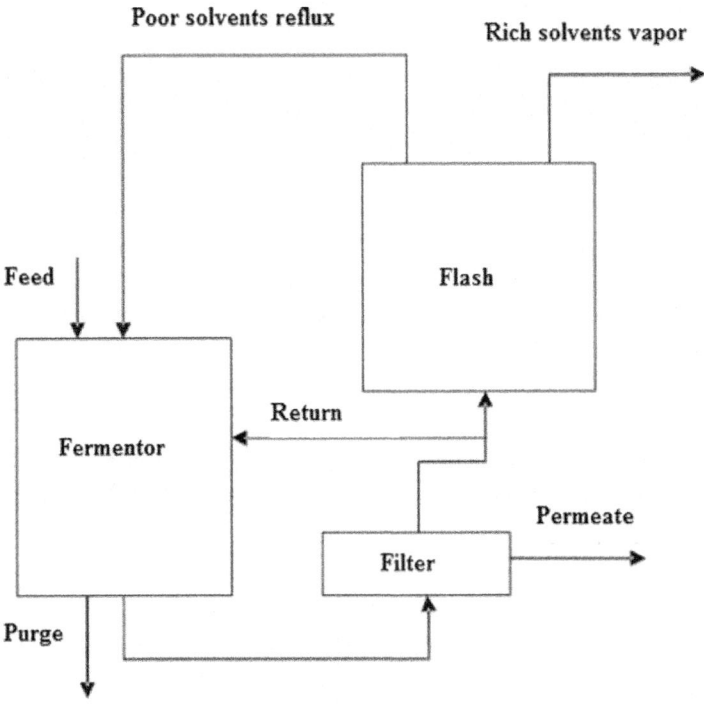

Figure 1: Schematic design of flash fermentation process.

Similarly, Liu et al. [19] propose a mathematical model to simulate the process consisting of a fermentor, gas striping, and a purification process for the condensate from gas striping. The objective was to simulate a process to produce 150.000 tons of butanol per year with purity of 99 wt%, and to evaluate the energy demand of all parts of this process. The authors concluded that ABE fermentation has lesser liquid fuel production (energy basis) using corn as a substrate than the ethanol production process. However, this scenario could change very quickly with the development of the process and genetic engineering.

Clearly, the use of *in situ* separation techniques in butanol production is promising for industrial applications by decreasing the product inhibition problems of the fermentation process. Furthermore, it can be considered a pre-separation process and decrease the quantity of butanol to be purified. The use of mathematical models to simulate the behavior of a fermentation process linked to any of these separation processes is important to predict behavior and production costs. With this it is possible to represent experimentally the best theoretical conditions and predict the adjustment necessary to scale up the process to industrial applications.

CONCLUDING REMARKS

In this review, biobutanol production was discussed in terms of microorganisms, substrates, types of bioreactors used in the process, separation techniques with special attention to *in situ* separation leading to decrease butanol inhibition and, the development of mathematical models to represent the *in situ* separation techniques. The engineering of microorganisms for butanol production has been reported over recent years. These changes in the bacterium decrease acetone production as well as increasing the resistance of microorganisms to high concentrations of butanol with consequent improvement in the use of sugar as a substrate for biobutanol production. Concerning substrates, there is a trend to use lignocellulosic materials, but the inhibitors generated during hydrolysis imposed difficulties for the industrial usage of this raw material for biobutanol production. The *in situ* separations associated with low energy expenditure during the removal of biobutanol are the technologies that will predominate in the future.

ACKNOWLEDGEMENTS

The authors thank CAPES for the scholarships and SCIT-RS for the financial support of this work.

AUTHORS' CONTRIBUTION

All authors participated in the compilation of data, discussion, data interpretation and manuscript drafting. All authors read and approved the final manuscript.

REFERENCES

1. Sorensen B: A sustainable energy future: construction of demand and renewable energy supply scenarios. *Int J Energy Res* 2008,32:436–470.

2. Cheng CL, Che PY, Chen BY, Lee WJ, Chien LJ, Chang JS: High yield bio-butanol production by solvent-producing bacterial microflora. *Bioresour Technol* 2012, 113:58–64.

3. Ezeji TC, Qureshi N, Blaschek HP: Microbial production of a biofuel (acetone–butanol–ethanol) in a continuous bioreactor: impact of bleed and simultaneous product removal. *Bioprocess Biosyst Eng* 2012. doi:10.1007/s00449–012–0766–5 doi:10.1007/s00449-012-0766-5

4. Ranjan A, Moholkar VS: Biobutanol: science, engineering, and economics. *Int J Energ Res* 2012, 36:277–323.

5. Knoshaug EP, Zhang M: Butanol tolerance in a selection of

microorganisms. *Appl Biochem Biotechnol* 2009, 153:13–20.

6. Oudshoorn A, van der Wielen LAM, Straathof AJJ: Adsorption equilibria of bio-based butanol solutions using zeolite. *Biochem Eng J* 2009, 48:99–103.

7. Cheng CL, Che PY, Chen BY, Lee WJ, Lin CY, Chang JS: Biobutanol production from agricultural waste by an acclimated mixed bacterial microflora. *Applied Energy* 2012. doi:10.1016/j.apenergy.2012.05.042 doi:10.1016/j.apenergy.2012.05.042

8. Sullivan L, Cates MS, Bennett GN: Structural correlations of activity of Clostridium acetobutylicum ATCC 824 butyrate kinase isozymes. *Enzyme Microb Technol* 2010, 46:118–124.

9. Pfromm PH, Amanor-Boadu V, Nelson R, Vadlani P, Madl R: Bio-butanol vs. bio-ethanol: a technical and economic assessment for corn and switchgrass fermented by yeast or Clostridium acetobutylicum. *Biomass Bioenerg* 2010, 34:515–524.

10. Mariano AP, Qureshi N, Maciel Filho R, Ezeji TC: Bioproduction of butanol in bioreactors: new insights from simultaneous in situ butanol recovery to eliminate product toxicity. *Biotechnol Bioeng* 2011, 108:1757–1765.

11. Shinto H, Tashiro Y, Yamashita M, Kobayashi G, Sekiguchi T, Hanai T, Kuriya Y, Okamoto M, Sonomoto K: Kinetic modeling and sensitivity analysis of acetone–butanol–ethanol production. *J Biotechnol* 2007, 131:45–56.

12. Green EM: Fermentative production of butanol- the industrial perspective. *Curr Opin Biotechnol* 2011, 22:337–343.

13. Taylor G: Biofuels and the biorefinery concept. *Energy Policy* 2008, 36:4406–4409.

14. Qureshi N, Saha BC, Cotta MA: Butanol production from wheat straw hydrolysate using *Clostridium beijerinckii* . *Bioprocess Biosyst Eng* 2007, 30:419–427.

15. Mariano AP, Maciel Filho R, Ezeji TC: Energy requirements during butanol production and in situ recovery by cyclic vacuum.*Renew Energy* 2012, 47:183–187.

16. Vane LM: Separation technologies for the recovery and dehydration of alcohols from fermentation broths. *Biofuels Bioprod Bior* 2008, 2:553–588.

17. Mariano AP, Angelis DF, Maugeri F, Atala DIP, Maciel MRW, Maciel Filho R: An alternative process for butanol production: continuous flash fermentation. *Chem Prod Process Mode* 2008, 3:34.

18. Mariano AP, Costa CBB, Maciel MRW, Maugeri Filho F, Atala DIP, Angelis DF, Maciel FR: Dynamics and control strategies for a butanol fermentation process. *Appl Biochem Biotechnol* 2010, 160:2424–2448.

19. Liu J, Wu M, Wang M: Simulation of the process for producing butanol from corn fermentation. *Ind Eng Chem Res* 2009, 48:5551–5557.

20. Jurgens G, Survase S, Berezina O, Sklavounos E, Linnekoski J, Kurkijärvi A, Väkeva M, Heiningen AV, Granström T: Butanol production from lignocellulosics. *Biotechnol Lett* 2012, 34:1415–1434.

21. Qureshi N, Ezeji TC: Butanol, 'a superior biofuel' production from agricultural residues (renewable biomass): recent progress in technology. *Biofuels Bioprod Bior* 2008, 2:319–330.

22. Qureshi N: Agricultural residues and energy crops as potentially economical and novel substrates for microbial production of butanol (a biofuel. *CAB Reviews: Perspectives in Agriculture, Veterinary Science, Nutrition and Natural Resources* 2010, 5:59.

23. Jang YS, Lee J, Malaviya A, Seung DY, Cho JH, Lee SY: Butanol production from renewable biomass: rediscovery of metabolic pathways and metabolic engineering. *Biotechnol J* 2012, 7:186–198.

24. Huang H, Liu H, Gan YR: Genetic modification of critical enzymes and involved genes in butanol biosynthesis from biomass.*Biotechnol Adv* 2010, 28:651–657.

25. Gheshlaghi R, Scharer JM, Moo-Young M, Chou CP: Metabolic pathways of clostridia for producing butanol. *Biotechnol Adv* 2009,27:764–781.

26. Kumar M, Gayen K: Developments in biobutanol production: New insights. *Appl Energ* 2011, 88:1999–2012.

27. Dürre P: Fermentative production of butanol- the academic perspective. *Curr Opin Biotechnol* 2011, 22:331–336.

28. Jin C, Yao M, Liu H, Lee CF, Ji J: Progress in the production and application of n-butanol as a biofuel. *Renew Sustain Energy Rev* 2011, 15:4080–4106.

29. Moholkar VS, Ranjan A, Mayank R: Economics of biobutanol: a review. *Res J Pharm Biol Chem Sci* 2012, 3:901–913.

30. Liu Z, Ying Y, Li F, Ma C, Xu P: Butanol production by *Clostridium beijerinckii* ATCC 55025 from wheat bran. *J Ind Microbiol Biotechnol* 2010, 37:495–501.

31. Khanna S, Goyal A, Moholkar VS: Production of n-butanol from biodiesel derived crude glycerol using *Clostridium pasteurianum* immobilized on Amberlite. *Fuel* 2013, 112:557–561.

32. Gottumukkala LD, Parameswaran B, Valappil SK, Mathiyazhakan K: Biobutanol production from rice straw by a non acetone producing Clostridium sporogenes BE01. *Bioresour Technol* 2013, 145:182–187.

33. Chen WH, Chen YC, Lin JG: Evaluation of biobutanol production from non-pretreated rice straw hydrolysate under non-sterile environmental conditions. *Bioresour Technol* 2013, 135:262–268.

34. Napoli F, Olivieri G, Russo ME, Marzocchella A, Salatino P: Butanol production by *Clostridium acetobutylicum* in a continuous packed bed reactor. *J Ind Microbiol Biotechnol* 2010, 37:603–608.

35. Lépiz-Aguilar L, Rodríguez-Rodríguez CE, Arias ML, Lutz G, Ulate W: Butanol production by *Clostridium beijerinckii* BA101 using cassava flour as fermentation substrate: enzymatic versus chemical pretreatments. *World J Microbiol Biotechnol* 2011, 27:1933–1939.

36. Qureshi N, Saha BC, Dien B, Hector RE, Cotta MA: Production of butanol (a biofuel) from agricultural residues: Part I – Use of barley straw hydrolysate. *Biomass Bioenerg* 2010, 34:559–565.

37. Qureshi N, Saha BC, Hector RE, Dien B, Hughes S, Liu S, Iten L, Bowman MJ, Sarath G, Cotta MA: Production of butanol (a biofuel) from agricultural residues: Part II – Use of corn stover and switchgrass hydrolysates. *Biomass Bioenerg* 2010, 34:566–571.

38. Lu C, Zhao J, Yang ST, Wei D: Fed-batch fermentation for n-butanol production from cassava bagasse hydrolysate in a fibrous bed bioreactor with continuous gas stripping. *Bioresour Technol* 2012, 104:380–387.

39. Noomtim P, Cheirsilp B: Production of Butanol from Palm Empty Fruit Bunches Hydrolyzate by *Clostridium acetobutylicum* .*Energy Procedia* 2011, 9:140–146.

40. Ezeji TC, Qureshi N, Blaschek HP: Production of acetone butanol (AB) from liquefied corn starch, a commercial substrate, using*Clostridium beijerinckii* coupled with product recovery by gas stripping. *J Ind Microbiol Biotechnol* 2007, 34:771–777.

41. Abd-Alla MH, El-Enany AWE: Production of acetone-butanol-ethanol from spoilage date palm (Phoenix dactyliferaL.) fruits by mixed culture of *Clostridium acetobutylicum* and *Bacillus subtilis* . *Biomass Bioenerg* 2012, 42:172–178.

42. Wang Y, Blaschek HP: Optimization of butanol production from tropical maize stalk juice by fermentation with *Clostridium beijerinckii* NCIMB 8052. *Bioresour Technol* 2011, 102:9985–9990.

43. Sun Z, Liu S: Production of n-butanol from concentrated sugar maple hemicellulosic hydrolysate by *Clostridia acetobutylicum*ATCC824. *Biomass Bioenerg* 2012, 39:39–47.

44. Al-Shorgani NKN, Kalil MS, Yusoff WMW: Biobutanol production from rice bran and de-oiled rice bran by *Clostridium saccharoperbutylacetonicum* N1–4. *Bioprocess Biosyst Eng* 2012, 35:817–826.

45. Guo T, Sun B, Jiang M, Wu H, Du T, Tang Y, Wei P, Ouyang P: Enhancement of butanol production and reducing power using a two-stage controlled-pH strategy in batch culture of *Clostridium acetobutylicum* XY16. *World J Microbiol Biotechnol* 2012,28:2551–2558.

46. Ranjan A, Khanna S, Moholkar VS: Feasibility of rice straw as alternate substrate for biobutanol production. *Appl Energ* 2013,103:32–38.

47. Ranjan A, Moholkar VS: Comparative study of various pretreatment techniques for rice straw saccharification for the production of alcoholic biofuels. *Fuel* 2011, 112:567–571.

48. Zhang WL, Liu ZY, Liu Z, Li FL: Butanol production from corncob residue using Clostridium beijerinckii NCIMB 8052. *Lett Appl Microbiol* 2012, 55:240–246.

49. Xue C, Zhao J, Lu C, Yang ST, Bai F, Tang IC: High-titern-butanol production by *Clostridium acetobutylicum* JB200 in fed-batch fermentation with intermittent gas stripping. *Biotechnol Bioeng* 2012, 109:2746–2756.

50. Isar J, Rangaswamy V: Improved n-butanol production by solvent tolerant *Clostridium beijerinckii* . *Biomass Bioenerg* 2012, 37:9–15.

51. Lin Y, Wang J, Wang X, Sun XH: Optimization of butanol production from corn straw hydrolysate by *Clostridium acetobutylicum*using response surface method. *Chin Sci Bull* 2011, 56:1422–1428.

52. Qureshi N, Maddox IS: Reduction in Butanol Inhibition by Perstraction: Utilization of Concentrated Lactose/Whey Permeate by*Clostridium acetobutylicum* to Enhance Butanol Fermentation Economics. *Food Bioprod Process* 2005, 83:43–52.

53. Li SY, Srivastava R, Suib SL, Li Y, Parnas RS: Performance of batch, fed-batch, and continuous A–B–E fermentation with pH-control. *Bioresour Technol* 2011, 102:4241–4250.

54. Chen Y, Zhou T, Liu D, Li A, Xu S, Liu Q, Li B, Ying H: Production of Butanol from Glucose and Xylose with Immobilized Cells of Clostridium acetobutylicum. *Biotechnol Bioprocess Eng* 2013, 18:234–241.

55. Ezeji TC, Qureshi N, Blaschek HP: Acetone butanol ethanol (ABE) production from concentrated substrate: reduction in substrate inhibition by fed-batch technique and product inhibition by gas stripping. *Appl Microbiol Biotechnol* 2004, 63:653–658.

56. Bahl H, Andersch W, Braun K, Gottschalk G: Effect of pH and Butyrate Concentration on the Production of Acetone and Butanol by Clostridium acetobutylicum Grown in Continuous Culture. *Eur J Appl Microbiol Biotechnol* 1982, 14:17–20.

57. Jang YS, Park JM, Choi S, Choi YJ, Seung DY, Cho JH, Lee SY: Engineering of microorganisms for the production of biofuels and perspectives based on systems metabolic engineering approaches. *Biotechnol Adv* 2012, 30:989–1000.

58. García V, Päkkilä J, Ojamo H, Muurinen E, Keiski RL: Challenges in biobutanol production: How to improve the efficiency? *Renew Sustain Energy Rev* 2011, 15:964–980.

59. Lee J, Jang YS, Choi SJ, Im JA, Song H, Cho JH, Seung DY, Papoutsakis T, Bennett GN, Lee SY: Metabolic Engineering of Clostridium acetobutylicum ATCC 824 for Isopropanol-Butanol-Ethanol Fermentation. *Appl Environ Microbiol* 2012, 78:1416–1423.

60. Collas F, Kuit W, Clément B, Marchal R, López-Contreras AM, Monot F: Simultaneous production of isopropanol, butanol, ethanol and 2,3-butanediol by Clostridium acetobutylicum ATCC 824 engineered strains. *AMB Express* 2012, 2:45.

61. Lütke-Eversloh T, Bahl H: Metabolic engineering of *Clostridium acetobutylicum* : recent advances to improve butanol production. *Curr Opin Biotechnol* 2011, 22:634–647.

62. Jiang Y, Xu C, Dong F, Yang Y, Jiang W, Yang S: Disruption of the acetoacetate decarboxylase gene in solvent-producing *Clostridium acetobutylicum* increases the butanol ratio. *Metab Eng* 2009, 11:284–291.

63. Gu Y, Jiang Y, Wu H, Liu X, Li Z, Li J, Xiao H, Shen Z, Dong H, Yang Y, Li Y, Jiang W, Yang S: Economical challenges to microbial producers of butanol: feedstock, butanol ratio and titer. *Biotechnol J* 2011, 6:1348–1357.

64. Van der Merwe AB, Cheng H, Görgens JF, Knoetze JH: Comparison of energy efficiency and economics of process designs for biobutanol production from sugarcane molasses. *Fuel* 2013, 105:451–458.

65. Whitfield MB, Chinn MS, Veal MW: Processing of materials derived from sweet sorghum for biobased products. *In Crops Prod* 2012,37:362–375.

66. Adsul MG, Singhvi MS, Gaikaiwari SA, Gokhale DV: Development of biocatalysts for production of commodity chemicals from lignocellulosic biomass. *Bioresour Technol* 2011, 102:4304–4312.

67. Lee J, Seo E, Kweon DH, Park K, Jin YS: Fermentation of rice bran and defatted rice bran for butanol production using *Clostridium beijerinckii* NCIMB 8052. *J Microbiol Biotechnol* 2009, 19:482–490.

68. Kumar P, Barrett DM, Delwiche MJ, Stroeve P: Methods for pretreatment of lignocellulosic biomass for efficient hydrolysis and biofuel production. *Ind Eng Chem Res* 2009, 48:3713–3729.

69. Ooi BG, Rambo AL, Hurtado MA: Overcoming the Recalcitrance for the Conversion of Kenaf Pulp to Glucose via Microwave-Assisted Pre-Treatment Processes. *Int J Mol Sci* 2011, 12:1451–1463.

70. Thirmal C, Dahman Y: Comparisons of existing pretreatment, saccharification, and fermentation processes for butanol production from agricultural residues. *Can J Chem Eng* 2012, 90:745–761.

71. Qureshi N, Saha BC, Hector RE, Cotta MA: Removal of fermentation inhibitors from alkaline peroxide pretreated and enzymatically hydrolyzed wheat straw: production of butanol from hydrolysate using *Clostridium beijerinckiiin* batch reactors. *Biomass Bioenerg* 2008, 32:1353–1358.

72. Parekh M, Formanek J, Blaschek HP: Pilot-scale production of butanol by Clostridium beijerinckii BA101 using a low-cost fermentation medium based on corn steep water. *Appl Microbiol Biotechnol* 1999, 51:152–157.

73. Zheng YN, Li LZ, Xian M, Ma YJ, Yang JM, Xu X, He DZ: Problems with the microbial production of butanol. *J Ind Microbiol Biotechnol* 2009, 36:1127–1138.

74. Mariano AP, Costa CBB, Angelis DF, Maugeri Filho F, Atala DIP, Maciel MRW, Maciel FR: Dynamics of a continuous flash fermentation for butanol production. *Chem Eng Trans* 2010, 20:285–290.

75. Secuianu C, Feroiu V, Geana D: High-Pressure Vapor -Liquid Equilibria in the System Carbon Dioxide + 1-Butanol at Temperatures from (293.15 to 324.15) K. *J Chem Eng Data* 2004, 49:1635–1638.

76. Dhamole PB, Wang Z, Liu Y, Wang B, Feng H: Extractive fermentation with non-ionic surfactants to enhance butanol production.*Biomass Bioenerg* 2012, 40:112–119.

77. Nakayama S, Morita T, Negishi H, Ikegami T, Sakaki K, Kitamoto D: Candida krusei produces ethanol without production of succinic acid; a potential advantage for ethanol recovery by pervaporation membrane separation. *FEMS Yeast Res* 2008, 8:706–714.

78. Yen HW, Chen ZH, Yang IK: Use of the composite membrane of poly (ether-block-amide) and carbon nanotubes (CNTs) in a pervaporation system incorporated with fermentation for butanol production by *Clostridium acetobutylicum* . *Bioresour Technol* 2012, 109:105–109.

79. Ezeji TC, Qureshi N, Blaschek HP: Bioproduction of butanol from biomass: from genes to bioreactors. *Curr Opin Biotechnol* 2007,18:220–227.

80. Tanaka S, Tashiro Y, Kobayashi G, Ikegami T, Negishi H, Sakaki K: Membrane-assisted extractive butanol fermentation by*Clostridium saccharoperbutylacetonicum* N1–4 with 1-dodecanol as the extractant. *Bioresour Technol* 2012, 116:448–452.

81. Mariano AP, Costa CBB, Angelis DF, Maugeri Filho F, Atala DIP, Maciel MRW, Maciel FR: Optimisation of a continuous flash fermentation for butanol production using the response surface methodology. *Chem Eng Res Des* 2009, 88:562–571.

82. Mariano AP, Costa CBB, Angelis DF, Maugeri Filho F, Atala DIP, Maciel MRW, Maciel FR: Optimisation of a fermentation process for butanol production by particle swarm optimisation (PSO). *J Chem Technol Biotechnol* 2010, 85:934–949.

Chapter 7

FACILE APPROACH FOR THE SYNTHESIS OF RIVAROXABAN USING ALTERNATE SYNTHON: REACTION, CRYSTALLIZATION AND ISOLATION IN SINGLE POT TO ACHIEVE DESIRED YIELD, QUALITY AND CRYSTAL FORM

Anil C Mali[1], Dattatray G Deshmukh[1], Divyesh R Joshi[1], Hitesh D Lad[1], Priyank I Patel[1], Vijay J Medhane[2] and Vijayavitthal T Mathad[1]

[1]Department of Process Research and Development, Megafine Pharma (P) Ltd., 201, Lakhmapur, Dindori, Nashik 422 202, Maharashtra, India

[2] Department of Chemistry, Organic Chemistry Research Center, K. T. H. M College, Nashik 422 002, Maharashtra, India.

ABSTRACT

An efficient and high yielding process for the production of impurity free rivaroxaban (1), an anti-coagulant agent using alternate synthon is reported. The key components of the process involve; synthesis of 4-(4-aminophenyl)-3-morpholinone (5) using easily available inexpensive nitro aniline (17), condensation of 4-{4-[(5S)-5-(aminomethyl)-2-oxo-1,3-oxazolidin-3-yl]phenyl}morpholin-3-one hydrochloride (9) with alternate synthon 4-nitrophenyl 5-chlorothiophene-2-carboxylate (11) in dimethylsulfoxide (DMSO) as a solvent and triethylamine as a base and isolation of the rivaroxaban (1) by designing the crystallization in same reaction pot using specific combination of acetonitrile and methanol as anti-solvents to obtain highly pure rivaroxaban (1) with desired ploymorphic form with an overall yield of around 22% (Calculated from 17). The developed process avoids the use of hazardous chemicals, critical operations and tedious work-ups. Potential impurities arouse during the reaction at various stages and carry-over impurities from starting materials were controlled selectively by designing reaction conditions and tuning the crystallization parameters.

GRAPHICAL ABSTRACT

Facile Approach for the Synthesis of Rivaroxaban using Alternate Synthon: Reaction, Crystallization and Isolation in Single Pot to Achieve Desired Yield, Quality and Crystal form[#].

BACKGROUND

Rivaroxaban (**1**), chemically known as 5-chloro-N-({(5S)-2-oxo-3-[4-(3-oxomorpholin-4-yl)phenyl]-1,3-oxazolidin-5-yl}methyl)thiophene-2-carboxamide is an orally active direct factor Xa (FXa) inhibitor drug developed by Bayer and approved by United states Food and Drug Administration (USFDA) during July 2011 under the trade name of Xarelto [1, 2]. Rivaroxaban (**1**) used for the prevention and treatment of various thromboembolic diseases, in particular pulmonary embolism, deep venous thrombosis, myocardial infarction, angina pectoris, reocclusion and restenosis after angioplasty or aortocoronary bypass, cerebral stroke, transitory ischemic attacks, and peripheral arterial occlusive diseases [1–4].

Alexander and co-workers [3, 4] reported first synthesis of **1** using linear approach that involves condensation of morpholin-3-one (**2**) with fluoro nitrobenzene (**3**) using sodium hydride as a base in N-methylpyrrolidone (NMP) to get nitro morpholinone (**4**), nitro group of **4** was reduced using palladium on carbon (Pd–C) and hydrogen in tetrahydrofuran (THF) to achieve 4-(4-aminophenyl)-3-morpholinone (**5**) which was then condensed with 2-[(2S)-2-oxiranylmethyl]-1H-isoindole-1, 3(2H)-dione (**6**) in ethanol and water mixture to provide amino alcohol **7**. Cyclization of **7** using N,N'-carbonyldiimidazole (CDI) in presence of 4-dimethylaminopyridine (DMAP) in THF furnished 2-({(5S)-2-oxo-3-[4-(3-oxomorpholin-4-yl)phenyl]-1,3-oxazolidin-5-yl}methyl)-1H-isoindole-1,3(2H)-dione (**8**). Deprotection of **8** using methyl amine in ethanol followed by condensation of obtained **9** with 5-chlorothiophene-2-carbonyl chloride (**10**) in pyridine furnished rivaroxaban (**1**) with an overall yield of 4.5% starting from compound **2** (Scheme 1).

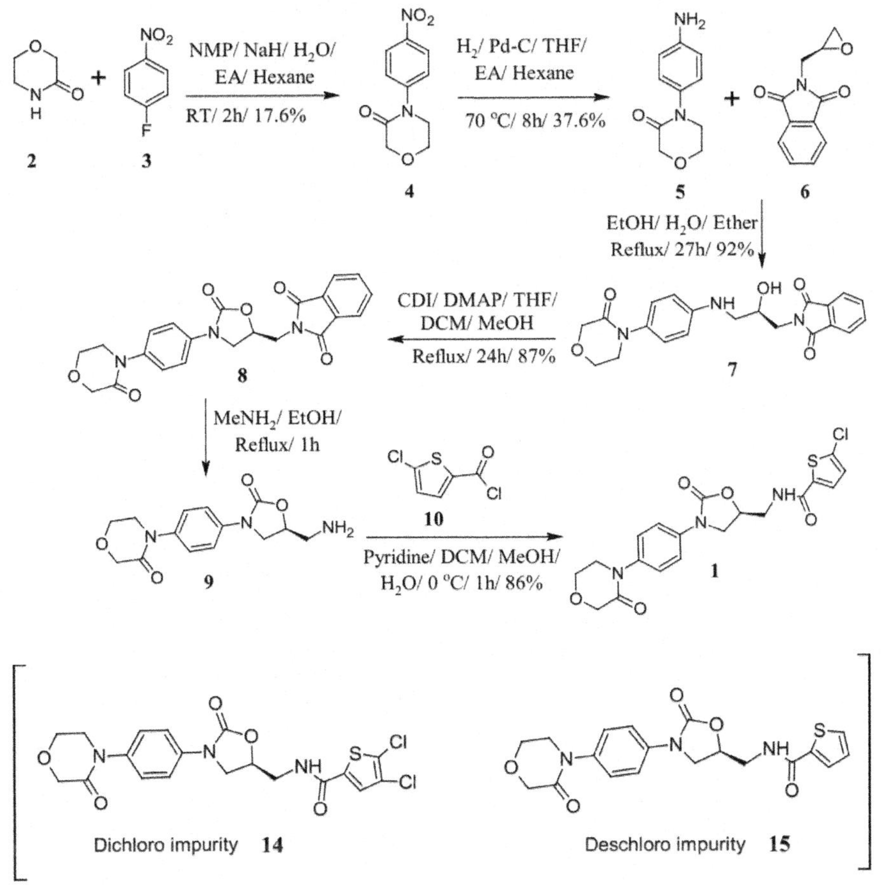

Scheme 1: Reported synthetic scheme for rivaroxaban (**1**) as per basic patent US 7576111.

The disclosed process has several disadvantages such as: (a) use of highly pyrophoric reagent like sodium hydride, (b) excess loading of expensive key raw material **6** and CDI that generate excess amount of by-product imidazole and makes the process economically and environmentally inefficient, (c) repeated filtrations involved for the isolation of **7** decreases the production throughput, (d) use of highly flammable solvent like diethyl ether for washing the compound **7** and hazard and toxic solvent like pyridine for the reaction at API stage makes the process unsafe, (e) purification of **1, 4, 5** and **8** by column chromatography makes process lengthy and time consuming, (f) incomplete conversion of intermediates leads to formation of many process related impurities, and (g) low overall yield of 4.5% starting from compound **2** which makes process less viable for commercial production.

Rafecas et al. [5] disclosed process for the preparation of rivaroxaban (1) and intermediates thereof (Scheme 2) with an overall yield of 5.93 with 98.4% purity by HPLC. Silvo et al. in 2011 reported the purification of rivaroxaban (1) by solvent mediated crystallization process [6] using various solvents and/or solvent combinations to provide maximum of around 99.70% purity by HPLC with a loss of around 10–20% yield making the process expensive. Several other processes reported in the literature also suffer disadvantages similar to as described above and thus makes the process unsuitable for large scale production [7–19].

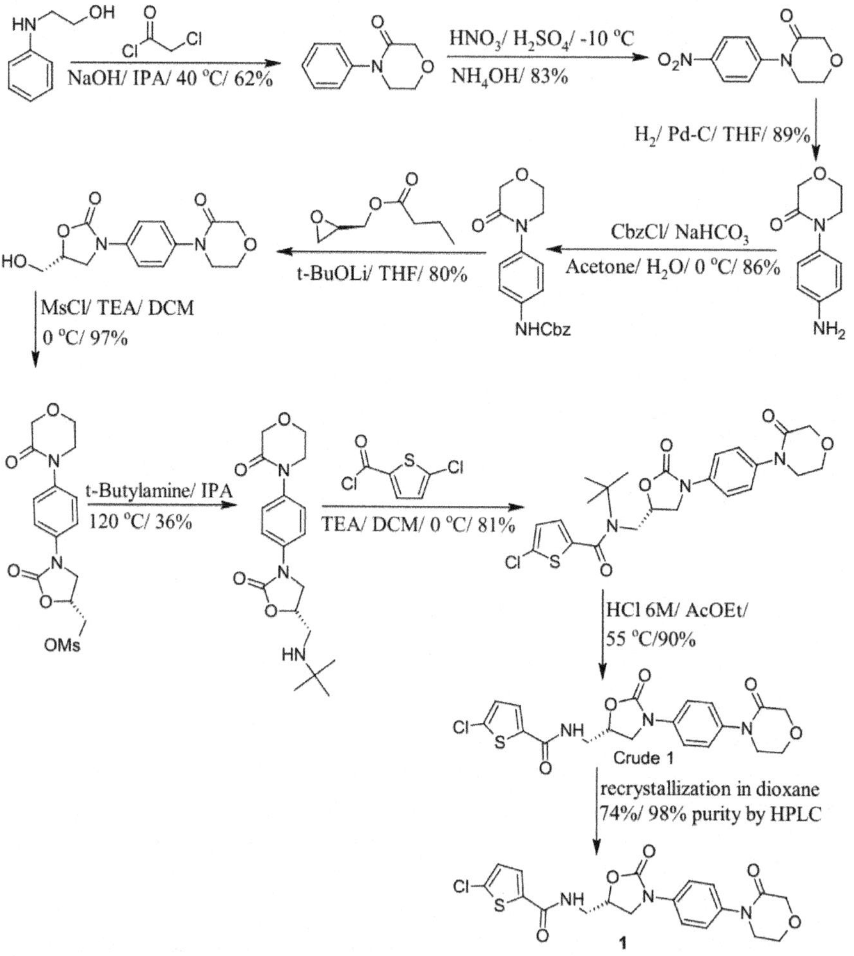

Scheme 2: Reported synthetic scheme for rivaroxaban (1) as per WO/2011 080341.

We hereby report an efficient, economic, scalable, impurity-free and production friendly process for the synthesis of rivaroxaban (1) which allows direct isolation of API from reaction mass without further purification complying with ICH quality [20] and desired polymorphic form [21,22]. Rivaroxaban (1) was obtained from 17 in seven steps with an overall yield of 22% and purity of 99.85% by HPLC using commercially available and less expensive raw materials and reagents.

Results and discussion

Process research and development of rivaroxaban mainly focused on following three aspects: (a) to develop an improved and high yielding process for the manufacture of 9 starting from 17 without the use of column chromatographic purification and avoiding the hazardous and unsafe chemical reagents and solvents, (b) to develop an efficient process for the manufacture of 1 starting from 9 which reduces the formation of impurities and provide high reaction yield and (c) to achieve crystallization and isolation of 1 with ICH quality and desired polymorphic form (modification-I) [21] directly from reaction mass from the same pot without subjecting the crude to a further purification.

Efficient Process for the Preparation of Compound 9

Critical and cost-contributing intermediate 5 was prepared [23] using easily available inexpensive nitro aniline (17) as a starting material. Boric acid catalyzed condensation of 17 with 2-(2-chloroethoxy) acetic acid in toluene followed by cyclization of the obtained chloroamide 18 in sodium hydroxide in the presence of phase transfer catalyst *tetra*-butyl ammonium bromide (TBAB) provided nitro compound 19. Catalytic reduction of 19 using Pd–C as catalyst and H_2 gas or ammonium formate as a source of hydrogen in methanol provided key intermediate 5(Scheme 3). However, another key intermediate compound 6 was prepared as per the literature procedure with little process improvements.

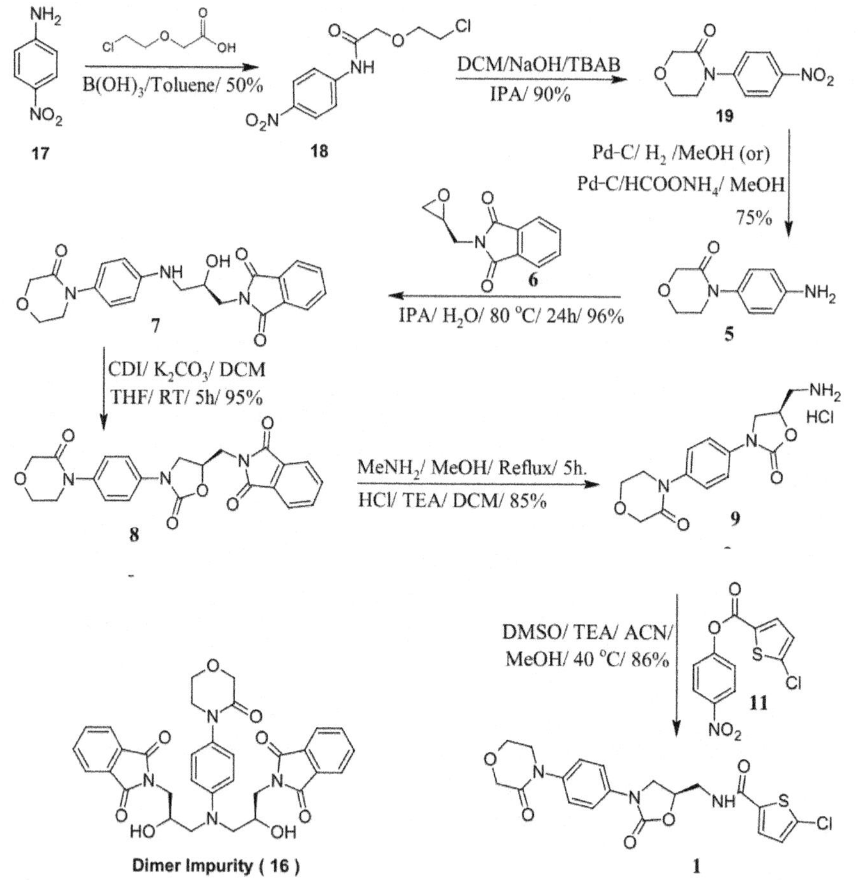

Scheme 3: Alternate and efficient synthesis of rivaroxaban (**1**).

Preparation of amino alcohol **7** (Scheme 3) by nucleophilic attack of **5** on oxirane **6** has been attempted using various solvents. Among several solvents screened such as methanol, ethanol, IPA, DMF, DMSO, toluene, water, acetonitrile, ethyl acetate, THF and their mixtures, the mixture of IPA and water found to be the suitable solvent for this reaction. It was noticed that oxirane **6** underwent degradation in other solvents except IPA–water resulting into unreacted **5** in the reaction. It was also noticed that excess use of oxirane **6** results into the formation of dimer impurity **16** (determined by LC–MS). Further, it has also been observed that oxirane **6** found to be stable in IPA–water mixture and thus reaction goes to completion with 1.05 mol equivalents of oxirane **6** at 80°C to provide **7** with negligible quantities of dimer **16**. The precipitated compound **7** was isolated by filtering and drying the product under vacuum at 60–65°C for 5–6 h with 95% yield and 97.0% purity by HPLC. During optimization, specific ratio of IPA and water (17:3) found to be optimum to provide excellent yield and quality of **7** (Table 1).

Table 1: Impact of IPA–water ratio on nucleophilic substitution reaction between **5** and **6** to give **7**

Expt. no.	Compd. 6 (mole. eq. w. r. to 5)	Volume w.r.to 5		Yield (%)	HPLC purity (%)	Content of 5 and 6	
		IPA	Water			5 (%)	6 (%)
1[a]	1.05	10	–	NI	62.00	12.00	10.0
2[b,c,d]	1.05	–	10	78.00	86.90	0.57	6.00
3[b,d]	1.05	9	1	90.00	94.38	1.70	2.00
4[b,d]	1.05	13.5	1.5	87.80	97.50	0.66	0.70
5	1.05	17	3	91.00	97.18	0.60	0.40

NI not isolated.

[a]Reaction not progressed.

[b]Reaction mass not stirrable.

[c]OXI (**6**) could not be removed in water.

[d]Thick reaction mass, unable to transfer from reaction flask to centrifuge.

Amino alcohol **7** obtained was then treated with CDI to achieve oxazolidinone **8** without further purification. The reaction was explored in various solvents (toluene, THF, ethyl acetate, 2-Me-THF, DCM and acetonitrile) using different bases (K_2CO_3, Na_2CO_3, DIPEA, TEA and $NaHCO_3$). During screening experiments incomplete conversion of **7** was observed in various solvents and different bases after prolonged maintenance at various temperature conditions resulting into the formation of many side products. However, reaction in DCM in the presence of K_2CO_3 as a base at 25–30°C proceeded smoothly with a reduced amount of impurities (~0.5% by HPLC). Work-up involves the filtration of the reaction mass to separate inorganic solids, concentration of the DCM and purification of obtained residue in THF to furnish **8** as a white crystalline solid with 99.8% purity by HPLC and 95% yield.

Synthesis of key intermediate **9** (Scheme 3) was then achieved by deprotection of **8** using 40% aqueous methyl amine solution in methanol at 60–65°C. After the reaction is completed the reaction mass was cooled gradually to 25–30°C and conc. HCl was added slowly to the mass to obtain HCl salt of **9** as white crystalline solid with 90% yield and 99.0% purity by HPLC.

Many by-products and analogous impurities formed during the deprotection reaction were identified by LC–MS in reaction mass as well as in the isolated solid (Figure 1). Based on the understanding of impurities purification has been established by recrystallization using methanol and DCM to eliminate these impurities. Traces of these impurities if left behind at this stage will be carried forward with **9** to impact the purity of rivaroxaban (**1**) thus it is necessary to eliminate at this step. It was also experienced that these impurities were found to be difficult to remove from the product and ultimately require column chromatographic purifications. The data of the pilot batches along with the impurity details are presented in Table 2.

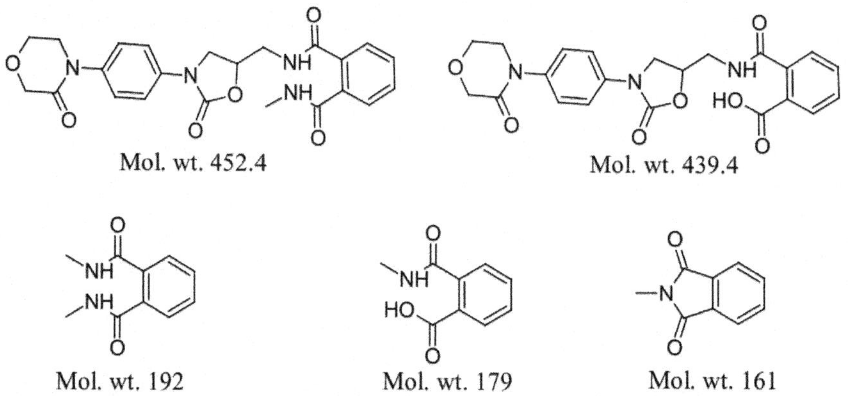

Mol. wt. 452.4   Mol. wt. 439.4

Mol. wt. 192   Mol. wt. 179   Mol. wt. 161

Figure 1: Identified impurities in the reaction mass of **9** by LC–MS. By-products and impurities formed during the deprotection of phthalimide group in **8** to form **9** by LC–MS in reaction mass as well as in the isolated solid of compound **9**.

Table 2: Yield and quality data of amine compound **9** from scale-up batches

No.	Comp 8 (kg)	MeNH₂ sol. (L)	MeOH (L)	Temp (°C)	Yield (%)	Quality details by HPLC (%)					
						9	8	7	6	5	16
1	2.88	2.93	28.8	60–65	87.5	99.76	ND	0.04	ND	ND	ND
2	2.87	2.92	28.7	60–65	86.0	99.80	ND	0.02	ND	ND	ND
3	2.90	2.95	29.0	60–65	85.0	99.86	0.01	0.02	ND	ND	ND

ND not detected.

Novel and Efficient Process for the Preparation of Rivaroxaban (1)

Our next objective was to establish the process for the condensation of pure **9** with 5-chlorothiophene-2-carboxylic acid or its derivatives to get rivaroxaban (**1**) with minimum or no impurities in the reaction mass itself. We observed that reaction of **9** with 5-chlorothiophene-2-carbonyl chloride (**10**) as pre reported was incomplete leading to presence of unreacted starting material with few degraded impurities both in the reaction mass and isolated product **1**. To overcome the issue we identified three alternative synthons (Scheme 4) such as active ester **11**, aldehyde **12** and alcohol **13** as the counterpart to react with **9** instead of **10** to explore improved or efficient reaction profile. During the exploration of the experiments to check the feasibility of each of these options interesting results were noticed which are tabulated in Table 3.

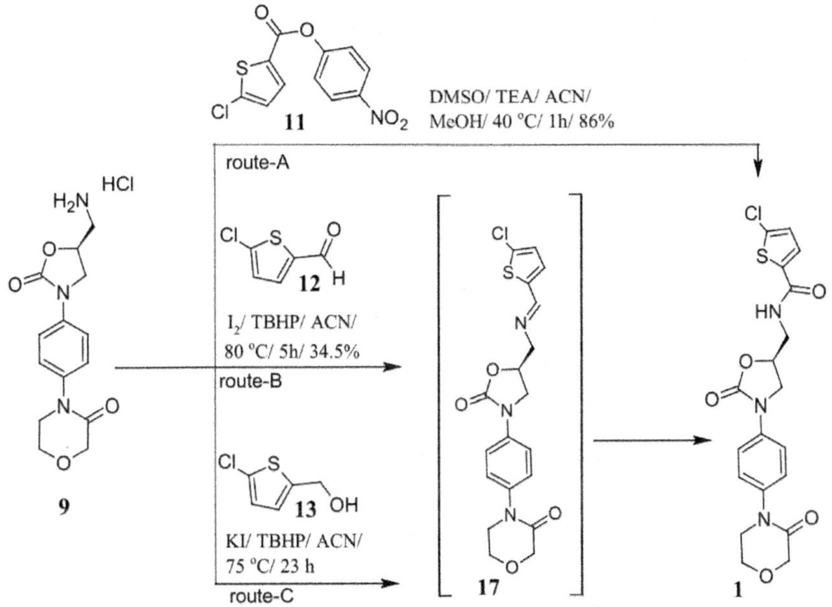

Scheme 4: Synthesis of **1** using coupling of different synthons **11**, **12** and **13** with **9**.

Table 3: Results obtained from different routes (A, B and C) for synthesis of **1**

Route	Compd. 9 (g)	Substrate (mol. eq.)	Solvents	Time (h)	Temp (°C)	HPLC purity (%)		Yield (%)
						Reaction mass	Isolated solid	
A	10.0	11 (1.05)	DMSO	1.0	40	78.0	99.85	86.0
B	10.0	12 (1.20)	ACN	5.0	80	60.8	91.0	34.5
C	10.0	13 (1.50)	ACN	23.0	75	NA	67.17	NI

NA not analyzed, *NI* not isolated.

Among the three synthons identified (**11–13**), active ester **11** as a substrate found to be the most promising synthon to react with **9** as the reaction underwent selectively and cleanly without formation of any impurities whereas in case of route-B and C, though oxidative amidation proceeded well via an imine intermediate (**17**, Scheme 4) but lead to the formation of many impurities with poor yields and hence discontinued. Thus route-A has been selected for further optimization.

Further, the situation was highly demanding to design the experiment for the synthesis of **1** which will directly yield ICH quality API from the reaction without any further purification of crude compound. In such a situation, the selection of suitable solvent and base was very important to achieve the target. Since the solubility of **1** is very low in most of the solvents except DMSO, DMF, NMP and acetic acid, we have selected DMSO and DMF as a reaction solvent for further screening based on the results of initial screen.

During the process optimization though the reaction of **9** with **11** progressed well with inorganic bases, filtration of the reaction mass to remove the inorganic base was not possible because the product is also precipitated out in the reaction mass as the reaction progresses. Alternatively, removing the inorganic base by adding water into the reaction mass and isolating the product found to possess several impurities which were difficult to remove even after repeated crystallizations. Thus we restricted our selection criteria to organic bases which are miscible with the selected solvents. Further, the solvents for the reaction were selected in which both organic base as well as rivaroxaban (**1**) is soluble.

Complete conversion **9** to **1** was observed when DMF or DMSO was used as a solvent in presence of triethylamine (TEA) as a base at 40°C. Though the reaction profile was same both in DMF and DMSO, we have chosen

DMSO as it is a class-3 solvent whereas DMF is class-2 solvent. At the end of the reaction, the reaction mass was in a complete dissolved state and anti-solvent mediated crystallization shall be designed. Accordingly MeOH, IPA, ACN, ethyl acetate, acetone, THF, water, DIPE and their mixtures were tried as anti-solvents to crystallize out the product with desired quality, yield and polymorph. The details of the same are summarized in Table 4.

Table 4: Trend data of yield and impurity profile of **1** in different reaction solvents/anti-solvents

Expt. no.	Reaction solvent	Base	Anti solvents	Yield (%)	HPLC purity (%)	Impurity profile by HPLC (%)				
						9	14	15	NP	11
1	DCM	TEA	–	NI	35.9	41	0.09	0.20	13.3	1.05
2	DMF	K$_2$CO$_3$	Water	60	98.84	0.56	0.02	0.10	0.16	0.12
3	DMF	TEA	Methanol	30	99.65	0.01	0.14	0.09	0.09	ND
4	DMF	TEA	Ethyl acetate	NI	99.62	0.07	0.14	0.08	0.02	ND
5	DMF	TEA	THF	20.0	99.70	0.07	0.05	0.08	0.02	ND
6	DMSO	TEA	Methanol	90.0	99.12	ND	0.14	0.22	0.14	ND
7	DMSO	TEA	IPA	NI	98.7	0.20	0.10	0.2	0.23	ND
8	DMSO	TEA	Acetic acid/water	75.0	97.1	0.04	0.13	0.10	1.78	ND
9	DMSO	TEA	Water	75.0	98.6	0.43	0.18	0.11	0.12	0.02
10	DMSO	TEA	Acetonitrile	69.1	99.85	0.01	0.06	0.03	0.01	0.02
11	DMSO	TEA	Acetonitrile	75.0	99.86	ND	0.07	0.02	0.02	ND
12	DMSO	DBU	Acetonitrile	75.1	99.70	ND	0.13	0.05	0.02	ND

NP Nitrophenol, *NI* not isolated, *ND* not detected.

When DMSO as solvent and methanol as anti-solvent are used the rivaroxaban **1** was precipitated out with a good yield but failed to eliminate dichloro impurity **14** and deschloro impurity **15**. When acetonitrile is used as an anti-solvent, impurities **14** and **15** were controlled well below 0.10% with good yield and purity of **1** (Scheme 1). Thus acetonitrile has been selected as an antisolvent for further optimization of the crystallization process. Alternatively source of these impurities (**14** and **15**) were further controlled in synthon **11**.

Mechanistically, the neucleophilic attack of amine **9** on active ester **11** was very facile because of the good leaving group (*p*-nitrophenol) on the electrophilic carbon of **11** that favored the progress of reaction without any side reaction. The by-product *p*-nitrophenol generated in the reaction was washed away with the mother liquors during the filtration. The content of *p*-nitrophenol in the pure **1** was measured using HPLC and found below 75 ppm in all the batches produced.

Crystallization and Isolation of 1 from the Reaction Pot

After the completion of reaction in DMSO at 40°C the temperature of the reaction mass was raised to 60°C and acetonitrile was added at same temperature and cooled gradually to 20–25°C to obtain the crystalline solid. The solid was filtered, washed with chilled acetonitrile and dried under vacuum to provide **1** with 75% yield and 99.8% purity with desired polymorph (Modification 1). Upon detailed analysis, it was learnt that the isolated solid was failed in residual solvent test by GC-HS with respect to acetonitrile content. The content of ACN by GC-HS was around 2,000 ppm against its acceptable limit of 410 ppm [20]. Different industrial techniques such as prolonged drying under vacuum, air tray drying, drying in RCVD (Rota Cone Vacuum Drier) at various temperatures, intermittent milling of the crystals followed by drying were tested to remove the acetonitrile content but none of these techniques were successful.

Observations of the crystallization process were thoroughly investigated to understand theory of nucleation and crystal growth in the reaction flask to overcome the issue of high residual content in the crystals. Detailed observation of the process provided the following insight on the crystallization viz.,

1.   *Solution* Reaction mass at 60°C to provide the clear solution indicating that the product is in the soluble state,

2.   *Saturation* Acetonitrile was added as anti-solvent at 60°C to achieve the saturated solution,

3.   *Super saturation* The saturated solution was gradually cooled to 20–25°C to achieve super saturation which lead to nucleation and crystal growth.

It was presumed that the nucleation and growth of crystals are occurring under acetonitrile environment. As the concentration of acetonitrile was more than that of DMSO which could get trapped into the crystal lattice during the primary nucleation and continued to do so as the crystal growth continues. Thus, we envisaged that if we limit the contact of acetonitrile during nucleation step by reducing its volume and/or by introducing the third anti solvent during crystal formation/nucleation process, the entrance of solvent into the crystal lattice may be arrested. Accordingly, different set of solvents were screened as a third solvent along with acetonitrile. As per our assumption, methanol has given an excellent result in controlling the residual solvent content as a third solvent among the many screened. Further the ratio of first anti solvent (ACN) and second anti solvent (methanol) played a tremendous impact on the quality and yield (Table 5).

Table 5: HPLC trend data, yield, and acetonitrile content by GC in **1**

Expt. no.	DMSO (Vol[a])	ACN (Vol[a])	MeOH (Vol[a])	Yield (%)	Content by HPLC (%)					ACN content by GC (ppm)
					1	14	15	NP	11	
1	5	10	–	69.1	99.85	0.06	0.03	0.02	ND	1,980
2	3	8	–	75	99.86	0.07	0.02	0.02	ND	753
3	4	10	–	71	99.51	ND	0.14	ND	ND	NA
4	4	6	4	71	99.51	0.15	0.03	ND	ND	580
5	4	4	6	75.0	99.56	0.06	ND	ND	ND	256
6	4	6	4	75.1	99.72	0.12	ND	0.02	ND	NA
7	4	6	9	80.2	99.68	0.05	ND	0.05	ND	337
8	4	4	6	87	99.79	0.04	0.03	0.03	ND	424
9	4	4	4	–	99.55	0.06	ND	0.01	ND	371
10	4	2	6	–	99.68	0.05	0.04	ND	ND	283
11	6	2	6	86.4	99.81	0.04	ND	0.04	ND	110
12	6	2	6	84.2	99.84	0.03	0.02	ND	ND	135

ND not detected, *NA* not analyzed.

[a]Volumes used with respect to wt of **9**.

Based on the data the solvent and anti-solvents ratio of 6:2:6 volumes of DMSO:ACN:MeOH was finalized (entry 11 and 12, Table 5) to achieve robust process to provide **1** with yield of 86% and purity of 99.8% by HPLC. The acetonitrile content was achieved below 300 ppm (GC-HS) in the final product. The trend data of content of impurities by HPLC and residual solvent content by GC-HS from pilot batches is provided in Table 6.

Table 6: Trend data of residual solvent, HPLC purity and yield of three pilot batches

Sr. no.	Solvent content by GC-HS in ppm			HPLC purity (%)					Yield (%)
	DMSO	Acetonitrile	Methanol	1	9	11	14	15	
1	1,068	238	352	99.81	0.01	ND	0.02	ND	85.2
2	991	226	368	99.81	0.01	ND	0.03	ND	85.3
3	969	241	407	99.83	0.01	ND	0.02	ND	85.9

ND not detected.

Further, to measure the extent of greenness of newly developed process, we have calculated process mass intensity (PMI), e-factor and atom economy (calculated for last four stages) using the standard calculations provided in ACS guideline [24] and compared with the originator's process. The details are provided in Table 7.

Table 7: PMI, e-factor, and atom economy of the developed process

Process	Theoretical PMI	Actual PMI	Aqueous PMI	Solvent PMI	e-factor	Atom economy
Megafine process	4.13	7.40	2.04	76.6	85.13	37.97
U.S. Pat. 7,576,111	2.24	28.14	25.0	415.3	455.95	44.91

By-products and impurities formed during the deprotection of phthalimide group in **8** to form **9** by LC–MS in reaction mass as well as in the isolated solid of compound **9**.

CONCLUSION

An efficient, economic, and production friendly process for the preparation of highly pure rivaroxaban via alternate synthon (**11**) is described. Reaction and isolation by crystallization have been designed in the same pot avoiding the work-up which normally involve extraction, distillation, separations etc., to provide the rivaroxaban (1) meeting the ICH quality and desired polymorphic form with an overall yield of 22% is starting from nitro aniline (17) over seven stages.

EXPERIMENTAL

All reagents, solvents, and processing aids are commercial products and were used as received. For reactions run of pilot scale, glass line reactors having variable rate agitation, a −10 to 150°C jacket temperature range were used for the reaction. ^{1}H NMR spectra was recorded in DMSO-D$_6$ and D$_2$O using Varian Gemini 400 MHz FT NMR spectrometer; the chemical shifts are reported in δ ppm relative to TMS. Related substance purity and residual solvent content in solid were monitored by high performance liquid chromatography (HPLC) on Agilent Technologies 1200 series. The gas chromatography on Agilent Technologies 7683B with head space was used for analyzing the residual solvents.

HPLC Method for Calculating the Chemical Purity, Chiral Purity and Assay

Related substances, assay and chiral purity of rivaroxaban (1) were estimated by a gradient HPLC analysis method developed at Megafine.

(a)   Related substances and assay of rivaroxaban were estimated using Zorbax SB-CN, (250 × 4.6 mm ID), 5 μ column; mobile phase A comprising a mixture of phosphate buffer (0.01 m potassium dihydrogen orthophosphate, 0.005 m 1-heptane sulphonic acid sodium salt, triethylamine, and adjust the pH 6.7 using orthophosphoric acid). Mobile phase B comprising a mixture of acetonitrile/water in the ratio 80:20 (v/v); gradient elution: time (min)/A (v/v): B (v/v), $T_{0.0}$/80:20, $T_{5.0}$/75:25, $T_{25.0}$/50:50, $T_{38.0}$/50:50, $T_{42.0}$/80:20, $T_{50.0}$/80:20; flow rate 1.5 ml/min column temperature 35°C wavelength 240 nm. The observed retention time of rivaroxaban under these chromatographic conditions is about 16.0 min.

(b)   Chiral purity was estimated using Chiralpak IE (250 × 4.6 mm ID), 5 μ column; mobile phase comprising a mixture of n-Hexane, ethanol, methanol, THF and ammonia in the ratio of 35:35:15:15:0.1 (v/v/v/v/v) respectively; flow rate 1.0 ml/min.; column temperature 35°C; wavelength 240 nm. The observed retention time of (S)-rivaroxaban is about 11.7 min and (R)-rivaroxaban is about 10 min.

GC Method for Calculating Residual Solvent Content

a. Residual solvents content was estimated using DB-624, 60 m, 0.25 mm ID, 1.4 μ film thickness fused silica capillary column; detector temperature 280°C; injector temperature 250°C; split ratio 10:1; column flow 0.8 ml/min; nitrogen used as carrier gas; head space conditions: oven temperature 90°C; loop temperature 115°C; Transfer line temperature 120°C. The observed elution order of solvents is methanol, isopropyl alcohol, acetonitrile, dichloromethane, and tetrahydrofuran at RT 9.1, 13.5, 13.9, 14.4 and 18.1, respectively.

b. DMSO was estimated by using HP-5, 30 m, 0.53 mm ID, 5.0 μ film thickness, Fused silica capillary column; detector temperature 280°C; injector temperature 250°C; split ratio 3:1; column flow 4 psi (at constant pressure).; nitrogen used as carrier gas; The observed retention time of DMSO is 5.5 min.

Preparation of 2-(2-chloroethoxy)-N-(4-nitrophenyl) Acetamide (18)

A solution of boric acid (13 g, 0.21 mol) and 4-nitroaniline (**17**, 50 g, 0.36 mol) in toluene (500 mL) were heated to 110–115°C to remove the water by azeotropic distillation. To this solution 2-(2-chloroethoxy)-acetic acid (50 g, 0.36 mol) was added at 50–60°C and reheated to 110–115°C for 18 h. Reaction mass was cooled and filtered. Filtrate was concentrated to get residue, residue was recrystallized using isopropyl alcohol (300 mL) to afford yellow solid of compound **18**. Yield: 47 g (50.0%). M.P.: 101–104°C. ^{1}H NMR (DMSO-d_6, 400 MHz): δ 10.38 (s, 1H), 8.22 (d, 2H), 7.91 (d, 2H), 4.22 (s, 2H), 3.82 (s, 4H). Chemical purity by HPLC: 98.5%.

Preparation of 4-(4-nitophenyl)-3-Morpholinone (19)

A solution of 2-(2-chloroethoxy)-N-(4-nitrophenyl) acetamide (**18**, 50 g, 0.19 mol), TBAB (3.5 g) and sodium hydroxide (12 g, 0.3 mol) in dichloromethane (500 mL) was stirred at 25–35°C for 3 h. After completion of reaction organic layer was washed with water (150 mL), organic layer was distilled out to provide residue which was crystallized in isopropyl alcohol (85 mL) to obtain yellow solid of compound **19**. Yield: 37.8 g (90.0%). M.P.: 148–152°C. ^{1}H NMR (DMSO-d_6, 400 MHz): δ 8.27 (d, 2H), 7.77 (d, 2H), 4.27 (s, 2H), 4.0 (t, 2H, 3.86 (t, 2H). Chemical purity by HPLC: 99.5%.

Preparation of 4-(4-aminophenyl) Morpholin-3-one (5)

A solution of 4-(4-nitophenyl)-3-morpholinone (19, 100 g, 0.45 mol) with Pd–C (10% w/v, 50% wet, 2 g) and ammonium formate (170 g) in methanol (1,000 mL) was heated to 55–60°C for 90 min, upon completion of reaction catalyst was filtered, filtrate was cooled to 5–10°C to precipitate compound 5, which was filtered and recrystallized in water to get white solid of compound 5. Yield: 65.0 g (75.0%). M.P.: 169–172°C. ^{1}H NMR (DMSO-d_6, 400 MHz): δ 6.94 (d, 2H), 6.54 (d, 2H), 5.15 (d, 2H), 4.27 (s, 2H), 3.90 (t, 2H), 3.56 (t, 2H). Purity by HPLC: 99.9%.Chemical purity by HPLC: 99.9%.

Preparation of 2-[(2R)-2-hydroxy-3-{[4-(3-oxomorpholin-4-yl) phenyl]amino}propyl]-1H-isoindole-1,3(2H)-dione (7)

A suspension of 4-(4-aminophenyl)morpholin-3-one (5, 100 g, 0.52 mol) and 2-[(2S)-oxiran-2-ylmethyl]-1H-isoindole-1,3(2H)-dione (6, 116.2 g, 0.57 mol) in isopropyl alcohol (1,700 mL) and water (300 mL) was refluxed for 24 h. After completion of reaction (by HPLC), reaction mass was cooled to 25–30°C, precipitated solid 7 was filtered, washed with isopropyl alcohol (100 mL) and dried the solid under vacuum (650–700 mm/Hg) at 50–55°C for 6 h to afford 7 as light yellow to off white colored solid. Yield: 196.8 g (96.0%). MS m/z: 395.9 (M$^+$ + 1). ^{1}H NMR (DMSO-d_6, 400 MHz): δ 2.98–3.04 (m, 1H), 3.13–3.19 (m, 1H), 3.68–3.57 (m, 4H), 4.02–3.91 (m, 3H), 4.13 (s, 2H), 5.16 (d, J = 5.2 Hz, 1H), 5.65 (t, J = 6.4 Hz, 1H), 6.61 (d, J = 8.8 Hz, 2H), 7.01 (d, J = 8.8 Hz, 2H), 7.82–7.88 (m, 4H). Purity by HPLC: 97.50%. Chemical purity by HPLC: 97.5%.

Preparation of 2-({(5S)-2-oxo-3-[4-(3-oxomorpholin-4-yl) phenyl]-1,3-oxazolidin-5-yl}methyl)-1H-isoindole-1,3(2H)-dione (8)

N–N-carbonyldiimidazole (CDI) (61.5 g, 0.38 mol) was added to a suspension of 7 (100 g, 0.25 mol) and potassium carbonate (34.5 g, 0.25 mol) in dichloromethane (1,000 mL) at 25–30°C. Reaction mixture was maintained at 25–30°C for 5–6 h until completion of the reaction (by HPLC). Inorganic solid was filtered and washed with dichloromethane (200 mL). Filtrate was collected and concentrated to obtain the residue. The residue was further slurried in tetrahydrofuran (500 mL) at 50–55°C and cooled to 25–30°C, filtered the solid, washed the solid with tetrahydrofuran (50 mL) to furnish 8 as light yellow to off white solid. Yield: 101.1 g (95.0%). MS m/z: 422.0 (M$^+$ + 1). ^{1}H NMR

(CDCl$_3$, 400 MHz): δ 3.75–3.72 (t, J = 5.2, 4.8 Hz, 2H), 4.0 (dt, 2H), 4.04 (t, 2H), 4.17 (dt, 2H), 4.33 (s, 2H), 5.02 (m, 1H), 7.35–7.32 (d, J = 12 Hz, 2H), 7.57–7.55 (d, J = 9.2 Hz, 2H), 7.77–7.74 (dd, 2H), 7.89–7.78 (dd, 2H). Purity by HPLC: 99.80%. Chemical purity by HPLC: 99.80%.

Preparation of 4-{4-[(5S)-5-(aminomethyl)-2-oxo-1,3-oxazolidin-3-yl]phenyl}morpholin-3-one hydrochloride (9)

40% Methylamine solution (102 mL) was added to the solution of **8** (100 g, 0.23 mol) in methanol (1,000 mL) at 25–30°C. Reaction mass was stirred at 60–65°C for 4–6 h (completion of reaction monitored by HPLC). Reaction mass was cooled to 25–30°C, pH of reaction mass was adjusted to 1–2 using concentrated hydrochloric acid, precipitated solid was filtered, and washed with methanol (100 mL) to obtain crude **9**. The obtained crude **9** was dissolved in mixture of methanol (800 mL) and dichloromethane (300 mL) by adjusting the pH 8–9 of reaction using triethylamine at 25–30°C to achieve clear solution. Reaction mass was acidified to pH 2–3 using concentrated hydrochloric acid to precipitate **9**. Precipitated solid was filtered, washed with methanol (150 mL) and dried to furnish pure **9** as white solid. Yield: 65.5 g (85.0%). MS m/z: 292.2 (M$^+$ + 1). ^{1}H NMR (D$_2$O, 400 MHz): δ 3.49 (d, 2H), 3.81–3.78 (t, J = 5.2, 4.8 Hz, 2H), 3.97 (q, 1H), 4.11–4.08 (t, J = 4.8, 5.2 Hz, 2H), 4.40–4.36 (t, J = 9.2, 6.8 Hz, 3H), 5.13 (m, 1H), 7.42–7.40 (d, J = 8.8 Hz, 2H), 7.61–7.58 (d, J = 8.8 Hz, 2H). Purity by HPLC: 99.80%. Chemical purity by HPLC: 99.80%.

Preparation of 5-chloro-N-({(5S)-2-oxo-3-[4-(3-oxomorpholin-4-yl)phenyl]-1,3-oxazolidin-5-yl}methyl)thiophene-2-carboxamide (1) as per route-A

4-Nitrophenyl 5-chlorothiophene-2-carboxylate (**11**, 95 g, 0.33 mol) was added to a solution of **9** (100 g, 0.30 mol) and triethylamine (46.2 g, 0.45 mol) in DMSO (600 mL) at 35–40°C. After completion of reaction (by HPLC) acetonitrile (200 mL) was added to the reaction mixture at 35–40°C. Reaction mixture was further heated up to 65°C, methanol (600 mL) was added to the reaction mixture at 60–65°C. Reaction mixture was cooled to 25–30°C, the obtained solid was filtered, washed with methanol (100 mL), and dried under vacuum (700 mm/Hg) for 4–5 h at 50–55°C to get pure rivaroxaban (**1**) as white to off white solid. Yield: 114.3 g (86.0%). MS m/z: 436.0 (M$^+$ + 1). ^{1}H NMR (DMSO-d_6, 400 MHz): δ 3.61–3.59 (t, J = 5.6 Hz, 2H), 3.72–3.69 (t, J = 5.2, 4.8 Hz, 2H), 3.86–3.82 (q, 1H), 3.97–3.95 (t, J = 4.8, 5.6 Hz, 2H), 4.21 (s, 2H), 4.16 (t, 1H), 4.87 (m, 1H), 7.2 (d, 1H), 7.42–7.38 (d, 2H), 7.57–

7.53 (d, 2H), 7.69 (d, 1H), 8.99–8.96 (t, $J = 4$, 6 Hz, 1H). ^{3}C NMR (DMSO-d_6, 400 MHz): δ 166.01, 160.85, 154.13, 138.48, 137.05, 136.51, 133.35, 128.46, 128.15, 125.95, 118.31, 71.38, 67.75, 63.50, 49.03, 47.44, 42.24. Elemental analysis: C, 52.5; H, 4.14; N, 9.65; S, 7.33. Purity by HPLC: 99.80%. Chiral purity by HPLC: 99.95%. Chemical purity by HPLC: 99.80%. Chiral purity by HPLC: 99.95%.

Residual Solvent by GC HS (ppm)

Methanol: 407, isopropyl alcohol: not detected, dichloromethane: not detected, tetrahydrofuran: not detected, acetonitrile: 120, DMSO: 510, triethylamine: 20.

AUTHOR'S CONTRIBUTIONS

All authors read and approved the final manuscript.

ACKNOWLEDGEMENTS

Authors thank the management of Megafine Pharma (P) Ltd. for the permission to publish this work. Authors also thank colleagues of the Analytical Research and Development team for their valuable contribution to this work.

REFERENCES

1.　Drug index. Xarelto (Rivaroxaban film coated oral tablet); Drug information, desription, user reviews, drug side effects, interactions—prescribing information at Rx List. http://www.rxlist.com/xarelto-drug.htm. Accessed June 2015

2.　Highlights of prescribing information of XARELTO, rivaroxaban, Extended-Release tablets. http://www.accessdata.fda.gov/drugsatfda_docs/label/2011/202439s001lbl.pdf. Accessed 2011

3.　Susanne R, Alexander S, Jens P, Thomas L, Josef P, Karl-Heinz S et al (2005) Discovery of the novel antithrombotic agent 5-chloro-N-({(5S)-2-oxo-3-[4-(3-oxomorpholin-4-yl)phenyl]-1,3-oxazolidin-5-yl}methyl) thiophene-2-carboxamide (BAY 59-7939): an oral, direct factor Xa inhibitor. J Med Chem 48:5900–5908

4.　Alexander S, Thomas L, Jens P, Susanne R, Elisabeth P et al (2009) Substituted Oxazolidinones and Their Use in the Field of Blood Coagulation. U.S. Patent 7,576,111

5. Rafecas JL, Comely AC, Ferrali A, Amelacortes C, Pasto AM (2011) Process for the preparation of rivaroxaban and intermediates thereof. PCT Int Appl WO2011080341 A1

6. Silvo Z, Anica P (2011) Processes for crystallization of rivaroxaban. PCT Int Appl WO2011012321 A1

7. Sturm H, Desouza D, Knepper K, Alert M (2011) Method for the preparation of rivaroxaban. PCT Int Appl WO2011098501 A1

8. Bodhauri P, Weeratunga G (2012) Processes for the preparation of rivaroxaban and intermediates thereof. PCT Int Appl WO2012051692 A1

9. Eva S, Gyorgyi KL, Balazs H, Balazs V, Gyorgy K, Gyorgy R et al (2012) Process for the preparation of a rivaroxaban and intermediates formed in said process. PCT Int Appl WO2012153155 A1

10. Rodriguez B, Olondriz M, Soldevilla L, Comas S (2012) Process for obtaining rivaroxaban and intermediate thereof. PCT Int Appl WO2012159992 A1

11. Singh PK, Hashmi MS, Sachdeva YP, Khanduri CH (2013) Process for the preparation of rivaroxaban and intermediates thereof. PCT Int Appl WO2013156936 A1

12. Halama A, Krulis R, Brichac J (2013) A process for the preparation of rivaroxaban based on saving of 1, 1'-carbonyl diimidazole. PCT Int Appl WO2013120464 A1

13. Dwivedi SD, Prasad A, Pal DR, Sharma MHP, Jain KN, Patel NB (2013) Processes and intermediates for preparing rivaroxaban. PCT Int Appl WO2013098833 A2

14. Rao MD, Reddy VB (2013) Improved process for preparing rivaroxaban using novel intermediates. PCT Int Appl WO2013164833 A1

15. Rao MD, Reddy VB (2013) A process for the preparation of 5-chloro-n-({(5s)-2-oxo-3-[4-(3-oxo-4-morpholinyl) phenyl]-1,3-oxazolidin-5-yl} methyl)-2-thiophene carboxamide. PCT Int Appl WO2013118130 A1

16. Keshav SP, Bhise SS, Kamble HH, Chaudhari G, Patil DS (2014) Process for the preparation of rivaroxaban. PCT Int Appl WO2014155259 A2

17. Pradhan NS, Patil NS, Walavalkar RR, Kulkarni NS, Pawar SB, Pawar TS (2014) Rivaroxaban intermediate and preparation thereof. PCT Int Appl WO2014102820 A2

18. Silvo Z, Primoz B (2014) A process for preparation of rivaroxaban. PCT Int Appl WO2014096214 A1

19. Pradhan NS, Patil NS, Walavalkar RR, Kulkarni NS, Pawar SB, Pawar TS (2014) Aldehyde derivative of substitute oxazolidinones. PCT Int Appl WO2014102822 A2

20. ICH Harmonized Tripartite Guideline. Impurities: guideline for residual solvents, Q3C (R5). http://www.ich.org/products/guidelines/quality/quality-single/article/impurities-guidline-for-residual-solvents.html. Accessed 3 Jan 2013

21. Grunenburg A, Lenz J, Braun GA, Keil B, Thomas CR (2010) Novel polymorphous form and the amorphous form of 5-chloro-n-({(5s)-2-oxo-3[4-(3-oxo-4-morpholinyl(-phenyl]-1,3-oxazolidine-5-yl}-methyl)-2-thiophene carboxamide. US Patent 20100152189 A1

22. Mathad VT (2013) Crystallization: theory and/vs. practice—few case studies', a presentation made in Crystallization-2013, held on 16–17 April 2013 by UBM/CPHI-India at Mumbai, India

23. Mathad VT, Joshi DR, Lad HD, Patel PI (2012) A process for production of 4-(4-aminophenyl)-3-morpholinone. IN patent 3478/MUM/2012

24. ACS GCI Pharmaceutical Roundtable. Green Chemistry Reporting Requirements for an Electronic Laboratory Notebook (ELN); http://portal.acs.org/portal/PublicWebsite/greenchemistry/industriainnovation/roundtableCNBP_026687. Accessed 2011

Chapter 8

ENHANCEMENT OF STABILITY OF A LIPASE BY SUBJECTING TO THREE PHASE PARTITIONING (TPP): STRUCTURES OF NATIVE AND TPP-TREATED LIPASE FROM THERMOMYCES LANUGINOSA

Mukesh Kumar[1] , Joyeeta Mukherjee[2] , Mau Sinha[1] , Punit Kaur[1] , Sujata Sharma[1] , Munishwar Nath Gupta[3] and Tej Pal Singh[1]

[1]Department of Biophysics, All India Institute of Medical Sciences, New Delhi, India

[2] Chemistry Department, Indian Institute of Technology Delhi, New Delhi, India

[3] Department of Biochemical Engineering and Biotechnology, Indian Institute of Technology Delhi, New Delhi, India.

ABSTRACT

Background

The lipase enzyme converts long chain acyltriglycerides into di- and monoglycerides, glycerol and fatty acids. The catalytic site in lipase is situated deep inside the molecule. It is connected through a tunnel to the surface of the molecule. In the unbound state under aqueous conditions, the tunnel remains closed. The tunnel can be opened when the enzyme is exposed to a lipid bilayer or a detergent or many hydrophobic/hydrophilic surfaces.

Results

In the present study, the lipase was subjected to three-phase partitioning (TPP) which consisted of mixing in *tert*-butanol and ammonium sulphate to the solution of lipase in the aqueous buffer. The enzyme formed an interfacial precipitate between the *tert*-butanol rich and water rich phases. The stability of the enzyme subjected to TPP was found to be higher (T_m of 80 °C) than the untreated enzyme (T_m of 77 °C). The activity of the enzyme subjected to TPP (3.3 U/mg) was nearly half of that of the untreated one (5.8 U/mg). However, the activity of the treated enzyme was higher (17.8 U/mg) than the untreated one (8.6 U/mg) when a detergent was incorporated in the assay buffer.

Conclusions

The structure determination showed that the substrate binding site in the treated enzyme was more tightly closed than that of the untreated protein.

BACKGROUND

Enzymes offer a more sustainable option for catalyzing chemical processes [1]. Hence, enhancing their stability and activity is a worthwhile goal. The present work describes efforts to understand how a simple process called three phase partitioning (TPP) resulted in altering both stability and activity of a lipase from *Thermomyces lanuginosus* (TLL). Lipases (triacylglycerol hydrolases; EC 3.1.1.3) have been extensively used in both aqueous media as well as non aqueous media [1–8]. One of the main reasons of their large-scale usage has been their broad specificity [8, 9]. Lately, this range has been further enlarged by the reports on the lipase catalysed promiscuous reactions [10–13]. The catalytic promiscuity refers to enzymes belonging to a particular class in the enzyme classification (EC) system catalyzing reactions of the type which are generally catalyzed by another class of enzymes. Lipases, classified as hydrolases, have been shown to catalyze many C–C bond formation reactions [10, 11, 14, 15]. It is believed that in such cases, substrates interact with the active site in a manner different from that observed by natural substrates [10].

The lipase from *Thermomyces lanuginosus* (TLL) is a glycosylated hydrolase which consists of 269 amino acid residues with a molecular weight of about 29 kDa. It has an optimum pH of 11–12 [16]. The structure of TLL was reported earlier in the native state which showed that the substrate binding cleft was in the closed form [17]. The structure determined in the presence of a detergent showed open conformation [18]. TLL is among the most frequently used lipases in biotechnology [19] and its applications include its use in catalysing C–C bond formation because of its promiscuous activity [20].

It was recently observed that TLL subjected to three phase partitioning under optimized conditions showed about 2-fold increase in the rates of hydrolytic activity of the lipase. It was also observed that the same lipase preparation showed 5-fold increase in the initial rates of a promiscuous reaction; the aldol reaction between 4-nitrobenzaldehyde and acetone carried out with 60 % acetone as the reaction medium (Scheme 1) [M. N. Gupta, unpublished results].

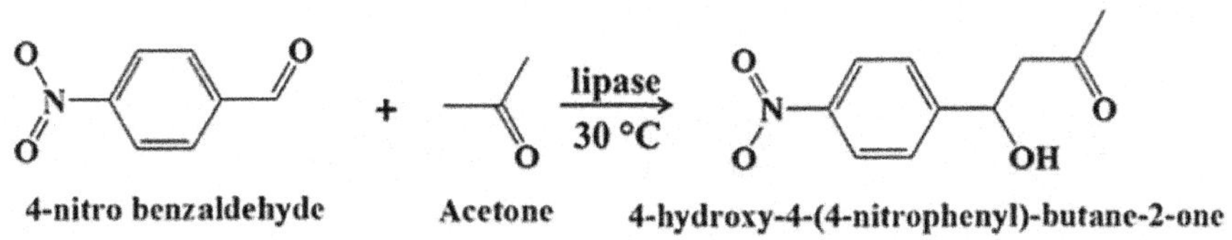

4-nitro benzaldehyde   **Acetone**   **4-hydroxy-4-(4-nitrophenyl)-butane-2-one**

Scheme 1: Aldol condensation reaction catalyzed by *Thermomyces lanuginosus* lipase.

TPP consists of precipitating a protein at the interface of two layers formed by the addition of *tert*-butanol and ammonium sulphate to a protein solution [21–26] (Fig. 1). It is essentially the precipitation of the protein by simultaneous addition of ammonium sulphate and *tert*-butanol at appropriate concentrations. However, unlike ammonium sulphate precipitation, this TPP often (though not always) results in subtle structural changes. This makes TPP valuable for wide range of applications such as protein purification, protein refolding and improving the catalytic efficiencies of the enzymes in both aqueous and non aqueous media [21–31]. The technique was introduced by Lovrein's group [21] who emphasized its importance in protein isolation and concentration [22]. The complexity of the mechanism by which it works was discussed adequately by Lovrein et al. [27], and more recently in the context of protein refolding [26]. There have been quite a few reports on the effect of TPP treatment on the structure and function of many enzymes and proteins [21–31]. In few cases, it has been clearly shown that the specific activity of the pure enzyme increased after being subjected to TPP [24, 25, 29]. Many years back, we had reported an X-ray crystallographic study comparing the structure of the untreated and TPP treated proteinase K [28]. The structural work indicated that the higher activity of Proteinase K, in both aqueous buffers and low water media primarily arose from greater accessibility of the substrate to the active site and overall higher molecular flexibility of Proteinase K as a result of the treatment. The effect of TPP treatment on alpha chymotrypsin was even more dramatic. The enzyme subjected to TPP showed significantly high activity in both aqueous and non-aqueous media [23]. When attempts were made to crystallize the alpha chymotrypsin subjected to TPP, selective autolysis generated a 14 amino acid peptide which was found to bind to the active site of the enzyme [32]. The enhanced protein flexibility has also been identified as one of the important causes of catalytic promiscuity [10]. One of our groups have recently reported that lyophilized preparation of subtilisin treated with 6 M urea (a highly flexible structure) showed quite high catalytic promiscuity with respect to the aldol condensation (Scheme 1) [33]. Hence, it was naturally of interest to investigate whether in the case of TLL subjected to TPP as well, the high initial rates of catalytic promiscuity arose from similar causes, that is, enhanced flexibility of the protein molecule especially around the active site.

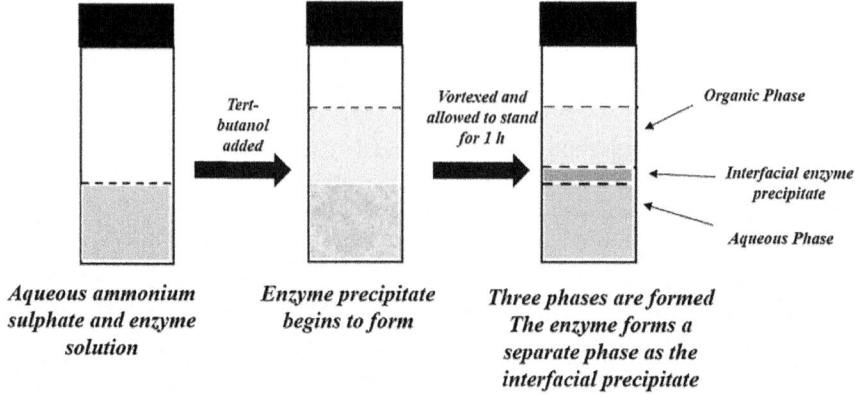

Figure. 1: Schematic diagram of the three phase partitioning of TLL.

In the present work, we report the protein structures based upon X-ray crystallographic studies of TLL including (1) native untreated form (TLL), (2) TLL after being subjected to three-phase partitioning (TPP-TLL). These structures are discussed alongwith the results from the biochemical studies. It was found that TPP of TLL, contrary to what was expected, resulted in the introduction of rigidity around the active site.

RESULTS AND DISCUSSION

Activity of TLL and TPP-TLL

TPP treatment has, in many cases, resulted in improvement of the catalytic rates of the various enzymes by the marginal to moderate degrees [23, 24, 28–31]. As indicated in the introduction, TLL subjected to TPP under optimized conditions showed an increase in the initial rates of hydrolytic activity from 8.6 to 17.8 U/mg protein (Table 1). The assay for measuring this activity is carried out in the presence of detergents [34]. Like many lipases the substrate binding site in the native TLL is shown to have a closed conformation such that the active site residues of the enzyme are inaccessible for the interaction with substrates [18]. The detergents are reported to cause an opening of the substrate binding site thereby rendering the active site accessible to the substrates [35]. In the present work both native TLL and TLL subjected to TPP were tested for enzymatic activity in the absence of detergents as well. The TPP-TLL showed that the specific activity of lipase decreased from 5.8 U/mg protein to 3.3 U/mg protein when the activities were measured in the absence of detergents

(Table 2). This is an unusual result as all the consequences of TPP treatment of enzymes reported so far in the literature either show no change or an increase in biocatalytic activity [28–31, 36]. Measurements of melting temperature (T_m) by recording the CD spectra of both untreated TLL and TLL subjected to TPP over a range of temperatures (Fig. 2) showed that the TPP-TLL had a T_m of 80 °C whereas the untreated enzyme had a T_m of 77 °C. The T_m values reflect how easy or difficult it is for a protein to unfold. The increase in the value of T_m indicated that the enzyme has become more rigid after the TPP-treatment.

Table 1: Three phase partitioning of *Thermomyces lanuginosus* lipase (TLL) at 25 °C

	Protein (mg/mL)	Activity (U/mL)	Specific activity (U/mg protein)	Fold Increase
Assay in the absence of Triton X-100				
TLL	17.5	102	5.8	–
TPP-TLL	0.22	0.72	3.3	0.57
Assay in the presence of Triton X-100				
TLL	17.5	151	8.6	–
TPP-TLL	0.22	3.91	17.8	2.1

The *Thermomyces lanuginosus* lipase was subjected to three phase partitioning as described in the materials and methods section. Experiments were carried out in triplicate and percentage error in each reading in a set was within 3 %. The activity was determined using pNPP assay and protein using bradford method

Table 2: The distances (Å) between atoms of residues from opposite sides of the substrate binding cleft indicating the very closed, closed and open states of substrate binding channel

Side 1	Side 2	TPP-TLL (PDB id: 4FLF)	TLL (PDB id:4ZGB)	Detergent-treated (PDB id: 1GT6)
86 ILE CD1	ILE259 CD2	3.99	5.48	19.35
87 GLU CG	LEU255 CD1	4.39	5.99	20.17
89TRP CZ3	VAL203 CG2	4.12	4.62	7.78
90ILECD1	VAL203 CG2	4.14	4.47	18.12
93LEUCD2	PRO207CG	4.21	5.12	9.03
95 PHE O	TYR213-OH	3.20	3.40	2.57
95 PHECZ	PRO208 CD	3.81	5.34	8.21

95 PHECE2	PHE211CE1	4.33	5.11	6.90
95PHE CD2	PHE211CZ	3.86	4.20	6.18
95PHE CG	PHE211CG	4.37	4.65	6.07
95PHE CD2	PHE211CD2	3.67	4.43	5.32
95PHE CD2	PHE211CD1	4.06	5.15	7.60
95PHE CD2	PHE211CE1	4.06	4.87	7.39
95PHE CD2	PRO208CD	4.04	5.56	7.35
95PHE CZ	PRO208 CD	3.81	5.34	7.45
95PHE CD1	PRO207CG	3.69	3.74	5.84
95PHE CE2	PRO208CG	4.70	6.30	8.39
97 LEUCG	TYR213 CZ	3.70	4.22	9.23

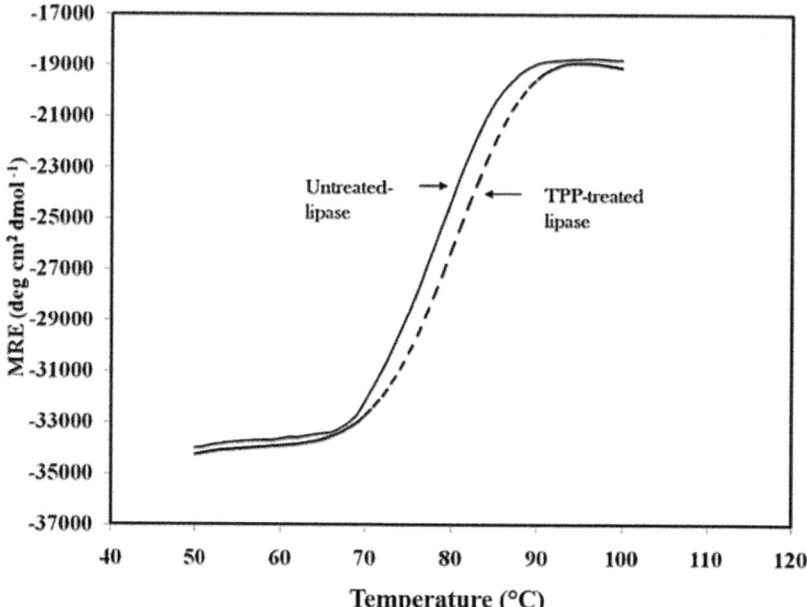

Figure. 2: Determination of the melting temperature of TPP-TLL using circular dichroism. Thermal denaturation curves were determined directly by monitoring the ellipticity changes at 222 nm. The samples with a concentration of 0.2 mg mL^{-1} were used. The temperature of sample solution was raised linearly by 1 °C min^{-1} from 50 to 100 °C. The heating curves were corrected for an instrumental baseline obtained by heating the buffer (10 mM sodium phosphate, pH 7.0) alone. The untreated lipase is shown with *solid line* and the TLL subjected to TPP is shown with *dashed line*.

However, when assayed in the presence of detergent triton X-100, the TPP-TLL showed an increase in the specific activity from 3.3 U/mg protein to 17.8 U/mg protein (Table 1). On the other hand, the native TLL showed an increase in the specific activity from 5.8 U/mg proteins to 8.6 U/mg proteins. This shows that the presence of detergent has a more marked effect on the specific activity of the TPP-TLL. In this case, the opening of the lid not only resulted in the more active "open" conformation (which is a well known phenomenon) but also presumably reversed the introduction of more rigidity in the active site by TPP treatment.

The lipase-detergent interactions have been exploited in multiple ways [8]. The most well known among these is via "interfacial activation". This leads to the movement of the molecular lid (present in several lipases including TLL) away from the active site access and makes the active site more accessible. This "lid opening" has also been achieved by immobilization [37, 38] and bioimprinting [39, 40].

Structure of the Untreated Native Lipase

The structure of the native TLL determined at 2.30 Å showed that protein chain adopted an α/β fold (Fig. 3). It consisted of 10 α-helices, $\alpha1$ (residues, Ser3–Ala20), $\alpha2$ (residues, Cys41–Lys46), $\alpha3$ (residues, Ile86–Gly91), $\alpha4$ (residues, Asp111–His135), $\alpha5$ (residues, Ser146–Arg160), $\alpha6$ (residues, GLy178–Gln188), $\alpha7$ (residues, Ile202–Leu206), $\alpha8$ (residues, Pro208–Gly212), $\alpha9$ (residues, Thr231–Asp234) and $\alpha10$ (Ile255–Leu259), and 10 β-strands, $\beta1$ (residues, Asp48–Ser58), $\beta2$ (residues, Val63–Asn71), $\beta3$ (residues, Lys74–Arg81), $\beta4$ (Asp96–Ile100), $\beta5$ (Cys107–His110), $\beta6$ (residues, Arg139–His145), $\beta7$ (residues, Asp165–Gly172), $\beta8$ (residues, Gly192–Thr199), $\beta9$ (residues, Glu219–Lys223) and $\beta10$ (residues, Asp234–Ile238). The active site residues in this lipase are Ser146, His258 and Asp201 and catalytic mechanism works in a manner similar to that of serine proteases [28]. The catalytic triad is located inside a deep cleft which is formed with the help of two segments, consisting of α-helix, $\alpha4$ and β-strands, $\beta4$ and $\beta5$ (Wall-1) and α-helices $\alpha6$, $\alpha7$ and $\alpha8$ (Wall-2). Several van der Waals interactions are observed between the two residues of two walls (Table 2). This represents the closed state of lipase whereby the active site residues are not freely accessible to interact with the substrates. In the closed state, the two sides are held together firmly with a number of attractive interactions between them (Fig. 4a). As seen from Fig. 4a, the inter-wall interactions include a large number of van der Waals contacts as a series of hydrophobic residues are lined up from both sides. Such a closed state corresponds to the inactive state of the enzyme. In order to open up the substrate binding cleft, the enzyme must be exposed to highly hydrophobic agents such as lipid aggregates or detergents. The substrate binding site has been shown to adopt an open conformation in the presence of detergents [41, 42]. Under the influence of aggregated lipid bilayer or detergents, Wall-2 moves away from Wall-1 so that the cleft adopts an open conformation as a result of which the diffusion of substrates can occur into the cleft leading to the required interactions with the active site residues. The opening of the cleft involves a large scale movement of Wall-2 which is often called as the lid. In the open state, the intra-Wall contacts are completely lost (Fig. 4b).

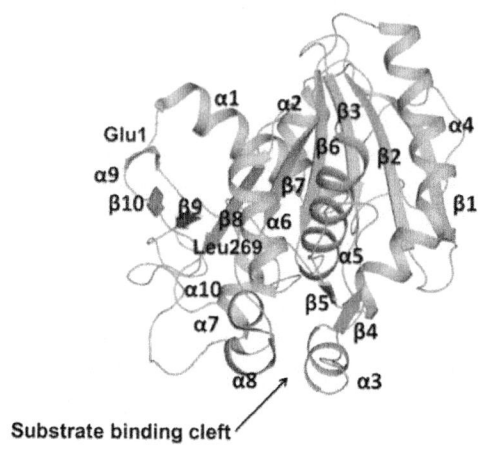

Figure. 3: Folding of TLL (PDB ID: 4ZGB) with secondary structure elements including α-helices and β-strands. The N-terminus (Glu1) and C-terminus (Leu269) as well as substrate binding cleft are also indicated [68].

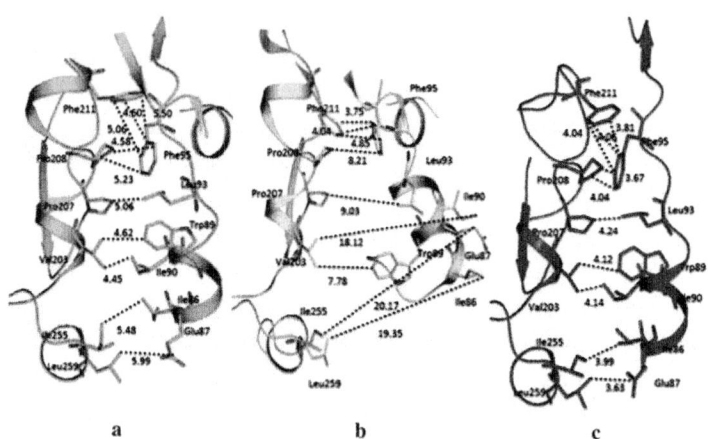

Figure. 4: The van der Waals contact distances between atoms of residues from two sides of the cleft in the **a** native untreated state (PDB ID: 4ZGB), **b** presence of a detergent (PDB ID: 1GT6) and **c** TPP-treated state (PDB ID:4FLF).

The closed and open states are two conditions for the inactive and active states of lipase (Fig. 5).

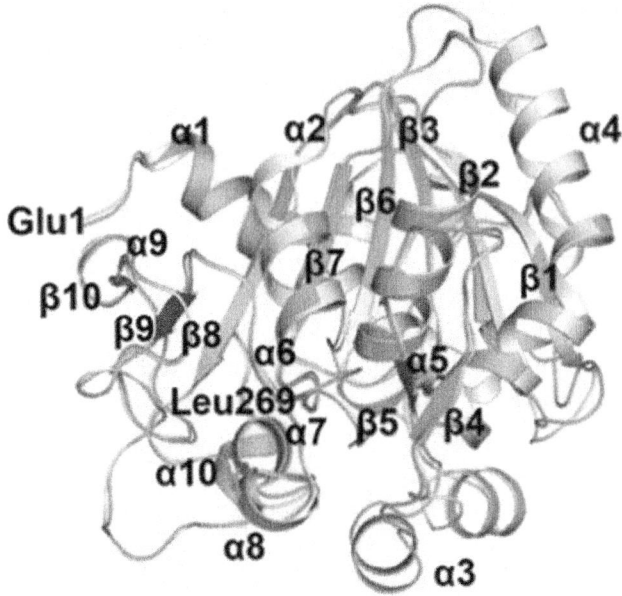

Figure. 5: The superimposition of TPP-treated lipase structure (PDB ID: 4FLF) (*cyan*) on that of lipase in the presence of detergent (PDB ID: 1GT6) (*grey*) are shown. The lid (*grey*) is open in the presence of detergent while it remains closed in the presence of organic solvent.

Structure of TLL Subjected to TPP

Crystal structure of TPP-TLL was determined at 2.15 Å resolution. The overall structure treated form is similar to that of untreated native protein with an r.m.s. shift of 0.7 Å for the Cα atoms. The exposure to the interface between the aqueous layer and the upper *tert*-butanol rich layer was expected to cause a conformational change in the cleft. However, the structure did not indicate the opening of the lid. This is easy to understand since it is well known that interfacial activation of the lipases requires interaction of the lipase in free solution with an interface [43,44]. In TPP treatment, the enzyme precipitates out of the solution and cannot interact with the *tert*-butanol rich layer (unlike the enzyme in free solution). In fact, rather than opening the lid, the TPP treatment resulted in making the active site less accessible due to enhanced molecular rigidity in that region. As a result, the lid further moved closer to the segment on the opposite side thus shortening the lengths of van der Waals contacts (Table 2). The TPP treatment, involving complex interactions of the

$SO_4{}^{2-}$ anions (kosmotropy, conformation tightening and electrostatic forces) simultaneously with interactions with *tert*-butanol induces a conformational change. Such conformational changes upon TPP treatment have been reported [22, 27, 28, 30, 45]. In the present structure of the TPP-TLL, as seen from Fig. 4c, the two walls moved closer to each other thus optimizing the van der Waals contacts between the two sides of the substrate binding cleft. In this case, the distances of hydrophobic contacts decreased at least by 10 %. This suggests that after the TPP treatment, the protein structure becomes more tight and hence more stable than the native state as observed in untreated TLL.

Enhancing the thermal stability of enzymes continues to be a useful goal in biocatalysis [46–48]. Many techniques including chemical crosslinking [49–51], immobilization on solid or soluble supports [52–54], protein engineering and directed evolution [55–58] have been employed for this purpose. In the present case, however, it was an incidental result of these studies wherein the aim was to improve both the natural and promiscuous activities of the lipase by subjecting it to TPP as seen with other enzymes [22, 23, 27–30, 45]. This is the first example where the TPP treatment seems to have resulted in the increase in the T_m. In cases reported so far, TPP treatment has resulted in a decrease of thermal stability [30]. The increase in the T_m of TLL upon TPP-treatment correlates well with the reduction in the activity. In this regard, it is noteworthy that the substrate binding site became more tight because the distances of the van der Waals contacts between the two sides of the substrate binding channel became considerably shorter than those observed in the untreated protein. This would have made it more difficult for the cleft to open when the substrate approached it. However, in the presence of detergent, the TLL subjected to TPP showed a higher activity because of the more favourable stereochemistry of the active site residues in the treated enzyme. The TLL subjected to TPP was less active and more stable which is good for its shelf life. On the other hand, in the presence of detergent, the TLL subjected to TPP showed activity more than two times higher than the untreated enzyme. An interesting aspect is that the TLL subjected to TPP in the absence of a detergent showed a decrease in the hydrolytic activity which involves hydrolysis of an analogue of a natural substrate, that is, *p*-nitrophenylpalmitate. However, it shows higher activity during catalysis of the promiscuous reaction, that is, aldol condensation. Broos [59] has provided convincing data which shows that changes in flexibility of the enzyme structure may have opposite consequences in terms of selectivity for natural and unnatural substrates. The promiscuous substrates are extreme examples of "unnatural substrates".

It is noteworthy that the TLL subjected to TPP showed 5-fold increase in the initial rates for the aldol condensation (in the presence of 40 % water in acetone as the medium which was also one of the substrates). This reaction was carried out in the absence of detergents, so this involved 'closed lid' structure of the enzyme. Also, one should not overlook the fact that one of the substrates in the promiscuous reaction was acetone. The presence of acetone in the medium was expected to influence the structure of the enzyme. The information about the change or its extent could not unfortunately be obtained by X-ray studies. The efforts to obtain the crystallization of TLL in the presence of acetone did not succeed.

EXPERIMENTAL SECTION

Materials

The lipase from *Thermomyces lanuginosus* was a kind gift from Novozyme A/S (Bagsvaerd, Denmark). *p*-Nitrophenyl palmitate (pNPP) was procured from Sigma-Aldrich Company (St. Louis, USA). All other chemicals used were of analytical grade.

Methods

Three Phase Partitioning (TPP) of Lipases

The protocol reported earlier was followed for carrying out the TPP treatment of TLL [60]. The solutions of TLL (2 mL, 20 mg/mL in 10 mM sodium phosphate buffer, pH 7.0 were saturated with varying concentrations of ammonium sulphate (w/v). This step was followed by the addition of 2 mL *tert*-butanol. The solutions were vortexed and allowed to stand at 25 °C for 1 h. Three phases (i.e., upper layer of *tert*-butanol, interfacial precipitate of protein and lower aqueous layer) were formed. The solutions were then centrifuged at 2000×g at 25 °C for 10 min. The lower aqueous and upper organic layer were separated using a pasteur pipette. The interfacial precipitates obtained were dissolved in 1 mL of 10 mM sodium phosphate buffer pH 7.0 and dialysed against the same buffer for 24 h with frequent changes of buffer. This was checked for lipase activity using 4-nitrophenyl palmitate (pNPP) as a substrate [34] and for amounts of protein using Bradford method [61]. The percentage activity and protein in the precipitates were calculated by taking the starting amount of activity and protein as 100 %. This preparation is referred to as TPP-TLL.

Assay for Lipase Activity

The hydrolytic activity of lipase was monitored by measuring the rate of hydrolysis of 4-nitrophenyl palmitate (pNPP) spectrophotometrically at

410 nm by following the procedure described earlier [34]. The reaction mixtures consisted of 1.8 mL of buffer (10 mM sodium phosphate, pH 7.0 containing 150 mM NaCl [and with or without 0.5 % (v/v) triton X-100]), 0.2 mL of enzyme solution in 10 mM sodium phosphate buffer, pH 7.0 and 20 µl of 50 mM substrate pNPP in acetonitrile. The mixtures were incubated for 30 min at 37 °C and after this the samples were kept in a domestic microwave oven along with a beaker containing a volume of water sufficient to make the total volume of the liquid in the cavity as 100 mL (the additional water absorbs a significant amount of microwave energy, this was done to avoid overheating of the sample) and irradiated for 30 s and read at 410 nm. One unit (U) of enzyme activity is defined as the amount of the enzyme that liberates 1 µmol of 4-nitrophenol per min at pH 7.0 and 37 °C.

Protein Estimation

Protein concentrations were determined according to the procedure described by Bradford [61]. Protein solutions (0.5 mL) were incubated with 4.5 mL of the dye reagent at 25 °C for 10 min and the absorbance of the solutions were read at 595 nm and bovine serum albumin was used as the standard protein.

Measurement of the Melting Temperature (T_m)

The melting temperature of TLL and TPP-TLL were determined by using circular dichroism studies. Thermal denaturation curves were determined directly by monitoring the ellipticity changes at 222 nm. The samples with a concentration of 0.2 mg mL^{-1} were used. The temperature of sample solution was raised linearly by 1 °C min^{-1} from 50 to 100 °C. The heating curves were corrected for an instrumental baseline obtained by heating the buffer (10 mM sodium phosphate, pH 7.0) alone. The melting temperature (T_m) was calculated from the first-order derivatives of the ellipticity-temperature plot.

Crystallization

The samples of both (1) non-treated lipase (TLL) and (2) the lipase subjected to three phase partitioning (TPP-TLL) were dissolved in solution containing 0.1 M HEPES buffer, 1 M NaCl pH 7.5 at a concentration of 30 mg/mL to make solutions (1) and (2), respectively. The drops of 10 µl of solutions (1) and (2) were prepared for the hanging drop vapour diffusion method. The protein drops were equilibrated against 1.6 M ammonium sulphate. The rectangular shaped crystals measuring up to dimensions of 0.4 × 0.3 × 0.3 mm^3 were obtained after 3 weeks from the drops of the solutions. The crystals from the two samples were washed with reservoir buffer before placing them into a fresh solution containing reservoir buffer and 25 % glycerol as a cryoprotectant for data collection at low temperatures.

Data Collection and Processing

Two X-ray intensity data sets were collected on the crystals of (1) TLL and (2) TPP-TLL with MAR CCD-225 Scanner (Mar research, Norderstedt, Germany) using the beamline, BM14 at the European Synchrotron Radiation Facility (ESRF), Grenoble, France. The reflections of both the data sets were indexed and scaled using the program, HKL2000 [62]. The summary of data collection statistics is given in Table 3.

Table 3: Crystallographic data collection and refinement statistics

Parameters	TLL	TPP-TLL
Resolution range (Å)	50.00–2.30	50.00–2.15
Space group	P6$_1$	P6$_1$
Unit cell parameters		
a = b (Å)	140.1	139.9
c (Å)	80.5	80.0
Number of molecules in the asymmetric unit	2	2
Total number of measured reflections	169,548	671,375
Number of unique reflections	37,653	46,024
Completeness (%)	98.8 (90.0)	99.2 (95.4)
I/σ (I)	26.8 (4.0)	33.0 (2.1)
Rsym (%)	6.9 (23.8)	8.6 (41.3)
Refinement		
R$_{cryst}$ (%)	18.2	22.7
R$_{free}$ (%)	20.6	26.8
R.m.s.d in bond lengths (Å)	0.02	0.02
R.m.s.d in bond angles (°)	2.2	2.1
R.m.s.d in torsion angles (°)	17.4	18.1
Ramachandran plot analysis		
Most favoured (%)	88.2	88.3
Additionally allowed (%)	10.1	9.5
Generously allowed region (%)	1.3	1.3
Disallowed region (%)	0.4	0.9

B factors (Å^2)		
Wilson B factor	37.6	37.8
Average B-factor for main chain atoms	41.6	36.7
Average B-factor for the side chain atoms and water oxygen atoms	42.1	38.7
Mean B-factor for all atoms	41.8	37.8
Number of protein atoms	4142	4142
Number of water oxygen atoms	245	288
Number of carbohydrate (NAG) residues	0	2
PDB ID	4ZGB	4FLF

The values in parentheses correspond to the values in the highest resolution shell

Structure Determinations and Refinements

The structure of TLL was determined with molecular replacement method [63] using the coordinates of lipase in its unbound state (PDB code: 1DT3) [18]. This produced a clear solution with two peaks. Initially, the structure was refined for 20 cycles using REFMAC 5.5 [64]. The coordinates of partially refined structure of TLL were used as the starting model for refining the TPP-TLL structure. The initial calculations for 25 cycles were carried out as the rigid body refinement. The electron density maps with $(2F_o-F_c)$ and (F_o-F_c) coefficients were calculated. The models were improved by manual model building with programs O [65] and Coot [66] using graphics workstations. After another 25 cycles of refinements, when the values of R_{cryst}/R_{free} factors dropped to less than 0.27/0.30, Fourier maps with $(2F_o-F_c)$ and (F_o-F_c) coefficients were calculated for both the structures. These maps were used for determining the positions of water oxygen atoms in both structures. The coordinates of 245 water oxygen atoms were obtained in the TLL structure while the coordinates of 288 water oxygen were determined in the TPP-TLL structure. These coordinates were included in the final cycles of refinements. The refinements finally converged to values of 0.221/0.270 and 0.227/0.268 for the R_{cryst}/R_{free} factors of the untreated and treated structures. The final refinement statistics are included in Table 3. The refined atomic coordinates of structures of TLL and TPP-TLL have been deposited in the protein data bank with accession codes of 4ZGB and 4FLF respectively.

CONCLUSIONS

The present studies did reveal some interesting information:

1. In the case of TLL, TPP treatment introduces a rigidity rather than flexibility around the active site region.

2. Upon opening up the lid by interfacial activation, TLL subjected to TPP assumes a structure which was more along the expected lines. As in the case of proteinase K [28], it might be more flexible in the state with an open lid than the native structure. This might have led to the significant increase in the enzyme activity.

Three phase partitioning has slowly emerged as a simple approach to subtly alter conformational flexibility. Even subtle changes in conformational flexibility are important for both natural and promiscuous reactions of enzymes and influence even their enantioselectivity [67].

The present work shows that TLL subjected to TPP behaves differently while it exists as a "closed lid" structure and as an "open lid" structure. In the "closed lid structure", TLL subjected to TPP has enhanced rigidity while the opening of the lid by well known interfacial activation removes this rigidity. The results are also useful in the context of understanding the role which conformational flexibility plays in catalytic promiscuity.

AUTHORS' CONTRIBUTIONS

The group at AIIMS (MK, MS, PK, SS and TPS) were involved in the crystallization of the enzyme and its complexes, interpretation of the X-ray data and discussion with the IITD group. JM carried out the experimental work consisting the activity assay, T_m measurements etc. MNG suggested probing TLL activity with or without TPP treatment and correlating the results of the X-ray studies with the biochemical implications. All authors read and approved the final manuscript.

ACKNOWLEDGEMENTS

This work was supported by Grants from the Department of Science and Technology (DST-SERB), Govt. of India (Grant No. SR/SO/BB-68/2010) to the IITD. The AIIMS group also thanks Department of Science and Technology (DST), New Delhi and Department of Biotechnology (DBT), New Delhi for financial support. TPS also thanks the Indian National Science Academy for the grant of position of INSA-Senior Scientist. JM thanks CSIR (Govt. of India) for the senior research fellowship.

REFERENCES

1. Fessner WD, Anthonsen T (2008) Modern biocatalysis: stereoselective and environmentally friendly reactions. Wiley, New York

2. Gupta MN (1992) Enzyme function in organic solvents. Eur J Biochem 203:25–32

3. Carrea G, Riva S (2000) Properties and synthetic applications of enzymes in organic solvents. Angew Chem Int Ed Engl 39:2226–2254

4. Gupta MN (ed) (2000) Methods in non-aqueous enzymology. Birkhäuser-Verlag, Basel

5. Halling PJ (2000) Biocatalysis in low-water media: understanding effects of reaction conditions. Curr Opin Chem Biol 4:74–80

6. Hudson EP, Eppler RK, Clark DS (2005) Biocatalysis in semi-aqueous and nearly anhydrous conditions. Curr Opin Biotechnol 16:637–643

7. Mattiasson B, Adlercreutz P (1991) Tailoring the microenvironment of enzymes in water-poor systems. Trends Biotechnol 9:394–398

8. Adlercreutz P (2013) Immobilisation and application of lipases in organic media. Chem Soc Rev 42:6406–6436

9. Kapoor M, Gupta MN (2012) Lipase promiscuity and its biochemical applications. Process Biochem 47:555–569

10. Khersonsky O, Tawfik DS (2010) Enzyme promiscuity: a mechanistic and evolutionary perspective. Annu Rev Biochem 79:471–505

11. Busto E, Gotor-Fernández V, Gotor V (2010) Hydrolases: catalytically promiscuous enzymes for non-conventional reactions in organic synthesis. Chem Soc Rev 39:4504–4523

12. Arora B, Pandey PS, Gupta MN (2014) Lipase catalyzed Cannizzaro-type reaction with substituted benzaldehydes in water. Tetrahedron Lett 55:3920–3922

13. Malhotra D, Mukherjee J, Gupta MN (2015) Sustainability of biocatalytic processes. In: Letcher T, Scott JL, Patterson DA (eds) Chemical processes for a sustainable future. Royal Society of Chemistry, Cambridge

14. Hult K, Berglund P (2007) Enzyme promiscuity: mechanism and applications. Trends Biotechnol 25:231–238

15. Arora B, Mukherjee J, Gupta MN (2014) Enzyme promiscuity: using the dark side of enzyme specificity in white biotechnology. Sustain Chem Process 2:25

16. Li N, Zong M-H, Ma D (2009) Unexpected reversal of the regioselectivity in *Thermomyces lanuginosus* lipase-catalyzed acylation of floxuridine. Biotechnol Lett 31:1241–1244

17. Derewenda U, Swenson L, Wei Y, Green R, Kobos PM, Joerger R, Haas MJ, Derewenda ZS (1994) Conformational lability of lipases observed in the absence of an oil-water interface: crystallographic studies of enzymes

from the fungi *Humicola lanuginosa* and *Rhizopus delemar*. J Lipid Res 35:524–534

18. Brzozowski AM, Savage H, Verma CS, Turkenburg JP, Lawson DM, Svendsen A, Patkar S (2000) Structural origins of the interfacial activation in *Thermomyces* (Humicola) lanuginosa lipase. Biochemistry 39:15071–15082

19. Fernandez-Lafuente R (2010) Lipase from *Thermomyces lanuginosus*: uses and prospects as an industrial biocatalyst. J Mol Catal B 62:197–212

20. Cai JF, Guan Z, He YH (2011) The lipase-catalyzed asymmetric C–C Michael addition. J Mol Catal B Enzym 68:240–244

21. Lovrein RE, Goldensoph C, Anderson P, Odegard B (1987) Three phase partitioning (TPP) via *t*-butanol: enzyme separation from crudes. In: Burgess R (ed) Protein purification, micro to macro. Marcel Dekker Inc., New York

22. Dennison C, Lovrien R (1997) Three phase partitioning: concentration and purification of proteins. Protein Expr Purif 11:149–161

23. Roy I, Gupta MN (2004) α-Chymotrypsin shows higher activity in water as well as organic solvents after three phase partitioning. Biocatal Biotransform 22:261–268

24. Narayan AV, Madhusudhan MC, Raghavarao KS (2008) Extraction and purification of Ipomoea peroxidase employing three-phase partitioning. Appl Biochem Biotechnol 151:263–272

25. Roy I, Sharma A, Gupta MN (2004) Obtaining higher transesterification rates with subtilisin Carlsberg in nonaqueous media. Bioorg Med Chem Lett 14:887–889

26. Raghava S, Barua B, Singh PK, Das M, Madan L, Bhattacharyya S, Bajaj K, Gopal B, Varadarajan R, Gupta MN (2008) Refolding and simultaneous purification by three-phase partitioning of recombinant proteins from inclusion bodies. Protein Sci 17:1987–1997

27. Lovrein RE, Conroy MJ, Richardson TI, Gregory RB (1995) Protein solvent interactions. Marcel Dekker Inc., New York

28. Singh RK, Gourinath S, Sharma S, Roy I, Gupta MN, Betzel C, Srinivasan A, Singh TP (2001) Enhancement of enzyme activity through three-phase partitioning: crystal structure of a modified serine proteinase at 1.5 A resolution. Protein Eng 14:307–313

29. Rather GM, Mukherjee J, Halling PJ, Gupta MN (2012) Activation of alpha chymotrypsin by three phase partitioning is accompanied by aggregation. PLoS One 7:e49241

30. Rather GM, Gupta MN (2013) Three phase partitioning leads to subtle structural changes in proteins. Int J Biol Macromol 60:134–140

31. Rather GM, Gupta MN (2013) Refolding of urea denatured ovalbumin with three phase partitioning generates many conformational variants. Int J Biol Macromol 60:301–308

32. Singh N, Jabeen T, Sharma S, Roy I, Gupta MN, Bilgrami S, Somvanshi RK, Dey S, Perbandt M, Betzel C, Srinivasan A, Singh TP (2005) Detection of native peptides as potent inhibitors of enzymes. Crystal structure of the complex formed between treated bovine alpha-chymotrypsin and an autocatalytically produced fragment, IIe-Val-Asn-Gly-Glu-Glu-Ala-Val-Pro-Gly-Ser-Trp-Pro-Trp, at 2.2 angstroms resolution. FEBS J 272:562–572

33. Mukherjee J, Mishra P, Gupta MN (2015) Urea treated subtilisin as a biocatalyst for transformations in organic solvents. Tetrahedron Lett 56:1976–1981

34. Jain P, Jain S, Gupta MN (2005) A microwave-assisted microassay for lipases. Anal Bioanal Chem 381:1480–1482

35. Mogensen JE, Sehgal P, Otzen DE (2005) Activation, inhibition, and destabilization of Thermomyces lanuginosus lipase by detergents. Biochemistry 44:1719–1730

36. Shah S, Gupta MN (2007) Obtaining high transesterification activity for subtilisin in ionic liquids. Biochim Biophys Acta 1770:94–98

37. Bastida A, Sabuquillo P, Armisen P, Fernández-Lafuente R, Huguet J, Guisán JM (1998) A single step purification, immobilization, and hyperactivation of lipases via interfacial adsorption on strongly hydrophobic supports. Biotechnol Bioeng 58:486–493

38. Fernandez-Lafuente R, Armisén P, Sabuquillo P, Fernández-Lorente G, Guisán JM (1998) Immobilization of lipases by selective adsorption on hydrophobic supports. Chem Phys Lipids 93:185–197

39. Yilmaz E (2002) Improving the application of microbial lipase by bio-imprinting at substrate-interfaces. World J Microbiol Biotechnol 18:37–40

40. Fishman A, Cogan U (2003) Bio-imprinting of lipases with fatty acids. J Mol Catal B 22:193–202

41. Yapoudjian S, Ivanova MG, Brzozowski AM, Patkar SA, Vind J, Svendsen A, Verger R (2002) Binding of Thermomyces (Humicola) lanuginosa lipase to the mixed micelles of cis-parinaric acid/NaTDC. Eur J Biochem 269:1613–1621

42. de Diego T, Lozano P, Gmouh S, Vaultier M, Iborra JL (2005) Understanding structure-stability relationships of Candida antartica lipase B in ionic liquids. Biomacromolecules 6:1457–1464

43. Desnuelle P (1961) Pancreatic lipase. Adv Enzymol 23:129–161

44. Brockman HL, Law JH, Kézdy FJ (1973) Catalysis by adsorbed enzymes. The hydrolysis of tripropionin by pancreatic lipase adsorbed to siliconized glass beads. J Biol Chem 248:4965–4970

45. Sharma A, Roy I, Gupta MN (2004) Affinity precipitation and macroaffinity ligand facilitated three-phase partitioning for refolding and simultaneous purification of urea-denatured pectinase. Biotechnol Prog 20:1255–1258

46. Santoro MM, Liu Y, Khan SM, Hou LX, Bolen DW (1992) Increased thermal stability of proteins in the presence of naturally occurring osmolytes. Biochemistry 31:5278–5283

47. Gupta MN (ed) (1993) Thermostability of enzymes. Springer, Heidelberg

48. Kuznetsova IM, Turoverov KK, Uversky VN (2014) What macromolecular crowding can do to a protein. Int J Mol Sci 15:23090–23140

49. Khare SK, Gupta MN (1988) A preparation of E. coli beta galactosidase. Appl Biochem Biotechnol 16:1–15

50. Tyagi R, Gupta MN (1998) Chemical modification and chemical crosslinking of protein/enzyme stabilization. Biochem (Moscow) 63:334–344

51. Yamazaki T, Tsugawa W, Sode K (1999) Increased thermal stability of glucose dehydrogenase by crosslinking chemical modification. Biotechnol Lett 21:199–202

52. Cao L (2005) Carrier bound immobilized enzymes: principles, application and design. Wiley-VCH Verlag GmbH and Co., Weinheim

53. Minteer SD (ed) (2011) Enzyme stabilization and immobilization: methods and protocols. Humana Press, New York

54. Guisan JM (ed) (2013) Immobilization of enzymes and cells. Humana Press, New York

55. Carey PR (ed) (1996) Protein engineering and design. Academic, New York

56. Nosoh Y, Sekiguchi T (1993) Protein engineering for thermostabilization. In: Gupta MN (ed) Thermostabilization of enzymes. Springer, Heidelberg, pp 182–204

57. Arnold FH, Georgiou G (2003) Directed enzyme evolution: screening and selection methods. Humana Press, Totowa

58. Arnold FH, Georgiou G (2003) Directed evolution library construction: methods and protocol. Humana Press, Totowa

59. Broos J (2002) Impact of the enzyme flexibility on the enzyme enantio-selectivity in organic media towards specific and non-specific substrates. Biocatal Biotransform 20:291–295

60. Mondal K, Roy I, Gupta MN (2006) Affinity-based strategies for protein purification. Anal Chem 78:3499–3504

61. Bradford MM (1976) A rapid and sensitive method for the quantitation of microgram quantities of protein utilizing the principle of protein-dye binding. Anal Biochem 72:248–254

62. Otwinowski Z, Minor W (1997) Denzo and scalepack. Meth Enzymol 276:307–326

63. Vagin A, Teplyakov A (2010) Molecular replacement with MOLREP. Acta Crystallogr D Biol Crystallogr 66:22–25

64. Murshudov GN, Skubák P, Lebedev AA, Pannu NS, Steiner RA, Nicholls RA, Winn MD, Long F, Vagin AA (2011) REFMAC5 for the refinement of macromolecular crystal structures. Acta Crystallogr D Biol Crystallogr 67:355–367

65. Jones TA, Zou JY, Cowan SW, Kjeldgaard M (1991) Improved methods for building protein models in electron density maps and the location of errors in these models. Acta Crystallogr A 47:110–119

66. Emsley P, Cowtan K (2004) Coot: model-building tools for molecular graphics. Acta Crystallogr D Biol Crystallogr 60:2126–2132

67. Mukherjee J, Gupta MN (2015) Increasing importance of protein flexibility in designing biocatalytic processes. Biotechnol Rep 6:119–123

68. DeLano WL (2005) The case for open-source software in drug discovery. Drug Discov Today 10:213–217

Chapter 9

ASSESSING THE MUTAGENICITY OF PROTIC IONIC LIQUIDS USING THE MINI AMES TEST

Joshua E. S. J. Reid[1,2], Neil Sullivan[3] , Lorna Swift[4] , Guy A. Hembury[5] , Seishi Shimizu[1] and Adam J. Walker[2]

[1]York Structural Biology Laboratory, Department of Chemistry, University of York, Heslington, York YO10 5DD, UK

[2] TWI Ltd., Granta Park, Great Abington, Cambridge CB21 6AL, UK

[3] The Durham Genome Centre, Park House, Station Road, Lanchester DH7 0EX, UK

[4] St. Saviour's and St. Olave's School, London SE1 4AN, UK

[5] University of Hull, Kingston-upon-Hull HU6 7RX, UK.

ABSTRACT

Background

Protic ionic liquids (PILs) have been suggested as "greener" alternatives to conventional solvents in various industrial applications. In order to assess their suitability for such purposes, a thorough evaluation of their toxicity, mutagenicity, carcinogenicity and environmental impact is crucial. Whilst some studies have been published concerning the biodegradability and toxicity towards microorganisms of a limited number of PILs, no data concerning the mutagenicity of any PIL exist within the literature. As part of our ongoing studies into the toxicity and environmental impact of PILs, we quantify herein the mutagenic potential of a range of PILs through the mini Ames test.

Results

In total, 16 PILs and two precursor amines were assessed based on the Ames test, using *Salmonella typhimurium* strains TA98 and TA100. The 16 PILs used in this study included both carboxylate and chloride anions, as well as secondary and tertiary ammonium cations. Our results show that out of the 16 PILs, 15 gave negative results to the mini Ames test, concluding that they are unlikely to be either mutagenic or carcinogenic. The PIL N,N-Dimethylethanolammonium Octanoate ([DMEtA][Oct]) was toxic to both test strains, and its mutagenic potential could not be assessed by the mini Ames test. The two precursor amines, diethanolamine and N,N-dimethylethanolamine, gave negative results to the mini Ames tests despite the suggestion from other mutagenicity tests of diethanolamine's suggested carcinogenicity.

Conclusions

15 PILs have been deemed likely to be neither mutagenic nor carcinogenic in accordance with the mini Ames test. We find that these results compare well to the relevant carboxylic acids and amines from the literature, suggesting that PILs exist as well solvated ions in these test conditions, similar to those of their precursors in the same test. From this, we caution the use of secondary ammonium cations in PILs, as certain secondary amines have been suggested to be potentially carcinogenic, despite their results from the mini Ames test.

BACKGROUND

The development of new chemical processes and applications, that are not only efficient but also pose a low hazard to both humans and the environment, forms a cornerstone of sustainable chemistry [1]. With the growing pressure from chemical legislation such as REACh in Europe, manufacturers are now strongly encouraged to find safer, alternative chemicals whenever possible. As a result, research into alternative, more sustainable and environmentally benign materials has grown considerably. Challenges such as reducing volatile organic compound (VOC) and greenhouse gas (GHG) emissions as well as decreasing the dependence of hazardous materials still persist for many chemical processes. Many common organic solvents of previous wide application, such as chloroform, 1,2-dichloroethane, benzene and hexamethylphosphoramide (HMPA), are now well-established as carcinogenic and their use has been restricted correspondingly. Replacing such materials with safer but equally effective solvents represents an ongoing challenge for the chemical industry, and a number of "neoteric" classes of solvent have thus emerged in recent years as potential alternatives—amongst them, supercritical fluids (SCFs) and ionic liquids (ILs) [3–5]. ILs, which are effectively low-melting salts, fall into two main chemical classes: aprotic ILs and protic ILs (PILs).

Typically, PILs are prepared from the reversible proton equilibrium between a Brønsted acid and an appropriate proton acceptor. This relatively quick and simple preparation is preferential to the multi-stage aprotic IL synthesis, as it avoids the use of stoichiometric quantities of highly toxic alkylation agents. The direct synthesis method can also minimize the presence of water in the product [6], known to have drastic effects on the IL's properties [7, 8]. This simple method for high purity preparation of PILs ultimately makes them much cheaper to produce than their aprotic counterparts [9, 10]. Even though they have been known for over a century [11], only very recently has their potential begun to be recognized [12–17]. The number of possible PIL structures is immense, arising from all the possible appropriate combinations of precursors. The physical and solvation properties of a PIL within the medium and at interfaces can therefore be modified with an appropriate selection of precursors. With such a wide variety of potential molecular structures, it is hard to generalise properties of protic ionic liquids. It is likely however that an extensive hydrogen bonding network exists in the condensed phase [18]. This has led to a variety of published case studies of their applications such as natural product extraction [19–24], desulfurization of fuel [25–27], absorbents for cooling-heating cycles [28], CO_2 capture [29, 30], anhydrous fuel cell electrolytes [31–33], catalysis [4, 34, 35], lubrication [36–40] and hydrometallurgy [41–44].

What makes PILs particularly attractive is their suggested low toxicity according to numerous in vitro case studies. Previously, a number of PILs (including primary, secondary and tertiary 2-hydroxylethylamines with carboxylate anions of various alkyl chain lengths) have been subject to tests to assess their toxicity towards a wide variety of organisms, such as soil microorganisms and terrestrial plants, marine bacteria, aquatic plants and rat leukaemia cells [45–47]. In these studies, the PILs generally displayed relatively high EC_{50}, implying that they are generally less toxic than the aprotic ILs also studied. Recently, quantitative structure activity relationship (QSAR) analysis has been used to successfully estimate the toxicity of some PILs in a number of tests [48]. While the toxicity of PILs have been explored for a number of structures, the mutagenicity of PILs has not been investigated. Mounting evidence suggest that the mutagenic potential of a substance can be used as an indicator of its carcinogenicity [49], and the strong correlation between mutagenicity and carcinogenicity of compounds has led to the development of the Ames test. By observing the effect of a chemical substance on the growth of particular strains of *Salmonella typhimurium* in specific test environments, it is possible to identify if a chemical substance can induce gene mutation within microorganisms. The results of these tests are used to calculate the mutagenic potential of a test material, which is utilised by regulatory agencies when assessing the risk associated with the use of chemical substances [49–53].

At present, we are only aware of one prior study into the mutagenic potential of ILs carried out by Docherty et al. [54]. Using the Ames test, they found that none of the ILs in their selection of imidazolium, pyridinium and quaternary ammonium bromide salts were mutagenic (Table 1). Unfortunately, there is no information relating the structure of the anion to mutagenic potential. This is an important issue to address in the interest of benign ionic liquid design for chemical processes. They summarize with the importance of the Ames test as a means for screening new, potentially environmentally benign green solvents. If we are to successfully apply these materials, we must understand any safety concerns they pose. To this end, we have screened 16 PILs for mutagenicity using the mini Ames test to better understand the potential risks associated with exposure to these materials. The precursor amines diethanolamine and N,N-dimethylethanolamine are also included in our mini Ames test screening, to better understand the relation of the precursor materials to the resulting PIL.

Table 1: Mutagenic Index (MI) for 10 aprotic ionic liquids as determined from the Ames test screening as reported by Docherty et al. [53]

Ionic liquid	MI (TA98)	MI (TA98 + S9)	MI (TA100)	MI (TA100 + S9)
1-Butyl-3-methyl-imidazolium bromide	0.86 ± 0.18	0.88 ± 0.20	1.15 ± 0.17	0.88 ± 0.15
1-Hexyl-3-methyl-imidazolium bromide	0.99 ± 0.37	0.84 ± 0.10	0.92 ± 0.05	1.21 ± 0.23
1-Octyl-3-methyl-imidazolium bromide	1.55 ± 0.28	0.85 ± 0.24	1.09 ± 0.13	1.13 ± 0.10
1-Butyl-3-methyl pyridinium bromide	1.07 ± 0.10	0.93 ± 0.13	1.39 ± 0.30	1.34 ± 0.04
1-Hexyl-3-methyl pyridinium bromide	1.16 ± 0.34	1.09 ± 0.13	1.27 ± 0.18	1.08 ± 0.19
1-Octyl-3-methyl pyridinium bromide	1.27 ± 0.42	1.01 ± 0.11	1.34 ± 0.28	0.89 ± 0.01
Tetramethyl-ammonium bromide	1.47 ± 0.26	0.90 ± 0.07	1.07 ± 0.13	1.16 ± 0.25
Tetraethyl-ammonium bromide	1.33 ± 0.28	1.74 ± 0.29	1.04 ± 0.03	1.26 ± 0.07

Tetrabutyl-ammonium bromide	1.07 ± 0.21	1.14 ± 0.35	1.19 ± 0.06	1.19 ± 0.02
Tetrahexyl-ammonium bromide	1.21 ± 0.80	0.70 ± 0.18	1.09 ± 0.08	0.82 ± 0.09

The results are from concentrations of 1 mg/plate, aside from Tetrahexylammonium bromide which was tested at 0.01 mg/plate as higher concentrations were toxic to the bacterial strains

EXPERIMENTAL SECTION

Materials

The names and abbreviations of the 16 PILs chosen for this study are shown in Table 2. The corresponding structures of all cations and anions of the PILs studied are shown in Fig. 1. The anion variation covers a number of common carboxylate structures as well as the chloride anion, while the cation selection includes both secondary and tertiary ammonium cations all featuring either alcohol or ether functional groups (Tables 3, 4).

Table 2: Names and abbreviations of the PILs used in this study

PIL name	Abbreviation
N,N-dimethylethanolammonium acetate	[DMEtA][OAc]
N,N-dimethylethanolammonium propionate	[DMEtA][Pr]
N,N-dimethylethanolammonium hexanoate	[DMEtA][Hex]
N,N-dimethylethanolammonium octanoate	[DMEtA][Oct]
N,N-dimethylethanolammonium glycolate	[DMEtA][Gly]
N,N-dimethylethanolammonium succinate	[DMEtA][Succ]
N,N-dimethylethanolammonium chloride	[DMEtA][Cl]
N-methyl-N-ethylethanolammonium acetate	[EMEtA][OAc]
N,N-diethylethanolammonium acetate	[DEEtA][OAc]
N-butyldiethanolammonium acetate	[BDEtA][OAc]
N-methyl-bis(2-methoxyethyl)ammonium acetate	[MDEOMA][OAc]
N-methylethanolammonium acetate	[MEtA][OAc]
Diethanolammonium acetate	[DEtA][OAc]
Diethanolammonium chloride	[DEtA][Cl]

| Bis(2-methoxyethyl)ammonium acetate | [DEOMA][OAc] |
| Bis(2-methoxyethyl)ammonium chloride | [DEOMA][Cl] |

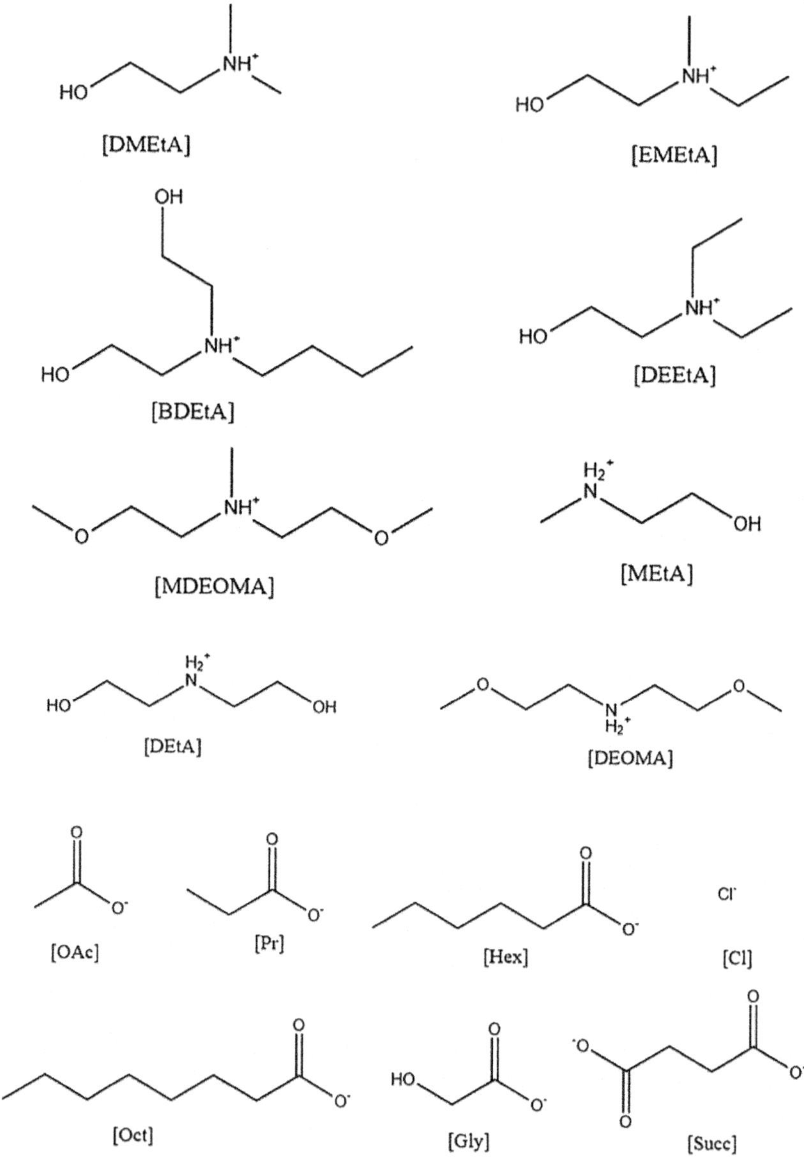

[DMEtA]

[EMEtA]

[BDEtA]

[DEEtA]

[MDEOMA]

[MEtA]

[DEtA]

[DEOMA]

[OAc]

[Pr]

[Hex]

[Cl]

[Oct]

[Gly]

[Succ]

Figure. 1: Structures and abbreviations of cations and anions used to prepare each PIL in this study.

Table 3: Mutagenicity Index (MI) and Mutagenic Activity Ratio (MAR) with the TA98 test strains of *S. typhimurium* at 5µL/plate both with and without S9 metabolic activation for all PILs and selected precursors in this study

PIL	TA 98		TA98	
	Mutagenicity Index (MI)		Mutagenic Activity Ratio (MAR)	
	–	S9	–	S9
[DMEtA][OAc]	0.95 ± 0.22	0.84 ± 0.43	−0.03 ± 0.008	−0.44 ± 0.144
[DMEtA][Pr]	0.59 ± 0.15	0.27 ± 0.17	−0.53 ± 0.094	−1.64 ± 1.126
[DMEtA][Hex]	0.23 ± 0.18	1.10 ± 0.26	−0.28 ± 0.193	0.06 ± 0.009
[DMEtA][Oct]	–	–	–	–
[DMEtA][Gly]	1.64 ± 0.15	0.52 ± 0.29	1.00 ± 0.149	−0.89 ± 0.181
[DMEtA][Succ]	0.82 ± 0.14	6.33 ± 2.50	−0.78 ± 0.084	1.78 ± 1.267
[DMEtA][Cl]	2.10 ± 0.26	0.56 ± 0.36	2.83 ± 0.237	−0.22 ± 0.151
[EMEtA][OAc]	0.95 ± 0.30	1.40 ± 0.46	−0.02 ± 0.004	0.33 ± 0.047
[DEEtA][OAc]	2.03 ± 0.08	3.05 ± 0.42	0.55 ± 0.015	2.39 ± 0.466
[BDEtA][OAc]	0.83 ± 0.21	1.00 ± 0.20	−0.18 ± 0.032	0.00 ± 0.000
[MDEOMA][OAc]	2.69 ± 0.87	0.63 ± 0.37	0.45 ± 0.161	−0.17 ± 0.084
[MEtA][OAc]	0.36 ± 0.39	0.90 ± 0.72	−0.15 ± 0.052	−0.06 ± 0.035
[DEtA][OAc]	0.88 ± 0.53	1.36 ± 0.62	−0.03 ± 0.010	0.22 ± 0.132
[DEtA][Cl]	1.00 ± 0.25	0.70 ± 0.69	0.00 ± 0.000	−0.17 ± 0.112
[DEOMA][OAc]	1.11 ± 0.35	1.30 ± 0.32	0.07 ± 0.031	0.50 ± 0.176
[DEOMA][Cl]	1.69 ± 0.67	0.96 ± 0.24	0.03 ± 0.010	−0.06 ± 0.016
Diethanolamine	0.48 ± 0.48	1.78 ± 0.54	−0.18 ± 0.083	0.39 ± 0.112
N,N-dimethyletha-nolamine	0.74 ± 0.46	0.17 ± 0.15	−0.32 ± 0.143	−1.94 ± 0.970

Table 4: Mutagenicity Index (MI) and Mutagenic Activity Ratio (MAR) with the TA100 test strains of *S. typhimurium* at 5µL/plate both with and without S9 metabolic activation for all PILs and selected precursors in this study

Code	TA 100		TA 100	
	Mutagenicity Index (MI)		Mutagenic Activity Ratio (MAR)	
	–	S9	–	S9
[DMEtA][OAc]	1.10 ± 0.02	1.13 ± 0.14	0.11 ± 0.003	0.15 ± 0.010
[DMEtA][Pr]	0.99 ± 0.08	1.45 ± 0.19	−0.01 ± 0.001	0.46 ± 0.220
[DMEtA][Hex]	0.07 ± 0.08	0.15 ± 0.03	−0.95 ± 0.149	−0.96 ± 0.091
[DMEtA][Oct]	–	–	–	–
[DMEtA][Gly]	0.74 ± 0.16	1.65 ± 0.10	−0.42 ± 0.024	0.35 ± 0.059
[DMEtA][Succ]	1.09 ± 0.05	0.99 ± 0.08	0.14 ± 0.003	−0.01 ± 0.001
[DMEtA][Cl]	1.11 ± 0.22	1.04 ± 0.08	0.15 ± 0.026	0.02 ± 0.003
[EMEtA][OAc]	0.95 ± 0.12	1.14 ± 0.05	−0.06 ± 0.003	0.16 ± 0.020
[DEEtA][OAc]	0.73 ± 0.06	0.74 ± 0.05	−0.69 ± 0.058	−0.64 ± 0.080
[BDEtA][OAc]	1.01 ± 0.10	1.30 ± 0.30	0.01 ± 0.001	0.32 ± 0.060
[MDEOMA][OAc]	1.09 ± 0.09	0.88 ± 0.33	0.09 ± 0.005	−0.09 ± 0.010
[MEtA][OAc]	0.96 ± 0.13	1.02 ± 0.22	−0.02 ± 0.002	0.01 ± 0.003
[DEtA][OAc]	0.68 ± 0.09	0.83 ± 0.06	−0.23 ± 0.037	−0.09 ± 0.011
[DEtA][Cl]	1.32 ± 0.09	1.11 ± 0.16	0.20 ± 0.009	0.06 ± 0.011
[DEOMA][OAc]	1.48 ± 0.10	0.85 ± 0.04	0.24 ± 0.028	−0.27 ± 0.025
[DEOMA][Cl]	1.21 ± 0.16	0.54 ± 0.11	0.14 ± 0.010	−0.85 ± 0.020
Diethanolamine	1.50 ± 0.17	0.51 ± 0.11	0.26 ± 0.032	−0.85 ± 0.217
N,N-dimethyletha-nolamine	0.66 ± 0.20	0.41 ± 0.08	−0.22 ± 0.052	−0.22 ± 0.048

The amines N-methyl-N-ethylethanolamine and N-methyl-bis(2-methoxyethyl)amine were purchased from Almac Sciences Ltd., Craigavon, UK. All other materials including carboxylic acids, secondary amines and tertiary amines were supplied by Sigma Aldrich Ltd., Gillingham, UK. All materials were of analytical grade and used without further purification.

Synthesis of PILs

All carboxylate PILs were prepared from the equimolar neutralization reaction between the corresponding precursor acid and amine as previously described [12]. The chloride PILs were prepared by bubbling an excess of dry HCl gas (generated from sodium chloride and sulfuric acid in a Kipps apparatus) through the appropriate amine, followed by removal of the excess acid *in vacuo*. Purity of 14 of the 16 PILs was assessed using ion chromatography to demonstrate the high purity achievable using this synthesis method. The results of this are recorded in the Additional file 1.

Mini Ames Test

The mini Ames test was carried out on the 16 PILs by Complement Genomics Ltd, The Durham Genome Centre, Park House, Station Road, Lanchester, Co. Durham, UK, DH7 0EX. The purpose of the mini-Ames test is the detection of mutations in specialised strains of *Salmonella typhimurium* that restore their ability to synthesise the amino acid histidine; hence the mutated bacteria are identified through their ability to grow in the absence of this essential amino acid. By using two different modified strains of *S. typhimurium*, TA98 and TA100, we can test for both frameshift and base pair mutations, respectively. The study of the development of mutations through the mini-Ames test can be applied to organic solvents to determine mutagenic potential, which can then be used as an indicator of likely carcinogenicity [53].

Because the mini-Ames test species are prokaryotic organisms and the tests are carried out in vitro, the results cannot provide direct information on the mutagenic and subsequent carcinogenic potency of a substance in mammals; however, despite the correlation not being absolute, many compounds that are positive in the mini-Ames test are also mammalian carcinogens.

Cultures of *Salmonella typhimurium* TA98 (to detect frameshift mutations) and TA100 (to detect missense mutations) strains were grown to late exponential phase of growth (approximately 10^9 cells per ml) at 37 °C with 25 µg ml^{-1} ampicillin to select for the R-factor (plasmids carrying antibiotic resistance genes [55]). The developed strains yielded spontaneous revertant colony plate counts which were found to be consistent with the laboratory's historical control data. Bacteria were exposed to the PILs both in the presence and absence of an S9 metabolic activation system, prepared from the livers of rats treated with the enzyme-inducing agent Aroclor 1254. The PILs were added directly to the test system in various concentrations; we exceeded recommended maximum test concentration for soluble non-cytotoxic substances (5µL/plate; OECD 471) and all calculations are based on test concentrations of 15μL/plate when possible. In addition, Positive control concentrations of 2-amino-fluorene in DMSO and sodium azide in water were used to demonstrate the effective performance of each assay used for TA98 and TA100, respectively.

The given bacterial culture (0.01 mL), either sterile buffer or the metabolic activation system (0.05 mL) and the test PIL (up to 0.015 mL) were pre-incubated for 20 min at 30 °C, after which the system was mixed with a top agar (2 mL) before being transferred onto the surface of a minimal agar plate. All plates were then incubated at 37 °C for 48 h, after which, the number of revertant colonies per plate was counted and compared to the control plates. Each assay was performed in triplicate.

By comparing the number of revertant colonies in the presence of the PIL at the highest measured dosage (R_{PIL}) to the number of spontaneous revertant colonies (R_s), the mutagenicity of each PIL was evaluated. By calculating the Mutagenicity Index (MI) (Eq. 1) for each PIL, the relative mutagenicity (according to each bacterial strain) can be compared for all PILs in this study [56].

$$MI = \frac{R_{PIL}}{R_s} \tag{1}$$

A further assessment of the data was the calculation of the Mutagenic Activity Ratio (MAR). Recommended by the US National Enforcement Investigations Centre, it has been suggested that a MAR ≥ 2.5 indicates a 95 % probability of likely carcinogenicity in test animals [56].

$$MAR = \frac{R_{PIL} - R_s}{R_H} \tag{2}$$

where R_H is the historical laboratory number for spontaneous revertant colonies. The complete data of all compounds screened with the mini Ames test in this study have been compiled into Additional file 1.

RESULTS AND DISCUSSION

In order for a compound to be classified as mutagenic according to the mini-Ames test, it must fulfil the following three criteria:

1. A significant correlation between test concentration of substance and the number of revertant colonies.

2. At least a twofold increase in the number of revertant colonies at the maximum test concentration.

3. The Mutagenicity Index (MI) and the Mutagenic Activity Ratio (MAR) of each substance must be greater than 2.0 and 2.5, respectively, at the maximum test concentration.

The MI and MAR results of each PIL in the TA98 strain and TA100 strain are summarised in Tables 3 and 4, respectively. Of the 16 screened PILs, 15 of them did not meet all three criteria. This means that these 15 PILs are likely to be non-mutagenic and non-carcinogenic as far as can be determined by the mini-Ames test. For some of the PILs, there are some test results where the MI and/or the MAR values are greater than the threshold as outlined in criteria point 3. However, because no PIL complied with criteria point 1, none of the PILs could be suggested to be mutagenic as a result of this study.

The PILs chosen for the purposes of this study were selected on the basis of assessing the effects of both anion and cation on the results of the mini-Ames test, which will be discussed in the following sections. The PILs selected for this work were based upon commercially available amine and acid precursors (or simple derivatives thereof), hence in many cases extensive toxicological data are available for these materials [57–60]. We hypothesise that the mutagenicity of PILs will be relatable to that of their precursors. Stange et al. has studied the binary system of triethylammonium methanesulfonate and water at various compositions using far infrared (FIR) spectroscopy and relation to density fluctuation theory (DFT) calculations of bond energy and frequencies [61]. They found that the PIL prefers to be in a solvent-separated ion pair structure as oppose to a contact-ion pair at around 80 % mole fraction of water. We expect that in similar aqueous conditions the PILs in this study would also exhibit the same solvent-separated ion pair structure. The conditions for the mini Ames test are designed to simulate the effects of these compounds at dilute concentrations. In this well solvated aqueous environment, we would expect the interactions between ions would be far less favourable than the interactions between ions. As a result, the mutagenic potential of a fully miscible PIL in the mini Ames test will be dependant more on the structure of the individual ions than the PILs ability to form aqueous macrostructures. To this end, we believe that the information of the Ames test results for the precursor materials used to produce a given PIL will be crucial in predicting the likely mutagenicity of a PIL.

Given that, at low concentrations and at physiological pH, the protonation state of the precursors and the PIL components would (assuming the PIL is fully water miscible) be identical. The mutagenic and carcinogenic potential of PILs might be expected to additively mirror that of its two molecular precursor chemicals.

Cations and Mutagenicity

Here we assess the effect of cation structure on the resulting mutagenicity of the PIL. This is made possible by our selection of PILs; 8 of which contain a common anion, acetate. By Increasing the length of the alkyl side chains on the cation ([DMEta][OAc] to [EMEtA][OAc] to [DEEtA][OAc])—and hence increasing lipophilicity—resulted in an increase in the MI with the TA98 strain both with and without the activating agent. This increase was consistent to the point where [DEEtA][OAc] passed the MI requirement for both TA98 tests (Fig. 2). However, as this PIL did not meet the remaining requirements as outlined above, it cannot be deemed mutagenic to this particular test strain. Interestingly, the PIL [BDEtA][OAc] did not show the same behaviour, despite having a longer aliphatic chain within its structure compared to all other hydroxyl functionalised cations in this study. This suggests that the use of hydroxyl functional groups can counteract the effects of an increased alkyl chain length on the resulting MI and MAR of the PIL. The incorporation of polar functional groups in IL cation structures has been linked to a reduction in observed toxicity, so it is reasonable that observed mutagenicity also decreases in a similar manner [62, 63].

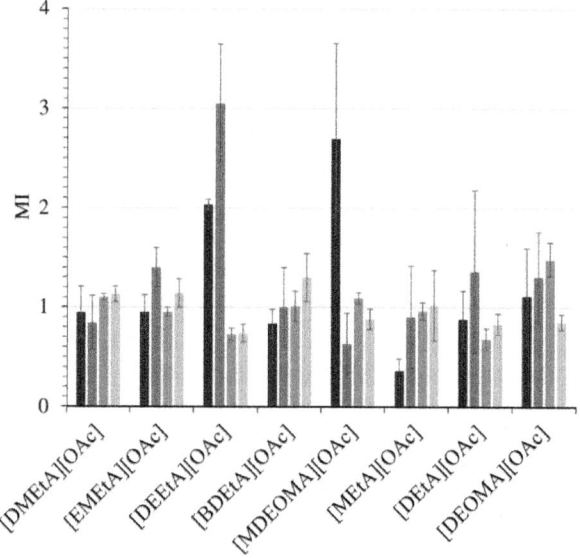

Figure. 2: Mutagenicity index (MI) of the 8 PILs with the acetate anion, showing the cation effect on the results of the mini-Ames test. [DEEtA][OAc] exhibited MI greater than 2 for both TA98 and TA98 + S9 tests, but lacked the necessary concentration dependence on number of revertant colonies necessary to be considered possibly mutagenic.

By substituting the hydroxyl component of the cation with a methoxy ether group, we increase the lipophilicity (by limiting the hydrogen bonding capability of the cation) while still incorporating oxygen into the PIL structure. This is a useful technique that is commonly utilised when designing readily biodegradable ILs, an important consideration to make in the design of new solvent technology [64, 65]. However, it has been suggested that methoxy functional groups can be mutagenic, as has been observed with methyl tertiary butyl ether in the unscheduled DNA synthesis assay test [66]. From our results, we observe a higher MI from the TA98 test for [MDEOMA][OAc] than compared to any other PIL (aside from [DEEtA][OAc]) but again lacked the necessary MAR to be deemed possibly mutagenic to this test strain. Aside from this result, [MDEOMA][OAc] has comparable MI (Fig. 2) and MAR (Fig. 3) values for all other tests to all hydroxyl functionalised cation PILs.

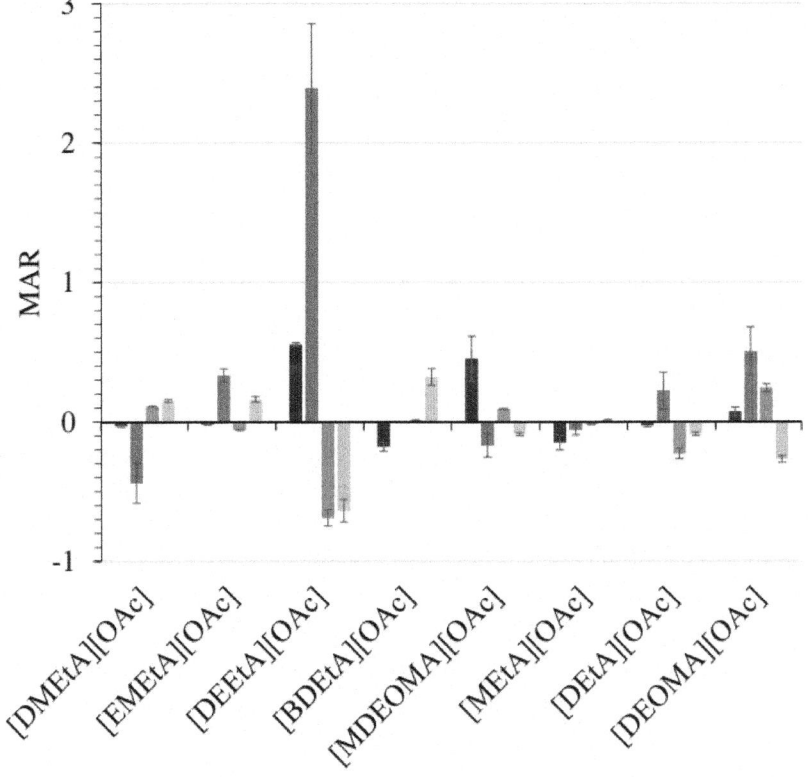

Figure. 3: Mutagenic Activity Ratio (MAR) of the 8 PILs with the acetate anion, showing the cation effect on the results of the mini-Ames test. [DEEtA][OAc] exhibited MAR greater than 2.5 for the TA98 + S9 tests, but not for any of the other test systems.

One of the questions we wish to address in this study is the difference in mutagenicity between secondary and tertiary ammonium PILs. Some secondary amines have been found to readily convert to N-nitrosamine compounds both in vivo and in vitro, which are virulently carcinogenic [67–69]. Additionally, diethanolamine has been determined to be possibly carcinogenic to humans (group 2B) by the International Agency for Research on Cancer [70]. This is despite the pass result given to previous Ames test investigations of diethanolamine [57]. Because of this concern, in addition to the PILs we also included the precursor amines diethanolamine and N,N-dimethylethanolamine into our mini Ames test screening. Both the precursor diethanolamine and the PIL [DEtA][OAc] showed that they are not likely to be either mutagenic according to the mini Ames test (Figs. 2, 3). We have already cautioned the use of results from the mini Ames test as a direct comparison to the mutagenicity potential of a compound. We conclude that more information is required before we can confirm the mutagenicity potential of diethanolammonium PILs. We remind here that the concentration of PILs in the test systems are typically less than 0.05 mol dm^{-3}. At these low concentrations, the structure of the individual ions will be more likely to govern mutagenicity rather than their tendency to form aqueous aggregates.

Variation of the Anion: Carboxylate Anions

Seven of the tested PILs were based on the *N,N*-dimethylethanolammonium cation, in combination with various anions (Figs. 4, 5). This facilitated a basic assessment of the effects of anion alkyl chain length and functionality on mutagenicity, as well as on any direct toxicity to the test organisms. The shorter-chain anions showed little correlation between chain length and either MI or MAR in either test strains. Indeed, it would have been extraordinary had any increase in mutagenicity directly attributable to such ubiquitous (and metabolically-essential) anions as acetate and succinate been indicated. *N,N*-Dimethylethanolammonium octanoate ([DMEtA][Oct]) was toxic to both strains at all concentrations, hence could not be screened for mutagenicity. The long aliphatic chain on the anion of this particular PIL is characteristic of anionic surfactants, which are obviously well established antimicrobial agents, hence this result was anticipated; similar effects having also been observed with other long chain alkyl-substituted ILs [71].

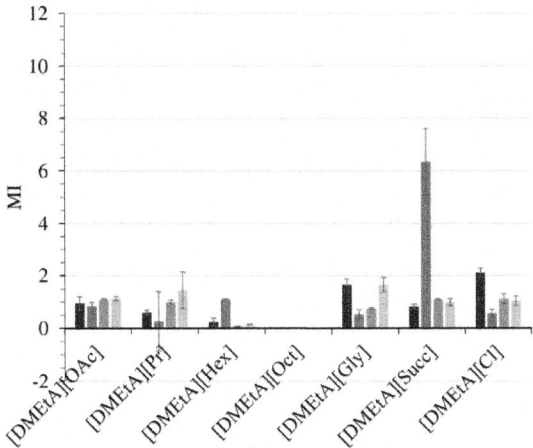

Figure. 4: Mutagenicity index (MI) of the 7 PILs with the N,N-dimethylethanolammonium cation, showing the cation effect on the results of the mini-Ames test. [DMEtA] [Oct] was toxic to both strains due to the long alkyl chain in the anion. The TA98 + S9 [DMEtA][Succ] result and the TA98 [DMEtA][Cl] had a fail result (greater than 2.0), but not for any of the other test systems.

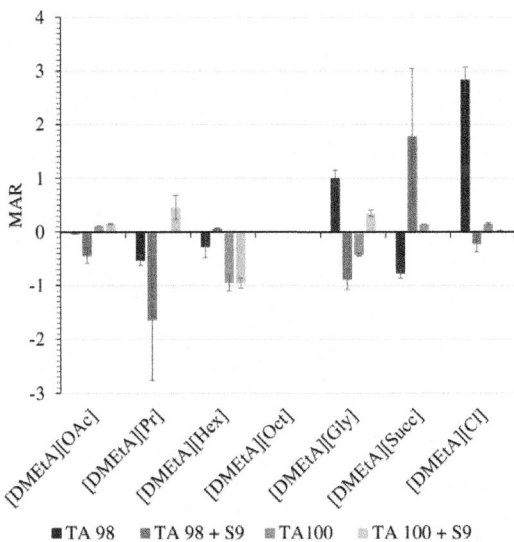

Figure. 5: Mutagenic Activity Ratio (MAR) of the 7 PILs with the N,N-dimethylethanolammonium cation, showing the cation effect on the results of the mini-Ames test. [DMEtA][Oct] was toxic to both strains due to the long alkyl chain in the anion. The TA98 [DMEtA][Cl] test had a fail result (MAR ≥ 2.5) but for none of the other test systems.

Variation of Anion: Chloride vs. Acetate

Unsurprisingly, only one of the test results showed any difference between acetate and chloride, this being a positive result [MI ≥ 2.0 (Fig. 6) and MAR ≥ 2.5 (Fig. 7)] seen with [DMEtA][Cl] with the TA98 strain alone, and not with [DMEtA][OAc]. However, this result lacked the linear relationship between concentration and number of revertant colonies necessary for it to be considered mutagenic. Again, we observe little difference in test results for all strains between secondary and tertiary ammonium cations, as well as between hydroxyl functionalised and methoxy functionalised cations.

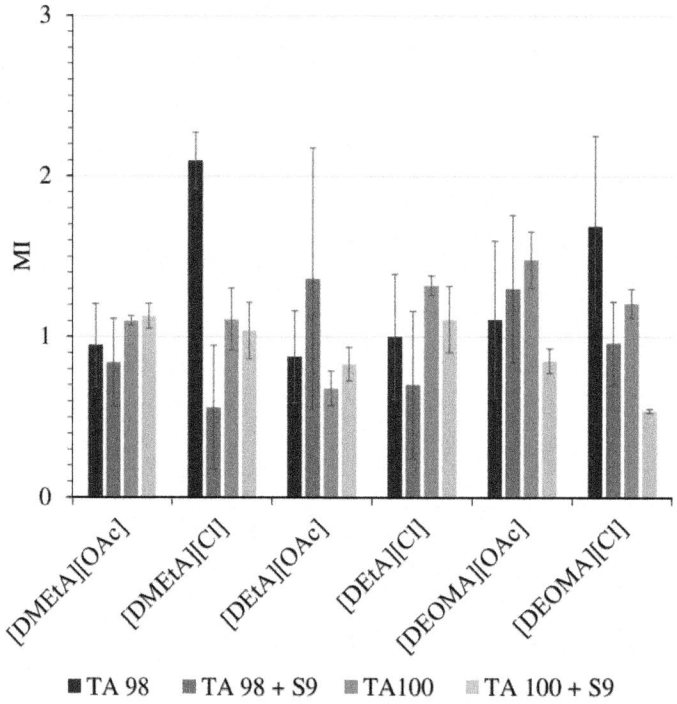

Figure. 6: Mutagenicity index (MI) of 6 PILs, featuring either an acetate or chloride anion, hydroxyl or ether functionalised groups on the cation and either secondary or tertiary ammonium cations. By changing between the acetate and chloride anion, there is little change in the MI. Secondary ammonium cation PILs show a comparable MI result to tertiary ammonium cation PILs in the mini Ames test. Methoxy ether functional groups also appear to have little impact on the MI of the PIL.

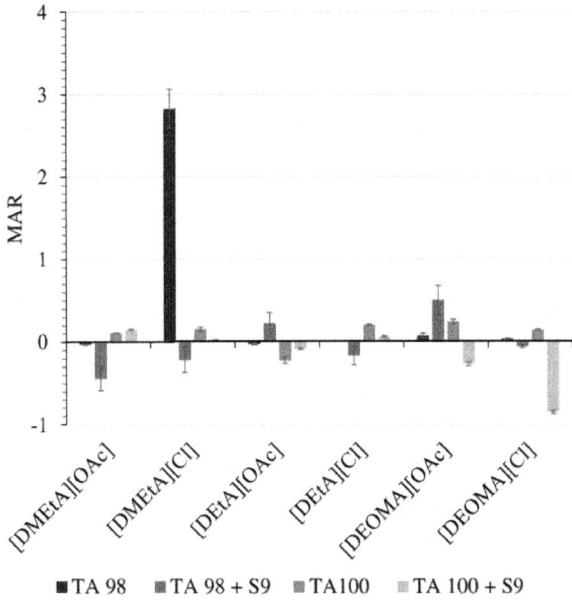

Figure. 7: Mutagenic Activity Ratio (MAR) of 6 PILs, featuring either an acetate or chloride anion, hydroxyl or ether functionalised groups on the cation and either secondary or tertiary ammonium cations. Aside from the TA 98 [DMEtA][Cl] result, all tests.

CONCLUSIONS

Of the 16 PILs tested, 15 were shown to be non-mutagenic in the mini-Ames test and may thus be considered unlikely to exhibit mutagenic or carcinogenic activity, within the accepted limitations of the test methodology. [DMEtA] [Oct] was toxic to both *S. typhimurium* strains in the mini Ames test due to the long alkyl chain on the carboxylate anion, and could therefore not be tested for mutagenicity by this test. All other PILs failed to meet all three of the requirements to be deemed potentially mutagenic or carcinogenic in accordance with the mini Ames test. However, because of the likely state of these PILs in physiological conditions being very similar to that of their precursors, we caution the use of secondary ammonium cation PILs as certain secondary amines. For example, it has been shown previously that diethanolamine have been deemed potentially carcinogenic to humans (group 2B). This is contradictory to our results for diethanolamine in the mini Ames test, which agree with prior Ames test results for diethanolamine. We propose that more in-depth mutagenicity studies into PILs should be carried out to either challenge or confirm the results of the mini Ames test. Ultimately, we conclude that the potential mutagenicity of carboxylate ammonium PILs can be treated as the same as that of their precursors due to the likely highly diluted state of both ions.

AUTHORS' CONTRIBUTIONS

JESJR drafted the manuscript with input from AJW and SS, LS and GAH prepared all PILs in this study, NS was responsible for screening of all compounds with the mini Ames test and SS and AJW supervised the entire study. All authors read and approved the final manuscript.

ACKNOWLEDGEMENTS

The authors gratefully acknowledge the financial support of the Engineering and Physical Sciences Research Council (EPSRC) through an industrial CASE award. All data supporting this study are provided as tables in the main paper and as supplementary information accompanying this paper. We gratefully acknowledge Complement Genomics Ltd. in undertaking mini Ames screening of all compounds within this study.

ADDITIONAL FILE

Additional file 1. The complete data of all compounds screened with the mini Ames test in this study.

Purity analysis using Ion Exchange Chromatography

Code	Name	Purity from Ion Exchange Chromatograph
[DMEtA][OAc]	dimethylethanolammonium acetate	99.16%
[DMEtA][Pr]	dimethylethanolammonium propionate	99.45%
[DMEtA][Hex]	dimethylethanolammonium hexanoate	98.84%
[DMEtA][Oct]	dimethylethanolammonium octanoate	99.37%
[DMEtA][Gly]	dimethylethanolammonium glycolate	99.93%
[DMEtA][Succ]	dimethylethanolammonium succinate	98.75%
[DMEtA][Cl]	dimethylethanolammonium Chloride	99.92%
[EMEtA][OAc]	ethyl-methylethanolammonium acetate	
[DEEtA][OAc]	diethylethanolammonium acetate	95.89%
[BDEtA][OAc]	butyldiethanolammonium acetate	95.32%
[MDEOMA][OAc]	N-methyl-bis(2-methoxyethyl)ammonium acetate	
[MEtA][OAc]	N-methylethanolammonium acetate	99.66%
[DEtA][OAc]	diethanolammonium acetate	94.42%
[DEtA][Cl]	diethanolammonium chloride	95.41%
[DEOMA][OAc]	bis(2-methoxyethyl)ammonium acetate	93.93%
[DEOMA][Cl]	bis(2-methoxyethyl)ammonium chloride	92.96%

Results from the Mini Ames test

Salmonella typhimurium TA 98 frameshift mutation results

Without S9

N,N-Dimethylethanolammonium acetate	Number of revetant colonies			Mean	STD	MI	MAR
Concentration / µL / plate	1	2	3				
0.17	11	6	18	11.67	6.03	0.95	-0.03
0.50	12	16	14	14.00	2.00	1.14	0.08
1.70	18	14	15	15.67	2.08	1.27	0.17
5.00	11	9	11	10.33	1.15	0.84	-0.10
15.00	8	13	14	11.67	3.21	0.95	-0.03
Controls							
2-AF	22	15	13	16.67	4.73	1.35	0.22
DMSO	4	10	14	9.33	5.03	0.76	-0.15
Water	11	14	7	10.67	3.51	0.86	-0.08
Spontaneous	11	11	15	12.33	2.31	1.00	0.00

With S9

N,N-Dimethylethanolammonium acetate	Number of revetant colonies			Mean	STD	MI	MAR
Concentration / µL / plate	1	2	3				
0.17	11	11	25	15.67	8.08	0.94	-0.17
0.50	11	7	16	11.33	4.51	0.68	-0.89
1.70	15	11	13	13.00	2.00	0.78	-0.61
5.00	13	8	12	11.00	2.65	0.66	-0.94
15.00	13	10	19	14.00	4.58	0.84	-0.44
Controls							
2-AF	72	80	64	72.00	8.00	4.32	9.22
DMSO	16	16	16	16.00	0.00	0.96	-0.11
Water	10	14	9	11.00	2.65	0.66	-0.94
Spontaneous	28	11	11	16.67	9.81	1.00	0.00

Without S9

N,N-Dimethylethanolammonium propionate	Number of revetant colonies			Mean	STD	MI	MAR
Concentration / µL / plate	1	2	3				
0.17	32	39	43	38.00	5.57	1.49	0.63
0.50	28	31	25	28.00	3.00	1.10	0.13
1.70	23	16	22	20.33	3.79	0.80	-0.26
5.00	21	15	12	16.00	4.58	0.63	-0.48
15.00	14	18	13	15.00	2.65	0.59	-0.53
Controls							
2-AF	30	21	21	24.00	5.20	0.94	-0.08
DMSO	19	22	14	18.33	4.04	0.72	-0.36
Water	229	176	171	192.00	32.14	7.53	8.33
Spontaneous	29	22		25.50	4.95	1.00	0.00

With S9

N,N-Dimethylethanolammonium propionate	Number of revetant colonies			Mean	STD	MI	MAR
Concentration / µL / plate	1	2	3				
0.17	5	11	4	6.67	3.79	0.49	-1.14
0.50	12	4	10	8.67	4.16	0.64	-0.81
1.70	4	9	9	7.33	2.89	0.54	-1.03
5.00	9	5	2	5.33	3.51	0.40	-1.36
15.00	4	6	1	3.67	2.52	0.27	-1.64
Controls							
2-AF	248	101	198	182.33	74.74	13.51	28.14
DMSO	17	21		19.00	2.83	1.41	0.92
Water	7	11	17	11.67	5.03	0.86	-0.31
Spontaneous	15	12		13.50	2.12	1.00	0.00

Without S9

N,N-Dimethylethanolammonium hexanoate	Number of revetant colonies			Mean	STD	MI	MAR
Concentration / µL / plate	1	2	3				
0.17	3	1	4	2.67	1.53	0.36	-0.23
0.50	4	8	5	5.67	2.08	0.77	-0.08
1.70	4	5	1	3.33	2.08	0.45	-0.20
5.00	2	2	4	2.67	1.15	0.36	-0.23
15.00	1	3	1	1.67	1.15	0.23	-0.28
Controls							
2-AF	9	13	9	10.33	2.31	1.41	0.15
DMSO	9	5	8	7.33	2.08	1.00	0.00
Water	4	2	7	4.33	2.52	0.59	-0.15
Spontaneous	5	3	14	7.33	5.86	1.00	0.00

With S9

N,N-Dimethylethanolammonium hexanoate	Number of revetant colonies			Mean	STD	MI	MAR
Concentration / µL / plate	1	2	3				
0.17	5	2	3	3.33	1.53	1.00	0.00
0.50	4	4	3	3.67	0.58	1.10	0.06
1.70	4	5	2	3.67	1.53	1.10	0.06
5.00	8	6	3	5.67	2.52	1.70	0.39
15.00	3	4	4	3.67	0.58	1.10	0.06
Controls							
2-AF	58	36	42	45.33	11.37	13.60	7.00
DMSO	5	6	6	5.67	0.58	1.70	0.39
Water	8	5	3	5.33	2.52	1.60	0.33
Spontaneous	2	4	4	3.33	1.15	1.00	0.00

Without S9

N,N-Dimethylethanolammonium glycolate	Number of revetant colonies			Mean	STD	MI	MAR
Concentration / µL / plate	1	2	3				
0.17	33	23	29	28.33	5.03	0.90	-0.15
0.50	48	31	48	42.33	9.81	1.35	0.55
1.70	42	44	47	44.33	2.52	1.41	0.65
5.00	49	47	68	54.67	11.59	1.74	1.17
15.00	43	53	58	51.33	7.64	1.64	1.00
Controls							
2-AF	21	25	15	20.33	5.03	0.65	-0.55
DMSO	19	22	19	20.00	1.73	0.64	-0.57
Water	20	41	42	34.33	12.42	1.10	0.15
Spontaneous	33	31	30	31.33	1.53	1.00	0.00

With S9

N,N-Dimethylethanolammonium glycolate	Number of revetant colonies			Mean	STD	MI	MAR
Concentration / µL / plate	1	2	3				
0.17	8	4	6	6.00	2.00	0.55	-0.83
0.50	4	9	4	5.67	2.89	0.52	-0.89
1.70	6	9	0	5.00	4.58	0.45	-1.00
5.00	6	5	8	6.33	1.53	0.58	-0.78
15.00	5	5	7	5.67	1.15	0.52	-0.89
Controls							
2-AF	227	219	161	202.33	36.02	18.39	31.89
DMSO	18	9	16	14.33	4.73	1.30	0.56
Water	9	19	22	16.67	6.81	1.52	0.94
Spontaneous	7	17	9	11.00	5.29	1.00	0.00

Without S9

N,N-Dimethylethanolammonium succinate	Number of revetant colonies			Mean	STD	MI	MAR
Concentration / µL / plate	1	2	3				
0.17	48	38	45	43.67	5.13	0.51	-2.10
0.50	84	98	73	85.00	12.53	0.99	-0.03
1.70	84	74	86	81.33	6.43	0.95	-0.22
5.00	62	57	60	59.67	2.52	0.70	-1.30
15.00	77	71	62	70.00	7.55	0.82	-0.78
Controls							
2-AF	3	6	8	5.67	2.52	0.07	-4.00
DMSO	3	6	5	4.67	1.53	0.05	-4.05
Water	89	108	64	87.00	22.07	1.02	0.07
Spontaneous	104	71	82	85.67	16.80	1.00	0.00

With S9

N,N-Dimethylethanolammonium succinate	Number of revetant colonies			Mean	STD	MI	MAR
Concentration / µL / plate	1	2	3				
0.17	6	4	2	4.00	2.00	2.00	0.33
0.50	16	12	5	11.00	5.57	5.50	1.50
1.70	8	14	13	11.67	3.21	5.83	1.61
5.00	3	6	8	5.67	2.52	2.83	0.61
15.00	22	4	12	12.67	9.02	6.33	1.78
Controls							
2-AF	190	232		211.00	29.70	105.50	34.83
DMSO	5	2	7	4.67	2.52	2.33	0.44
Water	8	6	5	6.33	1.53	3.17	0.72
Spontaneous	3	1	2	2.00	1.00	1.00	0.00

Without S9

N,N-Dimethylethanolammonium Chloride	Number of revetant colonies			Mean	STD	MI	MAR
Concentration / µL / plate	1	2	3				
0.17	57	64	67	62.67	5.13	1.21	0.55
0.50	90	84	44	72.67	25.01	1.41	1.05
1.70	65	58	76	66.33	9.07	1.28	0.73
5.00	61	68	51	60.00	8.54	1.16	0.42
15.00	112	115	98	108.33	9.07	2.10	2.83
Controls							
2-AF	16	18	16	16.67	1.15	0.32	-1.75
DMSO	31	24	52	35.67	14.57	0.69	-0.80
Water	61	72	55	62.67	8.62	1.21	0.55
Spontaneous	72	41	42	51.67	17.62	1.00	0.00

With S9

N,N-Dimethylethanolammonium Chloride	Number of revetant colonies			Mean	STD	MI	MAR
Concentration / µL / plate	1	2	3				
0.17	5	5	3	4.33	1.15	1.44	0.22
0.50	1	11		6.00	7.07	2.00	0.50
1.70	4	1	2	2.33	1.53	0.78	-0.11
5.00	9	4	3	5.33	3.21	1.78	0.39
15.00	1	3	1	1.67	1.15	0.56	-0.22
Controls							
2-AF	122	166	152	146.67	22.48	48.89	23.94
DMSO	66	53		59.50	9.19	19.83	9.42
Water	5	3	4	4.00	1.00	1.33	0.17
Spontaneous	4	3	2	3.00	1.00	1.00	0.00

Without S9

N-Ethyl-N-Methylethanolammonium acetate	Number of revetant colonies			Mean	STD	MI	MAR
Concentration / µL / plate	1	2	3				
0.17	5	8	2	5.00	3.00	0.75	-0.08
0.50	3	7	6	5.33	2.08	0.80	-0.07
1.70	5	9	1	5.00	4.00	0.75	-0.08
5.00	7	6	6	6.33	0.58	0.95	-0.02
15.00	5	7	7	6.33	1.15	0.95	-0.02
Controls							
2-AF	5	7	9	7.00	2.00	1.05	0.02
DMSO	9	5	7	7.00	2.00	1.05	0.02
Water	1	5	7	4.33	3.06	0.65	-0.12
Spontaneous	10	5	5	6.67	2.89	1.00	0.00

With S9

N-Ethyl-N-Methylethanolammonium acetate	Number of revetant colonies			Mean	STD	MI	MAR
Concentration / µL / plate	1	2	3				
0.17	4	6	6	5.33	1.15	1.07	0.06
0.50	10	4	9	7.67	3.21	1.53	0.44
1.70	8	6	5	6.33	1.53	1.27	0.22
5.00	7	5	9	7.00	2.00	1.40	0.33
15.00	7	8	6	7.00	1.00	1.40	0.33
Controls							
2-AF	98	72	92	87.33	13.61	17.47	13.72
DMSO	6	8	9	7.67	1.53	1.53	0.44
Water	5	2	3	3.33	1.53	0.67	-0.28
Spontaneous	1	8	6	5.00	3.61	1.00	0.00

Without S9

N,N-Diethylethanolammonium acetate	Number of revetant colonies			Mean	STD	MI	MAR
Concentration / µL / plate	1	2	3				
0.17	5	10	5	6.67	2.89	0.63	-0.20
0.50	5	11	12	9.33	3.79	0.88	-0.07
1.70	11	9	10	10.00	1.00	0.94	-0.03
5.00	13	15	14	14.00	1.00	1.31	0.17
15.00	22	21	22	21.67	0.58	2.03	0.55
Controls							
2-AF	18	8	16	14.00	5.29	1.31	0.17
DMSO	12	14	15	13.67	1.53	1.28	0.15
Water	10	10	11	10.33	0.58	0.97	-0.02
Spontaneous	12	10	10	10.67	1.15	1.00	0.00

With S9

N,N-Diethylethanolammonium acetate	Number of revetant colonies			Mean	STD	MI	MAR
Concentration / µL / plate	1	2	3				
0.17	10	5	7	7.33	2.52	1.05	0.06
0.50	5	7	5	5.67	1.15	0.81	-0.22
1.70	6	5	3	4.67	1.53	0.67	-0.39
5.00	1	14	6	7.00	6.56	1.00	0.00
15.00	20	26	18	21.33	4.16	3.05	2.39
Controls							
2-AF	71	57	49	59.00	11.14	8.43	8.67
DMSO	9	6	4	6.33	2.52	0.90	-0.11
Water	6	5	5	5.33	0.58	0.76	-0.28
Spontaneous	9	6	6	7.00	1.73	1.00	0.00

Without S9

N-Butyldiethanolammonium acetate	Number of revetant colonies			Mean	STD	MI	MAR
Concentration / µL / plate	1	2	3				
0.17	29	30		29.50	0.71	1.34	0.38
0.50	21	25	26	24.00	2.65	1.09	0.10
1.70	25	26		25.50	0.71	1.16	0.18
5.00	19	18	26	21.00	4.36	0.95	-0.05
15.00	16	17	22	18.33	3.21	0.83	-0.18
Controls							
2-AF	41	42	46	43.00	2.65	1.95	1.05
DMSO	22	27	21	23.33	3.21	1.06	0.07
Water	41	36	39	38.67	2.52	1.76	0.83
Spontaneous	27	24	15	22.00	6.24	1.00	0.00

With S9

N-Butyldiethanolammonium acetate	Number of revetant colonies			Mean	STD	MI	MAR
Concentration / µL / plate	1	2	3				
0.17	1	0	5	2.00	2.65	0.40	-0.50
0.50	13	5	2	6.67	5.69	1.33	0.28
1.70	5	5	1	3.67	2.31	0.73	-0.22
5.00	4	9	9	7.33	2.89	1.47	0.39
15.00	5	7	3	5.00	2.00	1.00	0.00
Controls							
2-AF	214	229	126	189.67	55.64	37.93	30.78
DMSO	1	5	5	3.67	2.31	0.73	-0.22
Water	5	7	5	5.67	1.15	1.13	0.11
Spontaneous	5	5	5	5.00	0.00	1.00	0.00

Without S9

N-Methyl-bis(2-methoxyethyl)ammonium acetate	Number of revetant colonies			Mean	STD	MI	MAR
Concentration / µL / plate	1	2	3				
0.17	10	10	11	10.33	0.58	1.94	0.25
0.50	14	9	9	10.67	2.89	2.00	0.27
1.70	7	14	15	12.00	4.36	2.25	0.33
5.00	13	10	20	14.33	5.13	2.69	0.45
15.00	0	0	0	0.00	0.00	0.00	-0.27
Controls							
2-AF	14	13	7	11.33	3.79	2.13	0.30
DMSO	7	5	6	6.00	1.00	1.13	0.03
Water	23	27	19	23.00	4.00	4.31	0.88
Spontaneous	10	2	4	5.33	4.16	1.00	0.00

With S9

N-Methyl-bis(2-methoxyethyl)ammonium acetate	Number of revetant colonies			Mean	STD	MI	MAR
Concentration / µL / plate	1	2	3				
0.17	6	2	7	5.00	2.65	1.88	0.39
0.50	3	1		2.00	1.41	0.75	-0.11
1.70	6	0	8	4.67	4.16	1.75	0.33
5.00	4	1	0	1.67	2.08	0.63	-0.17
15.00	6	2	6	4.67	2.31	1.75	0.33
Controls							
2-AF	62	67	95	74.67	17.79	28.00	12.00
DMSO	9	10	12	10.33	1.53	3.88	1.28
Water	3	4	2	3.00	1.00	1.13	0.06
Spontaneous	2	2	4	2.67	1.15	1.00	0.00

Without S9

N-Methylethanolammonium acetate	Number of revetant colonies			Mean	STD	MI	MAR
Concentration / µL / plate	1	2	3				
0.17	2	3	6	3.67	2.08	0.79	-0.05
0.50	2	4	5	3.67	1.53	0.79	-0.05
1.70	3	5	2	3.33	1.53	0.71	-0.07
5.00	1	4	2	2.33	1.53	0.50	-0.12
15.00	2	2	1	1.67	0.58	0.36	-0.15
Controls							
2-AF	6	6	5	5.67	0.58	1.21	0.05
DMSO	10	8	17	11.67	4.73	2.50	0.35
Water	3	3	6	4.00	1.73	0.86	-0.03
Spontaneous	8	4	2	4.67	3.06	1.00	0.00

With S9

N-Methylethanolammonium acetate	Number of revetant colonies			Mean	STD	MI	MAR
Concentration / µL / plate	1	2	3				
0.17	11	14		12.50	2.12	3.75	1.53
0.50	5	4	8	5.67	2.08	1.70	0.39
1.70	3	3	5	3.67	1.15	1.10	0.06
5.00	1	0	6	2.33	3.21	0.70	-0.17
15.00	5	2	2	3.00	1.73	0.90	-0.06
Controls							
2-AF	155	150	161	155.33	5.51	46.60	25.33
DMSO	9	11	13	11.00	2.00	3.30	1.28
Water	3	4	5	4.00	1.00	1.20	0.11
Spontaneous	0	6	4	3.33	3.06	1.00	0.00

Without S9

Diethanolammonium chloride	Number of revetant colonies			Mean	STD	MI	MAR
Concentration / µL / plate	1	2	3				
0.17	6	1	5	4.00	2.65	0.75	-0.07
0.50	4	3	5	4.00	1.00	0.75	-0.07
1.70	6	4	4	4.67	1.15	0.88	-0.03
5.00	3	7	6	5.33	2.08	1.00	0.00
15.00	3	7	6	5.33	2.08	1.00	0.00
Controls							
2-AF	13	24	28	21.67	7.77	4.06	0.82
DMSO	7	10	8	8.33	1.53	1.56	0.15
Water	5	6	6	5.67	0.58	1.06	0.02
Spontaneous	6	5	5	5.33	0.58	1.00	0.00

With S9

Diethanolammonium chloride	Number of revetant colonies			Mean	STD	MI	MAR
Concentration / µL / plate	1	2	3				
0.17	3	4	8	5.00	2.65	1.50	0.28
0.50	8	2	2	4.00	3.46	1.20	0.11
1.70	4	2	5	3.67	1.53	1.10	0.06
5.00	4	8	7	6.33	2.08	1.90	0.50
15.00	4	2	1	2.33	1.53	0.70	-0.17
Controls							
2-AF	102	126	117	115.00	12.12	34.50	18.61
DMSO	3	6	4	4.33	1.53	1.30	0.17
Water	7	3	11	7.00	4.00	2.10	0.61
Spontaneous	6	4	0	3.33	3.06	1.00	0.00

Without S9

Diethnaolammonium acetate	Number of revetant colonies			Mean	STD	MI	MAR
Concentration / µL / plate	1	2	3				
0.17	7	7	9	7.67	1.15	1.44	0.12
0.50	8	3	6	5.67	2.52	1.06	0.02
1.70	9	4	4	5.67	2.89	1.06	0.02
5.00	3	5	5	4.33	1.15	0.81	-0.05
15.00	6	5	3	4.67	1.53	0.88	-0.03
Controls							
2-AF	17	20	17	18.00	1.73	3.38	0.63
DMSO	7	18	12	12.33	5.51	2.31	0.35
Water	5	6	2	4.33	2.08	0.81	-0.05
Spontaneous	10	2	4	5.33	4.16	1.00	0.00

With S9

Diethnaolammonium acetate	Number of revetant colonies			Mean	STD	MI	MAR
Concentration / µL / plate	1	2	3				
0.17	2	4	4	3.33	1.15	0.91	-0.06
0.50	5	4	3	4.00	1.00	1.09	0.06
1.70	5	4	0	3.00	2.65	0.82	-0.11
5.00	2	2	1	1.67	0.58	0.45	-0.33
15.00	5	8	2	5.00	3.00	1.36	0.22
Controls							
2-AF	149	157	149	151.67	4.62	41.36	24.67
DMSO	14	14	9	12.33	2.89	3.36	1.44
Water	4	2	2	2.67	1.15	0.73	-0.17
Spontaneous	5	4	2	3.67	1.53	1.00	0.00

Without S9

Bis(2-methoxyethyl)ammonium chloride	Number of revetant colonies			Mean	STD	MI	MAR
Concentration / µL / plate	1	2	3				
0.17	4	7	9	6.67	2.52	0.80	-0.08
0.50	35	17	9	20.33	13.32	2.44	0.60
1.70	18	13	6	12.33	6.03	1.48	0.20
5.00	7	12	8	9.00	2.65	1.08	0.03
15.00	12	9	6	9.00	3.00	1.08	0.03
Controls							
2-AF	16	16	11	14.33	2.89	1.72	0.30
DMSO	14	23	23	20.00	5.20	2.40	0.58
Water	24	23	32	26.33	4.93	3.16	0.90
Spontaneous	8	7	10	8.33	1.53	1.00	0.00

With S9

Bis(2-methoxyethyl)ammonium chloride	Number of revetant colonies			Mean	STD	MI	MAR
Concentration / µL / plate	1	2	3				
0.17	1	9	6	5.33	4.04	0.67	-0.44
0.50	3	7	6	5.33	2.08	0.67	-0.44
1.70	3	2	10	5.00	4.36	0.63	-0.50
5.00	5	6	11	7.33	3.21	0.92	-0.11
15.00	7	6	10	7.67	2.08	0.96	-0.06
Controls							
2-AF	95	80	82	85.67	8.14	10.71	12.94
DMSO	13	12	16	13.67	2.08	1.71	0.94
Water	8	6	5	6.33	1.53	0.79	-0.28
Spontaneous	6	9	9	8.00	1.73	1.00	0.00

Without S9

Bis(2-methoxyethyl)ammonium acetate	Number of revetant colonies			Mean	STD	MI	MAR
Concentration / µL / plate	1	2	3				
0.17	20	17	18	18.33	1.53	1.49	0.30
0.50	18	16	8	14.00	5.29	1.14	0.08
1.70	11	12	11	11.33	0.58	0.92	-0.05
5.00	15	19	6	13.33	6.66	1.08	0.05
15.00	13	20	8	13.67	6.03	1.11	0.07
Controls							
2-AF		18	8	16	14.00	5.29	1.14	0.08
DMSO		18	16	20	18.00	2.00	1.46	0.28
Water		14	15	9	12.67	3.21	1.03	0.02
Spontaneous		12	15	10	12.33	2.52	1.00	0.00

With S9

Bis(2-methoxyethyl)ammonium acetate	Number of revetant colonies			Mean	STD	MI	MAR
Concentration / µL / plate	1	2	3				
0.17	16	6	16	12.67	5.77	1.27	0.44
0.50	7	10	9	8.67	1.53	0.87	-0.22
1.70	9	6	6	7.00	1.73	0.70	-0.50
5.00	7	10	11	9.33	2.08	0.93	-0.11
15.00	8	17	14	13.00	4.58	1.30	0.50
Controls							
2-AF		71	57	49	59.00	11.14	5.90	8.17
DMSO		14	10	14	12.67	2.31	1.27	0.44
Water		3	11	6	6.67	4.04	0.67	-0.56
Spontaneous		8	11	11	10.00	1.73	1.00	0.00

Without S9

N,N-Dimethylethanolamine	Number of revetant colonies			Mean	STD	MI	MAR
Concentration / µL / plate	1	2	3				
0.17	7	10	8	8.33	1.53	1.32	0.10
0.50	10	9	7	8.67	1.53	1.37	0.12
1.70	10	8	8	8.67	1.15	1.37	0.12
5.00	7	2	5	4.67	2.52	0.74	-0.08
15.00	7	3	4	4.67	2.08	0.74	-0.08
Controls							
2-AF		9	8	10	9.00	1.00	1.42	0.13
DMSO		3	4	4	3.67	0.58	0.58	-0.13
Water		9	2	6	5.67	3.51	0.89	-0.03
Spontaneous		9	7	3	6.33	3.06	1.00	0.00

With S9

N,N-Dimethylethanolamine	Number of revetant colonies			Mean	STD	MI	MAR
Concentration / µL / plate	1	2	3				
0.17	5	4		4.50	0.71	0.39	-1.19
0.50	3	10	12	8.33	4.73	0.71	-0.56
1.70	8	10	12	10.00	2.00	0.86	-0.28
5.00	7	7	7	7.00	0.00	0.60	-0.78
15.00	2	3	1	2.00	1.00	0.17	-1.61
Controls							
2-AF		92	86	109	95.67	11.93	8.20	14.00
DMSO		5	3	4	4.00	1.00	0.34	-1.28
Water		5	5	6	5.33	0.58	0.46	-1.06
Spontaneous		12	9	14	11.67	2.52	1.00	0.00

Without S9

Diethanolamine	Number of revetant colonies			Mean	STD	MI	MAR
Concentration / µL / plate	1	2	3				
0.17	9	8	13	10.00	2.65	1.43	0.15
0.50	6	6	9	7.00	1.73	1.00	0.00
1.70	3	7	4	4.67	2.08	0.67	-0.12
5.00	10	7	11	9.33	2.08	1.33	0.12
15.00	5	3	2	3.33	1.53	0.48	-0.18
Controls							
2-AF	7	6	7	6.67	0.58	0.95	-0.02
DMSO	3	10	6	6.33	3.51	0.90	-0.03
Water	7	8	9	8.00	1.00	1.14	0.05
Spontaneous	9	9	3	7.00	3.46	1.00	0.00

With S9

Diethanolamine	Number of revetant colonies			Mean	STD	MI	MAR
Concentration / µL / plate	1	2	3				
0.17	3	8	2	4.33	3.21	1.44	0.22
0.50	0	0	5	1.67	2.89	0.56	-0.22
1.70	5	7	9	7.00	2.00	2.33	0.67
5.00	5	7	8	6.67	1.53	2.22	0.61
15.00	5	7	4	5.33	1.53	1.78	0.39
Controls							
2-AF	49	51	37	45.67	7.57	15.22	7.11
DMSO	4	3	4	3.67	0.58	1.22	0.11
Water	0	5	0	1.67	2.89	0.56	-0.22
Spontaneous	5	2	2	3.00	1.73	1.00	0.00

Salmonella typhimurium TA100 missense mutation results

Without S9

N,N-Dimethylethanolammonium acetate	Number of revetant colonies			Mean	STD	MI	MAR
Concentration / µL / plate	1	2	3				
0.17	158	163	159	160.00	2.65	1.09	0.09
0.50	168	142	183	164.33	20.74	1.12	0.12
1.70	155	130	130	138.33	14.43	0.94	-0.06
5.00	139	154	125	139.33	14.50	0.95	-0.05
15.00	157	166	163	162.00	4.58	1.10	0.11
Controls							
Sodium Azide	934	907	777	872.67	83.94	5.95	5.04
Water	170	128	168	155.33	23.69	1.06	0.06
Spontaneous	147	146	147	146.67	0.58	1.00	0.00

With S9

N,N-Dimethylethanolammonium acetate	Number of revetant colonies			Mean	STD	MI	MAR
Concentration / µL / plate	1	2	3				
0.17	188	172	161	173.67	13.58	1.00	0.00
0.50	162	157	131	150.00	16.64	0.87	-0.17
1.70	176	168	166	170.00	5.29	0.98	-0.02
5.00	163	154	139	152.00	12.12	0.88	-0.15
15.00	198	192	173	187.67	13.05	1.08	0.11
Controls							
Water	177	189	174	180.00	7.94	1.04	0.05
Spontaneous	188	189	142	173.00	26.85	1.00	0.00

Without S9

N,N-Dimethylethanolammonium propionate	Number of revetant colonies			Mean	STD	MI	MAR
Concentration / µL / plate	1	2	3				
0.17	181	205	179	188.33	14.47	1.03	0.03
0.50	206	205	172	194.33	19.35	1.06	0.07
1.70	174	214	203	197.00	20.66	1.07	0.09
5.00	192	177	169	179.33	11.68	0.98	-0.03
15.00	173	200	174	182.33	15.31	0.99	-0.01
Controls							
Sodium Azide	1012	1194	908	1038.00	144.76	5.65	5.93
Water	190	173	181	181.33	8.50	0.99	-0.02
Spontaneous	169	199	183	183.67	15.01	1.00	0.00

With S9

N,N-Dimethylethanolammonium propionate	Number of revetant colonies			Mean	STD	MI	MAR
Concentration / µL / plate	1	2	3				
0.17	127	126	119	124.00	4.36	0.87	-0.13
0.50	111	134	138	127.67	14.57	0.90	-0.10
1.70	150	120	143	137.67	15.70	0.97	-0.03
5.00	153	91	162	135.33	38.66	0.95	-0.05
15.00	316	128	172	205.33	98.33	1.45	0.44
Controls							
Water	130	168	120	139.33	25.32	0.98	-0.02
Spontaneous	122	132	172	142.00	26.46	1.00	0.00

Without S9

N,N-Dimethylethanolammonium hexanoate	Number of revetant colonies			Mean	STD	MI	MAR
Concentration / µL / plate	1	2	3				
0.17	204	176	155	178.33	24.58	1.20	0.21
0.50	171	143	102	138.67	34.70	0.93	-0.07
1.70	130	126	144	133.33	9.45	0.90	-0.10
5.00	66	61	44	57.00	11.53	0.38	-0.63
15.00	9	12	12	11.00	1.73	0.07	-0.95
Controls							
Sodium Azide	490	470	506	488.67	18.04	3.29	2.36
Water	147	183	147	159.00	20.78	1.07	0.07
Spontaneous	129	172	144	148.33	21.83	1.00	0.00

With S9

N,N-Dimethylethanolammonium hexanoate	Number of revetant colonies			Mean	STD	MI	MAR
Concentration / µL / plate	1	2	3				
0.17	185	162	153	166.67	16.50	1.05	0.06
0.50	177	169	185	177.00	8.00	1.12	0.13
1.70	174	144	152	156.67	15.53	0.99	-0.01
5.00	92	93	86	90.33	3.79	0.57	-0.49
15.00	27	23	23	24.33	2.31	0.15	-0.96
Controls							
Water	156	167	139	154.00	14.11	0.97	-0.03
Spontaneous	156	164	155	158.33	4.93	1.00	0.00

Without S9

N,N-Dimethylethanolammonium glycolate	Number of revetant colonies			Mean	STD	MI	MAR
Concentration / µL / plate	1	2	3				
0.17	222	292	380	298.00	79.17	1.28	0.46
0.50	179	181	199	186.33	11.02	0.80	-0.32
1.70	179	188	135	167.33	28.36	0.72	-0.45
5.00	175	159	213	182.33	27.74	0.79	-0.34
15.00	161	176	179	172.00	9.64	0.74	-0.42
Controls							
Sodium Azide	830	910	1056	932.00	114.59	4.02	4.86
Water	189	203	191	194.33	7.57	0.84	-0.26
Spontaneous	186	306	204	232.00	64.71	1.00	0.00

With S9

N,N-Dimethylethanolammonium glycolate	Number of revetant colonies			Mean	STD	MI	MAR
Concentration / µL / plate	1	2	3				
0.17	106	99	94	99.67	6.03	1.32	0.18
0.50	129	85	105	106.33	22.03	1.41	0.22
1.70	268	296	308	290.67	20.53	3.86	1.55
5.00	238	175	218	210.33	32.19	2.79	0.97
15.00	148	115	109	124.00	21.00	1.65	0.35
Controls							
Water	59	61	41	53.67	11.02	0.71	-0.16
Spontaneous	67	80	79	75.33	7.23	1.00	0.00

Without S9

N,N-Dimethylethanolammonium succinate	Number of revetant colonies			Mean	STD	MI	MAR
Concentration / µL / plate	1	2	3				
0.17	164	281	271	238.67	64.86	1.06	0.09
0.50	292	287	134	237.67	89.81	1.05	0.08
1.70	260	265	194	239.67	39.63	1.06	0.09
5.00	208	266	207	227.00	33.78	1.00	0.01
15.00	246	240	252	246.00	6.00	1.09	0.14
Controls							
Sodium Azide	972	1048	836	952.00	107.41	4.21	5.04
Water	290	254	257	267.00	19.97	1.18	0.28
Spontaneous	240	229	209	226.00	15.72	1.00	0.00

With S9

N,N-Dimethylethanolammonium succinate	Number of revetant colonies			Mean	STD	MI	MAR
Concentration / µL / plate	1	2	3				
0.17	230	181	227	212.67	27.47	1.36	0.41
0.50	212	193	183	196.00	14.73	1.25	0.29
1.70	214	164	259	212.33	47.52	1.36	0.40
5.00	215	203	207	208.33	6.11	1.33	0.37
15.00	166	168	130	154.67	21.39	0.99	-0.01
Controls							
Water	226	219	213	219.33	6.51	1.40	0.45
Spontaneous	147	171	151	156.33	12.86	1.00	0.00

Without S9

N,N-Dimethylethanolammonium chloride	Number of revetant colonies			Mean	STD	MI	MAR
Concentration / µL / plate	1	2	3				
0.17	224	206	267	232.33	31.34	1.20	0.27
0.50	241	226	227	231.33	8.39	1.19	0.26
1.70	249	188	253	230.00	36.43	1.19	0.25
5.00	24	239	237	166.67	123.56	0.86	-0.19
15.00	234	239	171	214.67	37.90	1.11	0.15
Controls							
Sodium Azide	740	786	652	726.00	68.09	3.75	3.70
Water	228	257	238	241.00	14.73	1.24	0.33
Spontaneous	182	155	244	193.67	45.63	1.00	0.00

With S9

N,N-Dimethylethanolammonium chloride	Number of revetant colonies			Mean	STD	MI	MAR
Concentration / µL / plate	1	2	3				
0.17	93	74	54	73.67	19.50	1.08	0.04
0.50	90	90	74	84.67	9.24	1.25	0.12
1.70	79	87	72	79.33	7.51	1.17	0.08
5.00	67	60	72	66.33	6.03	0.98	-0.01
15.00	78	57	78	71.00	12.12	1.04	0.02
Controls							
Water	79	84	65	76.00	9.85	1.12	0.06
Spontaneous	69	62	73	68.00	5.57	1.00	0.00

Without S9

N-Ethyl-N-Methylethanolammonium acetate	Number of revetant colonies			Mean	STD	MI	MAR
Concentration / µL / plate	1	2	3				
0.17	167	193	155	171.67	19.43	1.04	0.04
0.50	173	140	104	139.00	34.51	0.84	-0.18
1.70	74	166	166	135.33	53.12	0.82	-0.21
5.00	201	185	180	188.67	10.97	1.14	0.16
15.00	150	165	156	157.00	7.55	0.95	-0.06
Controls							
Sodium Azide	960	1600	1272	1277.33	320.03	7.73	7.72
Water	196	163	175	178.00	16.70	1.08	0.09
Spontaneous	180	130	186	165.33	30.75	1.00	0.00

With S9

N-Ethyl-N-Methylethanolammonium acetate	Number of revetant colonies			Mean	STD	MI	MAR
Concentration / µL / plate	1	2	3				
0.17	165	152	159	158.67	6.51	0.98	-0.02
0.50	151	153	155	153.00	2.00	0.95	-0.06
1.70	145	163	160	156.00	9.64	0.96	-0.04
5.00	158	152	152	154.00	3.46	0.95	-0.06
15.00	169	210	172	183.67	22.85	1.14	0.16
Controls							
Water	174	186	168	176.00	9.17	1.09	0.10
Spontaneous	171	157	157	161.67	8.08	1.00	0.00

Without S9

N,N-Diethylethanolammonium acetate	Number of revetant colonies			Mean	STD	MI	MAR
Concentration / µL / plate	1	2	3				
0.17	342	332	346	340.00	7.21	0.93	-0.17
0.50	359	392	362	371.00	18.25	1.02	0.04
1.70	335	386	358	359.67	25.54	0.99	-0.04
5.00	384	340	344	356.00	24.33	0.98	-0.06
15.00	240	281	275	265.33	22.14	0.73	-0.69
Controls							
Sodium Azide	992	1408	1088	1162.67	217.82	3.19	5.54
Water	330	285	297	304.00	23.30	0.83	-0.42
Spontaneous	356	393	346	365.00	24.76	1.00	0.00

With S9

N,N-Diethylethanolammonium acetate	Number of revetant colonies			Mean	STD	MI	MAR
Concentration / µL / plate	1	2	3				
0.17	329	469	417	405.00	70.77	1.20	0.48
0.50	342	354	352	349.33	6.43	1.03	0.08
1.70	350	383	312	348.33	35.53	1.03	0.08
5.00	369	358	405	377.33	24.58	1.12	0.29
15.00	263	220	265	249.33	25.42	0.74	-0.64
Controls							
Water	325	301	261	295.67	32.33	0.88	-0.30
Spontaneous	316	310	387	337.67	42.83	1.00	0.00

Without S9

N-Butyldiethanolammonium acetate	Number of revetant colonies			Mean	STD	MI	MAR
Concentration / µL / plate	1	2	3				
0.17	203	176	148	175.67	27.50	0.99	-0.01
0.50	153	167	197	172.33	22.48	0.97	-0.03
1.70	262	188	224	224.67	37.00	1.27	0.33
5.00	214	229	226	223.00	7.94	1.26	0.32
15.00	176	206	153	178.33	26.58	1.01	0.01
Controls							
Sodium Azide	553	526	533	537.33	14.01	3.04	2.50
Water	233	229	210	224.00	12.29	1.27	0.33
Spontaneous	178	168	185	177.00	8.54	1.00	0.00

With S9

N-Butyldiethanolammonium acetate	Number of revetant colonies			Mean	STD	MI	MAR
Concentration / µL / plate	1	2	3				
0.17	271	83	200	184.67	94.93	1.27	0.28
0.50	261	235	173	223.00	45.21	1.53	0.56
1.70	123	180	137	146.67	29.70	1.01	0.01
5.00	150	235	227	204.00	46.94	1.40	0.42
15.00	159	181	228	189.33	35.25	1.30	0.32
Controls							
Water	245	145	294	228.00	75.94	1.57	0.59
Spontaneous	117	124	195	145.33	43.15	1.00	0.00

Without S9

N-Methyl-bis(2-methoxyethyl)ammonium acetate	Number of revetant colonies			Mean	STD	MI	MAR
Concentration / µL / plate	1	2	3				
0.17	161	162	120	147.67	23.97	1.10	0.09
0.50	165	89	111	121.67	39.11	0.90	-0.09
1.70	174	179	153	168.67	13.80	1.25	0.24
5.00	154	135	105	131.33	24.70	0.98	-0.02
15.00	140	146	155	147.00	7.55	1.09	0.09
Controls							
Sodium Azide	1084	924	1012	1006.67	80.13	7.48	6.06
Water	126	101	114	113.67	12.50	0.84	-0.15
Spontaneous	153	126	125	134.67	15.89	1.00	0.00

With S9

N-Methyl-bis(2-methoxyethyl)ammonium acetate	Number of revetant colonies			Mean	STD	MI	MAR
Concentration / µL / plate	1	2	3				
0.17	115	111	94	106.67	11.15	1.00	0.00
0.50	103	104	66	91.00	21.66	0.85	-0.12
1.70	93	119	79	97.00	20.30	0.91	-0.07
5.00	73	121	101	98.33	24.11	0.92	-0.06
15.00	106	91	85	94.00	10.82	0.88	-0.09
Controls							
Water	105	102	83	96.67	11.93	0.90	-0.07
Spontaneous	141	109	71	107.00	35.04	1.00	0.00

Without S9

N-Methylethanolmmonium chloride	Number of revetant colonies			Mean	STD	MI	MAR
Concentration / µL / plate	1	2	3				
0.17	96	98	95	96.33	1.53	1.01	0.01
0.50	80	88	112	93.33	16.65	0.98	-0.01
1.70	66	100	115	93.67	25.11	0.99	-0.01
5.00	86	96	80	87.33	8.08	0.92	-0.05
15.00	82	96	97	91.67	8.39	0.96	-0.02
Controls							
Sodium Azide	856	880	801	845.67	40.50	8.90	5.21
Water	117	97	87	100.33	15.28	1.06	0.04
Spontaneous	89	113	83	95.00	15.87	1.00	0.00

With S9

N-Methylethanolmmonium chloride	Number of revetant colonies			Mean	STD	MI	MAR
Concentration / µL / plate	1	2	3				
0.17	76	65	49	63.33	13.58	1.09	0.04
0.50	79	58	56	64.33	12.74	1.11	0.05
1.70	87	65	103	85.00	19.08	1.47	0.19
5.00	64	82	74	73.33	9.02	1.26	0.11
15.00	60	79	38	59.00	20.52	1.02	0.01
Controls							
Water	74	77	65	72.00	6.24	1.24	0.10
Spontaneous	58	71	45	58.00	13.00	1.00	0.00

Without S9							
Diethanolammonium chloride	Number of revetant colonies			Mean	STD	MI	MAR
Concentration / µL / plate	1	2	3				
0.17	90	115	103	102.67	12.50	1.12	0.08
0.50	107	99	74	93.33	17.21	1.02	0.01
1.70	108	110	109	109.00	1.00	1.19	0.12
5.00	110	83	102	98.33	13.87	1.07	0.05
15.00	115	121	126	120.67	5.51	1.32	0.20
Controls							
Sodium Azide	348	318	474	380.00	82.78	4.15	2.00
Water	63	88	81	77.33	12.90	0.84	-0.10
Spontaneous	103	82	90	91.67	10.60	1.00	0.00

With S9							
Diethanolammonium chloride	Number of revetant colonies			Mean	STD	MI	MAR
Concentration / µL / plate	1	2	3				
0.17	67	63	73	67.67	5.03	0.95	-0.03
0.50	70	76	56	67.33	10.26	0.94	-0.03
1.70	64	68	76	69.33	6.11	0.97	-0.01
5.00	83	83	79	81.67	2.31	1.14	0.07
15.00	71	70	96	79.00	14.73	1.11	0.06
Controls							
Water	64	76	53	64.33	11.50	0.90	-0.05
Spontaneous	77	79	58	71.33	11.59	1.00	0.00

Without S9							
Diethanolammonium acetate	Number of revetant colonies			Mean	STD	MI	MAR
Concentration / µL / plate	1	2	3				
0.17	114	80	86	93.33	18.15	0.92	-0.06
0.50	116	113	86	105.00	16.52	1.03	0.02
1.70	100	107	83	96.67	12.34	0.95	-0.04
5.00	106	91	67	88.00	19.67	0.86	-0.10
15.00	81	59	68	69.33	11.06	0.68	-0.23
Controls							
Sodium Azide	1012	904	976	964.00	54.99	9.45	5.99
Water	73	98	111	94.00	19.31	0.92	-0.06
Spontaneous	99	97	110	102.00	7.00	1.00	0.00

With S9							
Diethanolammonium acetate	Number of revetant colonies			Mean	STD	MI	MAR
Concentration / µL / plate	1	2	3				
0.17	64	75	50	63.00	12.53	0.87	-0.07
0.50	66	66	59	63.67	4.04	0.88	-0.06
1.70	66	69	45	60.00	13.08	0.83	-0.09
5.00	55	58	59	57.33	2.08	0.79	-0.11
15.00	52	61	67	60.00	7.55	0.83	-0.09
Controls							
Water	51	76	53	60.00	13.89	0.83	-0.09
Spontaneous	76	68	73	72.33	4.04	1.00	0.00

Without S9							
Bis(2-methoxyethyl) ammonium chloride	Number of revetant colonies			Mean	STD	MI	MAR
Concentration / µL / plate	1	2	3				
0.17	102	97	92	97.00	5.00	1.04	0.02
0.50	74	104	90	89.33	15.01	0.95	-0.03
1.70	90	107	87	94.67	10.79	1.01	0.01
5.00	84	99	92	91.67	7.51	0.98	-0.01
15.00	109	123	108	113.33	8.39	1.21	0.14
Controls							
Sodium Azide	443	478	580	500.33	71.18	5.34	2.82
Water	98	73	106	92.33	17.21	0.99	-0.01
Spontaneous	100	112	69	93.67	22.19	1.00	0.00

With S9

Bis(2-methoxyethyl) ammonium chloride		Number of revetant colonies			Mean	STD	MI	MAR
Concentration / µL / plate		1	2	3				
	0.17	173	239	178	196.67	36.75	0.77	-0.43
	0.50	234	272	264	256.67	20.03	1.00	0.00
	1.70	283	277	253	271.00	15.87	1.06	0.11
	5.00	257	235	220	237.33	18.61	0.93	-0.13
	15.00	142	136	137	138.33	3.21	0.54	-0.85
Controls								
Water		244	262	241	249.00	11.36	0.97	-0.05
Spontaneous		283	227	258	256.00	28.05	1.00	0.00

Without S9

Bis(2-methoxyethyl) ammonium acetate		Number of revetant colonies			Mean	STD	MI	MAR
Concentration / µL / plate		1	2	3				
	0.17	90	95	78	87.67	8.74	1.23	0.11
	0.50	113	102	92	102.33	10.50	1.43	0.22
	1.70	101	81	84	88.67	10.79	1.24	0.12
	5.00	92	94	95	93.67	1.53	1.31	0.16
	15.00	113	91	112	105.33	12.42	1.48	0.24
Controls								
Sodium Azide		650	704	734	696.00	42.57	9.76	4.34
Water		90	95	104	96.33	7.09	1.35	0.17
Spontaneous		71	74	69	71.33	2.52	1.00	0.00

With S9

Bis(2-methoxyethyl) ammonium acetate		Number of revetant colonies			Mean	STD	MI	MAR
Concentration / µL / plate		1	2	3				
	0.17	259	275	284	272.67	12.66	1.06	0.11
	0.50	276	264	312	284.00	24.98	1.10	0.19
	1.70	296	251	304	283.67	28.57	1.10	0.18
	5.00	255	325	224	268.00	51.74	1.04	0.07
	15.00	210	243	207	220.00	19.97	0.85	-0.27
Controls								
Water		319	272		295.50	33.23	1.15	0.27
Spontaneous		246	265	263	258.00	10.44	1.00	0.00

Without S9

N,N-Dimethylethanolamine		Number of revetant colonies			Mean	STD	MI	MAR
Concentration / µL / plate		1	2	3				
	0.17	102	100	98	100.00	2.00	1.07	0.04
	0.50	101	116	96	104.33	10.41	1.11	0.07
	1.70	144	182	151	159.00	20.22	1.70	0.45
	5.00	157	184	184	175.00	15.59	1.87	0.56
	15.00	77	60	48	61.67	14.57	0.66	-0.22
Controls								
Sodium Azide		480	578	420	492.67	79.76	5.26	2.77
Water		98	73	106	92.33	17.21	0.99	-0.01
Spontaneous		100	112	69	93.67	22.19	1.00	0.00

Without S9

Diethanolamine		Number of revetant colonies			Mean	STD	MI	MAR
Concentration / µL / plate		1	2	3				
	0.17	100	123	81	101.33	21.03	1.34	0.18
	0.50	106	102	93	100.33	6.66	1.33	0.17
	1.70	91	93	94	92.67	1.53	1.22	0.12
	5.00	166	158	142	155.33	12.22	2.05	0.55
	15.00	113	128	100	113.67	14.01	1.50	0.26
Controls								
Sodium Azide		552	568	624	581.33	37.81	7.68	3.51
Water		90	103	108	100.33	9.29	1.33	0.17
Spontaneous		62	86	79	75.67	12.34	1.00	0.00

With S9							
N,N-Dimethylethanolamine	Number of revetant colonies			Mean	STD	MI	MAR
Concentration / µL / plate	1	2	3				
0.17	49	42	52	47.67	5.13	0.94	-0.02
0.50	40		46	43.00	4.24	0.85	-0.06
1.70	80	40	76	65.33	22.03	1.29	0.11
5.00	81	87	92	86.67	5.51	1.71	0.26
15.00	25	21	16	20.67	4.51	0.41	-0.22
Controls							
Water	67	54	47	56.00	10.15	1.11	0.04
Spontaneous	55	50	47	50.67	4.04	1.00	0.00

With S9							
Diethanolamine	Number of revetant colonies			Mean	STD	MI	MAR
Concentration / µL / plate	1	2	3				
0.17	197	269	194	220.00	42.46	0.92	-0.14
0.50	313	286	222	273.67	46.74	1.14	0.24
1.70	184	144	235	187.67	45.61	0.78	-0.37
5.00	239	176	157	190.67	42.92	0.80	-0.35
15.00	157	105	102	121.33	30.92	0.51	-0.85
Controls							
Water	321	275	209	268.33	56.30	1.12	0.21
Spontaneous	269	217	233	239.67	26.63	1.00	0.00

REFERENCES

1. Clark JH (1999) Green chemistry: challenges and opportunities. Green Chem 1:1–8

2. Postle M, Holmes P, Salado R, Thorell A, Tuffnell N, Guittat A, Fantke P, Rushton L (2011) Assessing the health and environmental impacts in the context of socio-economic analysis under REACH; Part 1: literature review and recommendations, European Chemicals Agency. Accessed 25th June 2015

3. Eckert CA, Knuston BL, Debendetti PG (1996) Supercritical fluids as solvents for chemical and materials processing. Nature 383:313–318

4. Hallett JP, Welton T (2011) Room-temperature ionic liquids: solvents for synthesis and catalysis. 2. Chem Rev 111:3508–3576

5. Clark JH, Tavener SJ (2007) Alternative solvents: shades of green. Org Process Res Dev 11:149–155

6. Burrell GL, Burgar IM, Separovic F, Dunlop NF (2010) Preparation of protic ionic liqids with minimal water content and ^{15}N NMR study of proton transfer. Phys Chem Chem Phys 12:1571–1577

7. Seddon KR, Stark A, Torres MJ (2000) Influence of chloride, water, and organic solvents on the physical properties of ionic liquids. Pure Appl Chem 72:2275–2287

8.   Reid JESJ, Walker AJ, Shimizu S (2015) Residual water in ionic liquids: clustered or dissociated? Phys Chem Chem Phys 17:14710–14718

9.   Chen L, Sharifzadeh M, Mac Dowell N, Welton T, Shah N, Hallett JP (2014) Inexpensize ionic liquids: [HSO$_4$]$^-$ based solvent production at bulk scale. Green Chem. 16:3098–3106

10. George A, Brandt A, Tran K, Zahari SMSNS, Klein-Marcuchamer D, Sun N, Sathitsuksanoh N, Shi J, Stavila V, Parthasarathi R, Singh S, Holmes BM, Welton T, Simmons BA, Hallett JP (2015) Design of low-cost ionic liquids for lignocellulosic biomass pretreatment. Green Chem. 17:1728–1734

11. Walden P (1914) Molecular weights and electrical conductivity of several fused salts. Bull Acad Imper Sci (St. Petersburg) 8:405–422

12. Walker AJ (2007) Protic ionic liquids and their potential industrial applications. Chim Oggi 5:17–19

13. Walker AJ, inventor; Innovia Films Limited, assignee (2004) Ionic liquids comprising nitrogen containing cations. European patent EP 1805131 B1

14. Angell CA, Byrne N, Belieres JP (2007) Parallel developments in aprotic and protic ionic liquids: physical chemistry and applications. Acc Chem Res 40:1228–1236

15. Greaves TL, Drummond CJ (2008) Protic ionic liquids: properties and applications. Chem Rev 108:206–237

16. Angell CA, Ansari Y, Zhao Z (2012) Ionic liquids: past, present and future. Faraday Discuss 154:9–27

17. MacFarlane DR, Tachikawa N, Forsyth M, Pringle JM, Howlett PC, Elliott GD, Davis JH Jr, Watanabe M, Simon P, Angell CA (2014) Energy applications of ionic liquids. Energy Environ Sci 7:232–250

18. Hunt PA, Ashworth CR, Matthews RP (2015) Hydrogen bonding in ionic liquids. Chem Soc Rev 44:1257–1288

19. Sanders MW, Wright L, Tate L, Fairless G, Crowhurst L, Bruce NC, Walker AJ, Hembury GA, Shimizu S (2009) Unexpected preferential dehydration of artemisinin in ionic liquids. J Phys Chem A 133:10143–10145

20. Chi YS, Zhang ZD, Li CP, Liu QS, Yan PF, Welz-Biermann U (2011) Microwave-assisted extraction of lactones from Ligusticum chuanxiong Hort. using protic ionic liquids. Green Chem 13:666–670

21. Goujon N, Wang X, Rajkowa R, Byrne N (2012) Regenerated silk fibroin using protic ionic liquids solvents: towards an all-ionic-liquid process for producing silk with tunable properties. Chem Commun 48:1278–1280

22. Idris A, Vijayaraghavan R, Patti AF, MacFarlane DR (2014) Distillable protic ionic liquids for keratin dissolution and recovery. ACS Sustain Chem Eng. 2:1888–1894

23. Achinivu EC, Howard RM, Li G, Gracz H, Henderson WA (2014) Lignin extraction from biomass with protic ionic liquids. Green Chem 16:1114–1119

24. Verdía P, Brandt A, Hallett JP, Ray MJ, Welton T (2014) Fractionation of lignocellulosic biomass with the ionic liquid 1-butylimidazolium hydrogen sulfate. Green Chem 16:1617–1627

25. Li Z, Li C, Chi Y, Wang A, Zhang Z, Li H, Liu Q, Welz-Biermann U (2012) Extraction process of dibenzothiophene with new distillable amine-based protic ionic liquids. Energy Fuels 26:3723–3727

26. Lü H, Wang S, Deng C, Ren W, Guo B (2014) Oxidative desulfurization of model diesel via dual activation by a protic ionic liquid. J Hazard Mater 279:220–225

27. Li Z, Xu J, Li C (2015) Extraction process of sulfur compounds from fuels with protic ionic liquids. RSC Adv. 5:15892–15897

28. Yokozeki A, Shiflett MB (2007) Vapor-liquid equilibria of ammonia + ionic liquid mixtures. Appl Energy 84:1258–1273

29. Mumford KA, Pas SJ, Linseisen T, Statham TM, Nicholas NJ, Lee A, Kezia K, Vijayraghavan R, MacFarlane DR, Stevens GW (2015) Evaluation of the protic ionic liquid, N, N-dimethyl-aminoethylammonium formate for CO2 capture. Int J Greenhouse Gas Control 32:129–134

30. Vijayraghavan R, Pas SJ, Izgorodina EI, MacFarlane DR (2013) Diamino protic ionic liquids for CO_2 capture. Phys Chem Chem Phys 15:19994–19999

31. Nakamoto H, Watanabe M (2007) Bronsted acid-base ionic liquids for fuel cell electrolytes. Chem Comm. 24:2539–2541

32. Lee SY, Ogawa A, Kanno M, Nakamoto H, Yasuda T, Watanbe M (2010) Nonhumidified intermediate temperature fuel cells using protic ionic liquids. J Am Chem Soc 132:9764–9773

33. Snyder J, Fujita T, Chen MW, Erlebacher J (2010) Oxygen reduction in nanoporous metal-ionic liquid composite electrocatalysts. Nat Mater 9:904–907

34. Olivier-Bourbigou H, Manga L, Morvan D (2010) Ionic liquids and catalysis: recent progress from knowledge to applications. Appl Catal A-Gen. 373:1–56

35. Clark JH, Farmer TJ, Macquarrie DJ, Sherwood J (2013) Using metrics and sustainability considerations to evaluate the use of bio-based and non-renewable Brønsted acidic ionic liquids to catalyse Fischer esterification reactions. Sustain Chem Process. 1:23–35

36. Qu J, Truhan JJ, Dai S, Luo H, Blau PJ (2006) Ionic liquids with ammonium cations as lubricants or additives. Tribol Lett 22:207–214

37. Kondo H (2008) Protic ionic liquids with ammonium salts as lubricants for magnetic thin film media. Tribol Lett 31:211–218

38. Zhao Q, Zhao G, Zhang M, Wang X, Liu W (2012) Tribological behavior of protic ionic liquids with dodecylamine salts of dialkyldithiocarbamate as additives in lithium complex grease. Tribol Lett 48:133–144

39. Espinosa T, Jimenez M, Sanes J, Jimenez AE, Inglesias M, Bermudez MD (2014) Ultra-low friction with a protic ionic liquid boundary film at the water-lubricated sapphire-stainless steel interface. Tribol Lett 53:1–9

40. Kondo H, Ito M, Hatsuda K, Yun K, Watanabe M (2013) Novel ionic lubricants for magnetic thin film media. IEEE Trans Magnet. 49:3756–3759

41. Luo H, Yu M, Dai S (2007) Solvent extraction of Sr^{2+} and Cs^+ based on hydrophobic protic ionic liquids. Z Naturforsch A. 62:281–291

42. Katsuta S, Yoshimoto Y, Okai M, Takeda Y, Bessho K (2011) Selective extraction of palladium and platinum from hydrochloric acid solutions by trioctylammonium-based mixed ionic liquids. Ind Eng Chem Res 50:12735–12740

43. Katsuta S, Okai M, Yoshimoto Y, Kudo Y (2012) Extraction of gallium(III) from hydrochloric acid solutions by trioctylammonium-based mixed ionic liquids. Anal Sci 28:1009–1012

44. Watanabe Y, Araki Y, Katsuta S (2014) Extraction behaviour of gold (III) from hydrochloric acid solutions by protic ionic liquids. Bunseki Kagaku 63:563–567

45. Peric B, Martí E, Sierra J, Cruañas R, Inglesias M, Garau MA (2011) Terrestrial ecotoxicity of short aliphatic protic ionic liquids. Ecotoxicol Environ Saf 30:2802–2809

46. Peric B, Sierra J, Martí E, Cruañas R, Garau MA, Arning J, Bottin-Weber U, Stolte S (2013) (Eco)toxicity and biodegradability of selected protic and aprotic ionic liquids. J Hazard Mater 261:99–105

47. Peric B, Sierra J, Martí E, Cruañas R, Garau MA (2014) A comparative study of the terrestrial ecotoxicity of selected protic and aprotic ionic liquids. Chemosphere 108:418–425

48. Peric B, Martí E, Sierra J, Cruañas R, Inglesias M, Garau MA (2015) Quantitative structure-activity relationship (QSAR) prediction of (eco) toxicity of short aliphatic protic ionic liquids. Ecotoxicol Environ Saf 115:257–262

49. Ames BN, McCann J, Yamasaki E (1975) Methods for detecting carcinogens and mutagens with salmonella-mammalian-microsome mutagenicity test. Mutat Res 31:347–363

50. McCann J, Choi E, Yamasaki E, Ames BN (1975) Detection of carcinogens as mutagens in salmonella microsome test—assay of 300 chemicals. Proc Nat Acad Sci USA 72:5135–5139

51. Maron DM, Ames BN (1983) Revised methods for the salmonella mutagenicity test. Mutat Res 113:173–215

52. Mortelmans K, Zeiger E (2000) The Ames Salmonella/microsome mutagenicity assay. Mutat Res-Fund Mol M. 455:29–60

53. Maron DM, Katzenellenbogen J, Ames BN (1981) Compatability of organic-solvents with the salmonella-microsome test. Mutat Res 88:343–350

54. Docherty KM, Hebbeler SZ, Kupla CF Jr (2006) An assessment of ionic liquid mutagenicity using the Ames test. Green Chem 8:560–567

55. McCann J, Spingarn N, Kobori J, Ames BN (1975) Detection of carcinogens as mutagens: bacterial tester strains with R Factor plasmids. Proc Natl Acad Sci USA 72:979–983

56. Kirkland DJ (1989) Guidelines for mutagenicity testing part II. Statistical evaluation of mutagenicity test data, 1st edn. Cambridge University Press, Cambridge

57. Haworth S, Lawlor T, Mortelmans K, Speck W, Zeiger E (1983) Salmonella mutagenicity test results for 250 chemicals. Environ Mol Mutagen 5:3–142

58. Zeiger E, Anderson B, Haworth S, Lawlor T, Mortelmans K, Speck W (1987) Salmonella mutagenicity tests. 3. Results from the testing of 255 chemicals. Environ Mol Mutagen 9:1–109

59. Zeiger E, Anderson B, Haworth S, Lawlor T, Mortelmans K (1992) Salmonella mutagenicity tests. 5. Results from the testing of 311 chemicals. Environ Mol Mutagen 19:2–141

60. Fujita H, Sumi C, Sasaki M (1992) Mutagenicity test of food additives with Salmonella typhimurium TA97 and TA102. Ann Rep Tokyo Metrop Res Lab Public Health. 43:219–227

61. Stange P, Fumino K, Ludwig R (2013) Ion speciation of protic ionic liquids in water: transition from contact to solvent-separated ion pairs. Angew Chem Int Ed 52:2990–2994

62. Stolte S, Matzke M, Arning J, Böschen A, Pitner W, Welz-Biermann U, Jastorff B, Ranke J (2007) Effects of different head groups and functionalised side chains on the aquatic toxicity of ionic liquids. Green Chem 9:1170–1179

63. Pham TPT, Cho C, Yun Y (2010) Environmental fate and toxicity of ionic liquids: a review. Water Res 44:352–372

64. Coleman D, Gathergood N (2010) Biodegradation studies of ionic liquids. Chem Soc Rev 39:600–637

65. Jordan A, Gathergood N (2015) Biodegradation of ionic liquids—a critical review. Chem. Soc, Rev

66. Zhou W, Yuan D, Huang G, Zhang H, Ye S (2000) Mutagenicity of methyl tertiary butyl ether. J Environ Pathol Toxicol Oncol 19:35–39

67. Greenbla M, Mirvish S, So BT (1971) Nitrosamine studies—induction of lung adenomas by concurrent administration of sodium nitrate and secondary amines in swiss mice. J Natl Cancer Inst 46:1029–1034

68. Hawkswor GM, Hill MJ (1971) Bacteria and N-nitrosation of secondary amines. Brit J Cancer. 25:520–526

69. Szylit O, Ducluzeau R, Champ M, Klein D (1976) La formation des nitrosamines dans le tube digestif. Ann Nutr Aliment. 30:805–812

70. Some chemicals present in industrial and consumer products, food and drinking-water, IARC monographs on the evaluation of carcinogenic risks to humans; International Agency for Research on Cancer; World Health Organisation. 2012. http://monographs.iarc.fr/ENG/Monographs/vol101/mono101.pdf. Accessed 26 June 2015

71. Garcia MT, Gathergood N, Scammells PJ (2005) Biodegradable ionic liquids—Part II. Effect of the anion and toxicology. Green Chem 7:9–14

CITATION

CHAPTER 1

M.A. Martin-Luengo, M. Yates, M. Ramos, F. Plou, J. L. Salgado, A. Civantos, J.L. Lacomba, G. Reilly, C. Vervaet, E. Sáez Rojo, A.M. Martínez Serrano, M. Diaz, L. Vega Argomaniz, L. Medina Trujillo, S. Nogales and R. Lozano Pirrongell (2012). Sustainable Materials and Biorefinery Chemicals from Agriwastes, Resource Management for Sustainable Agriculture, Dr. Vikas Abrol (Ed.), ISBN: 978-953-51-0808-5, InTech, DOI: 10.5772/50551.

CHAPTER 2

Paula Saavalainen, Satish Kabra, Esa Turpeinen, et al., "Sustainability Assessment of Chemical Processes: Evaluation of Three Synthesis Routes of DMC," Journal of Chemistry, vol. 2015, Article ID 402315, 12 pages, 2015. doi:10.1155/2015/402315

CHAPTER 3

Mohamed Samer (2015). Biological and Chemical Wastewater Treatment Processes, Wastewater Treatment Engineering, Associate Prof. Mohamed Samer (Ed.), ISBN: 978-953-51-2179-4, InTech, DOI: 10.5772/61250.

CHAPTER 4

Mahesh K. Potdar, Geoffrey F. Kelso, Lachlan Schwarz, Chunfang Zhang and Milton T. W. Hearn, "Recent Developments in Chemical Synthesis with Biocatalysts in Ionic Liquids," Molecules 2015, 20(9), 16788-16816; doi:10.3390/molecules200916788.

CHAPTER 5

Rinkoo Devi Gupta, "Recent advances in enzyme promiscuity," Sustainable Chemical Processes20164:2, DOI: 10.1186/s40508-016-0046-9.

CHAPTER 6

Luiz J Visioli, Heveline Enzweiler, Raquel C Kuhn, Marcio Schwaab and Marcio A Mazutt, "Recent advances on biobutanol production," Sustainable Chemical Processes20142:15, DOI: 10.1186/2043-7129-2-15.

CHAPTER 7

Anil C Mali, Dattatray G Deshmukh, Divyesh R Joshi, Hitesh D Lad, Priyank I Patel, Vijay J Medhane and Vijayavitthal T Mathad, "Facile approach for the synthesis of rivaroxaban using alternate synthon: reaction, crystallization and isolation in single pot to achieve desired yield, quality and crystal form," Sustainable Chemical Processes20153:11, DOI: 10.1186/s40508-015-0036-3.

CHAPTER 8

Mukesh Kumar, Joyeeta Mukherjee, Mau Sinha, Punit Kaur, Sujata Sharma, Munishwar Nath Gupta and Tej Pal Singh, "Enhancement of stability of a lipase by subjecting to three phase partitioning (TPP): structures of native and TPP-treated lipase from Thermomyces lanuginose," Sustainable Chemical Processes20153:14, DOI: 10.1186/s40508-015-0042-5.

CHAPTER 9

Joshua E. S. J. Reid, Neil Sullivan, Lorna Swift , Guy A. Hembury , Seishi Shimizu and Adam J. Walker, "Assessing the mutagenicity of protic ionic liquids using the mini Ames test," Sustainable Chemical Processes20153:17, DOI: 10.1186/s40508-015-0044-3.

INDEX